Intelligent Systems Reference Library

Volume 278

The aim of this series is to publish a Reference Library, including novel advances and developments in all aspects of Intelligent Systems in an easily accessible and well structured form. The series includes reference works, handbooks, compendia, textbooks, well-structured monographs, dictionaries, and encyclopedias. It contains well integrated knowledge and current information in the field of Intelligent Systems. The series covers the theory, applications, and design methods of Intelligent Systems. Virtually all disciplines such as engineering, computer science, avionics, business, e-commerce, environment, healthcare, physics and life science are included. The list of topics spans all the areas of modern intelligent systems such as: Ambient intelligence, Computational intelligence, Social intelligence, Computational neuroscience, Artificial life, Virtual society, Cognitive systems, DNA and immunity-based systems, e-Learning and teaching, Human-centred computing and Machine ethics, Intelligent control, Intelligent data analysis, Knowledge-based paradigms, Knowledge management, Intelligent agents, Intelligent decision making, Intelligent network security, Interactive entertainment, Learning paradigms, Recommender systems, Robotics and Mechatronics including human-machine teaming, Self-organizing and adaptive systems, Soft computing including Neural systems, Fuzzy systems, Evolutionary computing and the Fusion of these paradigms, Perception and Vision, Web intelligence and Multimedia.

Indexed by SCOPUS, DBLP, zbMATH, SCImago.

All books published in the series are submitted for consideration in Web of Science.

Yen-Wei Chen · Lanfen Lin · Rahul Kumar Jain

Recent Advances in Deep Learning for Medical Image Analysis

Paradigms and Applications

Springer

Yen-Wei Chen
College of Information Science
and Engineering
Ritsumeikan University
Ibaraki, Osaka, Japan

Rahul Kumar Jain
College of Information Science
and Engineering
Ritsumeikan University
Ibaraki, Osaka, Japan

Lanfen Lin
College of Computer Science
and Technology
Zhejiang Key Laboratory of Multi-omics
Precision Diagnosis and Treatment of Liver
Diseases
Zhejiang University
Hangzhou, Zhejiang, China

ISSN 1868-4394 ISSN 1868-4408 (electronic)
Intelligent Systems Reference Library
ISBN 978-3-031-94790-2 ISBN 978-3-031-94791-9 (eBook)
https://doi.org/10.1007/978-3-031-94791-9

This Springer imprint is published by the registered company Springer Nature Switzerland AG
The registered company address is: Gewerbestrasse 11, 6330 Cham, Switzerland

Preface

In recent years, deep learning has emerged as a transformative technology across various fields, with medical image analysis standing out as one of its most impactful applications. By leveraging the power of deep learning, researchers and practitioners have been able to address longstanding challenges in medical imaging, enabling more accurate diagnostics, efficient workflows, and deeper insights into complex diseases. However, the field of medical image analysis is evolving rapidly, with new advances and methodologies continually redefining the possibilities. This book, *Recent Advances in Deep Learning for Medical Image Analysis*, aims to provide a comprehensive overview of the latest developments in this dynamic area of research.

This book is designed for researchers, students, and professionals in both computer science and healthcare domains who seek to deepen their understanding of cutting-edge techniques in medical image analysis using deep learning. This book is organized into two parts: Part I, *Deep Learning Fundamentals and Paradigms* (Chaps. 1–3), and Part II, *Advanced Deep Learning for Medical Image Analysis* (Chaps. 4–10) as shown in below:

Part I. *Deep Learning Fundamentals and Paradigms*

1. **Deep Convolutional Neural Networks (CNNs)** introduces the cornerstone of deep learning, discussing its role in medical image processing.
2. **Deep CNNs for Image Classification, Object Detection, and Segmentation** covers the core tasks essential for medical image understanding.
3. **Attention and Transformer Networks** explores how attention mechanisms and transformer architectures are reshaping the field.

Part II. *Advanced Deep Learning for Medical Image Analysis*

4. **Transformer-Based Approaches for Medical Image Analysis** delves deeper into transformer-based solutions for advanced image analysis.
5. **Deep Learning Networks for 3D Medical Image Analysis** focuses on the challenges and innovations in analyzing three-dimensional medical data.
6. **Multimodal Deep Learning for Medical Image Analysis** examines techniques that integrate diverse data modalities for enriched medical insights.

7. **Semi-Supervised Learning for Medical Image Analysis** discusses leveraging unlabeled data to improve model performance.
8. **Domain Adaptation and Generalization for Medical Image Analysis** addresses the critical issue of ensuring robust performance across diverse datasets such as multi-center datasets.
9. **Deep Learning Models for Medical Image Translation** highlights generative methods and their applications in creating synthetic medical images for diagnosis and treatments.
10. **Foundation Models for Medical Image Analysis** explores how large-scale pre-trained models can be adapted for medical imaging tasks.

The chapters cover a wide range of topics, providing insights into both foundational and cutting-edge methodologies for medical image analysis. They also highlight the authors' recent research achievements, offering a well-balanced combination of theoretical knowledge, practical implementations, and real-world case studies.

We hope this book will serve as a valuable resource for understanding the latest advancements in deep learning and inspire new ideas and collaborations in the field of medical image analysis.

We extend our heartfelt gratitude to all the contributors—Shiao Xie, Gang Zhan, Hongyi Wang, Yang Ai, Shuyi Ouyang, Huiming Huang, Ziwei Niu, Yulin Yang, Jihong Hu, Xiaoyu Shi, and Shurong Chai—from Ritsumeikan University, Japan, and Zhejiang University, China, for their efforts and significant contributions to each chapter.

Osaka, Japan — Yen-Wei Chen
Hangzhou, China — Lanfen Lin
Osaka, Japan — Rahul Kumar Jain
April 2025

Contents

About the Authors

Prof. Yen-Wei Chen received his B.E. degree in 1985 from Kobe University, Kobe, Japan. He received his M.E. degree in 1987 and his D.E. degree in 1990, both from Osaka University, Osaka, Japan. From 1991 to 1994, he was a research fellow at the Institute of Laser Technology, Osaka. From October 1994 to March 2004, he was an associate professor and a professor in the Department of Electrical and Electronic Engineering, University of the Ryukyus, Okinawa, Japan. He is currently a professor at the College of Information Science and Engineering, Ritsumeikan University, Japan. He is an associate editor for the *International Journal of Image and Graphics* (IJIG), and the *International Journal of Knowledge-Based and Intelligent Engineering Systems*. His research focuses on computer vision, deep learning, and medical image analysis. He has published more than 300 research papers in these fields.

Prof. Lanfen Lin received her B.S. and Ph.D. degrees from Northwestern Polytechnical University in 1990 and 1995, respectively. She held a postdoctoral position with the Department of Computer Science and Technology, Zhejiang University, China, from January 1996 to December 1997. She was an associate professor from 1998 to 2005. Now she is a full professor and the vice-director of the Artificial Intelligence Institute in Zhejiang University. Her research interests include computer vision, medical image processing, and intelligent manufacturing. She has published more than 200 research papers in these fields.

Dr. Rahul Kumar Jain received his Ph.D. degree from Ritsumeikan University, Shiga, Japan, in 2022. He has been an intern trainee at Tiwaki Co., Ltd., Japan, since 2019. He is now working as a senior researcher at the College of Information Science and Engineering, Ritsumeikan University, Japan. His research interests include computer vision, deep learning, and image processing as well as the applications of artificial intelligence in areas including engineering, science, computer science, healthcare, and so on.

Part I
Deep Learning Fundamentals and Paradigms

Chapter 1
Deep Convolutional Neural Networks (CNNs)

1.1 Introduction

In recent years, Artificial Intelligence (AI) has gained significant attention in advancing automation across various domains. Terms such as Machine Learning (ML), Deep Learning (DL), Computer Vision (CV) and AI itself are frequently mentioned in various technical literature, research articles, and media. Nowadays, DL frameworks have emerged as the leading AI-based technology, dominating both academic research and industrial applications. They are widely employed for predictive tasks such as data analysis, text mining, image classification, image segmentation, object detection, human pose estimation, emotion detection, automatic vehicle driving, navigation and more. These AI related tasks are tremendously useful in various sectors in the modern era such as biomedicine, healthcare, Information and Communication Technology (ICT), digital technology and industrial automation. DL framework has a wide range of applications in medical imaging, proving invaluable support to healthcare professionals [1]. These algorithms excel in a range of medical tasks, including disease classification, organ segmentation (anatomy modeling), lesion and abnormal tissue detection, and the generation of contrast-enhanced CT and MRI images.

Over the past few years, alongside the advancement of traditional machine learning (ML)-based AI, there has been growing interest in data-driven DL technique. Traditional ML algorithms looks at input data and corresponding outputs to determine rules for learning from data to make predictions. Model based on these algorithms is considered "trained" when it is presented with examples relevant to a task, and it finds statistical structures within these examples. As shown in Fig. 1.1a, these algorithms rely on manual feature extraction, a process known as hand-crafted features. This involves manually selecting and transforming input raw data into a set of low-level or mid-level representative features. Feature engineering plays a critical role in traditional ML as its performance heavily depends on the quality of these features. Depending on the application, these features may include statistical measures, frequency components, texture information, and any others.

Y. Chen et al., *Recent Advances in Deep Learning for Medical Image Analysis*, Intelligent Systems Reference Library 278, https://doi.org/10.1007/978-3-031-94791-9_1

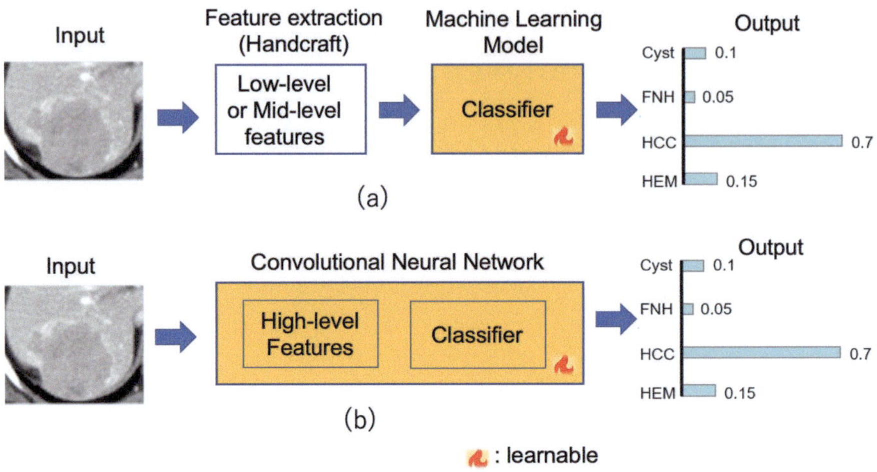

Fig. 1.1 Comparison of traditional machine learning and deep learning approaches

In contrast, DL is a data-driven technique that automates the feature extraction process. Unlike traditional methods, DL automatically extracts the ideal features from the data through several iterations of optimization as shown in Fig. 1.1b. This approach requires a substantial amount of training data to effectively learn and extract high-level feature representations. By minimizing manual intervention, DL enables the automatic extraction of intricate patterns in high-dimensional data, reducing dependence on human expertise and enhancing the ability to discover complex relationships within the data [2]. Figure 1.1 provides a conceptual overview of the key differences between traditional ML and DL techniques.

The development of DL models consists of two phases. The first phase is the training phase, during which the model learns from data. In a supervised training scheme, annotated or labeled (ground truth) training data is provided for the model learning. We can train a DL model for a specific task by computing a loss function (objective loss) that quantifies the difference between the predicted output and the ground truth label. The second phase is the testing or inference, where the trained model is evaluated on test data to assess its performance and generalization ability. Understanding how input samples (e.g., images or other data) are processed with the fundamental building blocks of a deep network is essential for designing effective DL models and gaining insights into the learning process [3]. The fundamental design and key components of DL models include multi-layer perceptron (MLPs), convolutional neural networks (CNNs) and some other functioning unit such as regularization techniques, activation functions, pooling operations, and loss functions.

The following subsections explore the fundamental building blocks of DL, which serve as essential components for constructing neural network architectures. These fundamental concepts provide insights into the functioning and underlying mechanisms of DL-based intelligent systems.

1.2 Multi-layer Perceptron (MLP)

A MLP network is a computational model inspired by the structure and functioning of the human brain. MLP is composed of layers of interconnected neurons or nodes, each layer having learnable parameters (weights) and performing operations on input data [4]. Figure 1.2 shows a simple structural representation of neural network layer. Following are the key components of MLP network:

Neurons (Nodes): Neurons are the fundamental processing units in a neural network. They receive input signals, perform computations, and produce output signals. Neurons allow a neural network to model relationships between input and output.

Layers: Neurons are organized into layers, with three main types: Input Layer (receives data), Hidden Layers (perform feature processing/pattern extraction), and Output Layer (generates predictions).

Weights: Neurons are connected to each other through weighted connections. Weights determine the strength of these connections and are updated during training according to specific tasks and datasets.

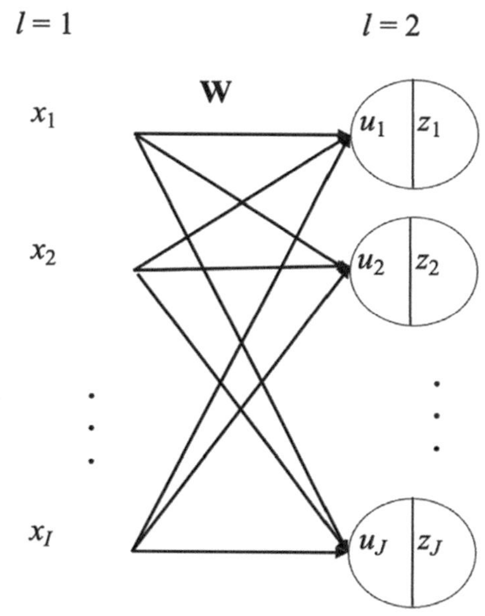

Fig. 1.2 A simple structural representation of neural network layer

Biases: Biases are added to the weighted sum before applying the activation function. They are useful in modeling complex functions, reducing prediction errors and helping the network to make more accurate predictions.

Activation Function: The activation function of a neuron determines its output based on the weighted sum of its inputs. Common activation functions include the Sigmoid, ReLU (Rectified Linear Unit) [5], and softmax functions. Activation functions introduce non-linearity, allowing the network to model complex relationships in the data.

Forward and Backward Pass: During the forward pass, data is passed through the neural network layers to generate a prediction. The backward pass (backpropagation) is the process of adjusting weights and biases to reduce the prediction error. Here the objective loss is calculated, i.e., the difference between the predicted and ground truth to optimize the network [6].

A mathematical formulation for the operation of a neuron in the Multi-Layer Perceptron (MLP) is given below:

$$\mathbf{u} = \mathbf{w} \cdot \mathbf{x} + \mathbf{b} \tag{1.1}$$

$$\mathbf{w} = \begin{bmatrix} w_{11} & w_{12} & \dots & w_{1n} \\ w_{21} & w_{22} & \dots & w_{2n} \\ \vdots & \vdots & \ddots & \vdots \\ w_{m1} & w_{m2} & \dots & w_{mn} \end{bmatrix} \tag{1.2}$$

Here, $\mathbf{w}$ is the weight matrix and $\mathbf{b}$ is the bias term. $\mathbf{x} = [x_1, x_2, \dots, x_I]^T$ and $\mathbf{u} = [u_1, u_2, \dots, u_J]^T$, where I and J are the number of input and output neurons, respectively. The weight matrix $\mathbf{w}$ determines the strength of the connection between the input and the neuron, while the bias $\mathbf{b}$ helps to shift the output, enabling the neuron to model more complex functions. The weighted sum (u_j) is then passed through an activation function to produce the final output. The output z_j can be given by $z_j = f(u_j)$, where $f(\cdot)$ is an activation function, which will be described in Sect. 1.3.6.

1.3 Convolution Neural Networks (CNNs)

CNNs are a class of DL models specifically designed for processing data with a grid-like topology, such as images. Leveraging the power of convolutional operations, CNNs can automatically and adaptively learn spatial hierarchies of features, making them highly effective for tasks like image recognition, object detection, and segmentation. CNNs have revolutionized the field of computer vision by achieving significant improvements in accuracy and performance over traditional ML methods.

1.3.1 Architecture

CNNs are primarily designed for visual tasks though it can also be applied to non-visual tasks such as time-series analysis, text processing, and graph data analysis. They are composed of several key components, including convolutional layers for feature extraction, pooling layers for downsampling, activation functions such as Rectified Linear Unit (ReLU) [5] layers for introducing nonlinearity, and Batch Normalization [7] layers to improve training stability. CNNs learn in a hierarchical manner, advancing from detecting basic features (e.g. edges, textures) to more complex features (e.g., object semantics). Throughout this learning process, CNNs optimize various parameters (weights) to identify relevant features.

1.3.2 Convolution Operation and Filter

In convolutional operations, CNNs utilize a set of learnable filters or kernels that are applied to the input data. These filters are rectangular sliding windows with numerical values representing weight coefficients. During the convolution operation, these filters slide over the input, conducting element-wise multiplications and summing the results. The weight values within these filters can either be initialized with small random values or derived from pretrained weights obtained from large datasets, and they are subsequently updated during the task-specific training process. CNNs use backpropagation to update these weights and recognize local patterns, minimizing the error between the predicted output and the ground truth. Figure 1.3 illustrates a simple example of the 3×3 convolution (or filtering) operation. In this example, kernel filters are applied to an input image to extract specific features. To detect edges, horizontal and vertical edge detection filters based on the Sobel operator are used. The process begins with the filter placed at the top-left corner of the input tensor. As the filter slides across the input tensor, it performs element-wise multiplication at each position. The resulting values are then summed to produce the corresponding entry in the output matrix. This convolution operation can be mathematically expressed as:

$$u(i, j) = \sum_{l=0}^{K-1} \sum_{m=0}^{K-1} w(l, m) \cdot x\left(i + l - \left\lfloor \frac{K}{2} \right\rfloor, j + m - \left\lfloor \frac{K}{2} \right\rfloor\right) \tag{1.3}$$

where $u(i, j)$ is the output value of the convolution operation at position (i, j). $w(l, m)$ is the kernel (filter) value at position (l, m). $x(i, j)$ is the input value (e.g., image or feature map) at position (i, j). K is the size of the convolution kernel, assuming a square kernel with dimensions $K \times K$. l is the row index of the kernel element, ranging from 0 to $K - 1$. m is the column index of the kernel element, ranging from 0 to $K - 1$. $\lfloor \frac{K}{2} \rfloor$ is the floor of $K/2$, used to center the kernel over the input pixel.

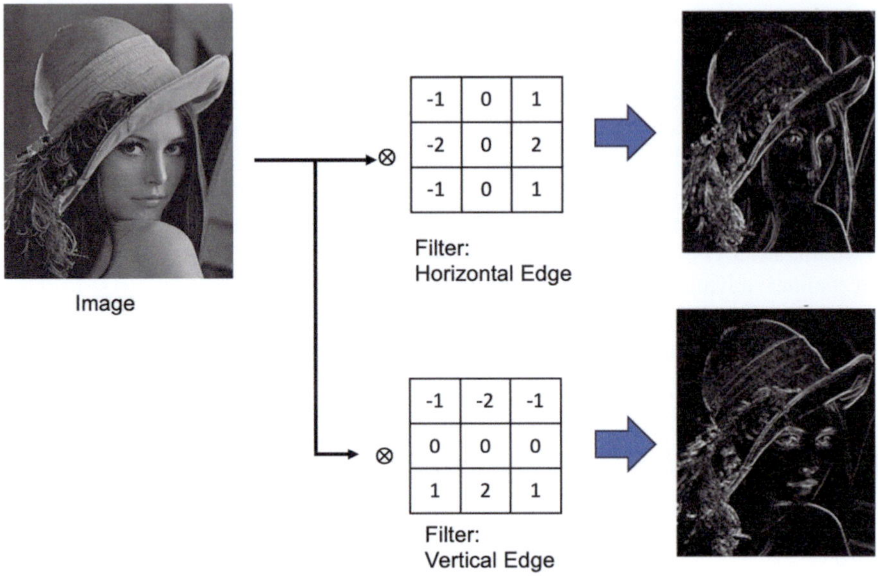

Fig. 1.3 An example of convolution operation using kernels

1.3.2.1 Stride

Stride refers to the step size by which the convolutional filter moves across the input data during the convolution operation. A stride of 1 means that the filter shifts by one pixel at a time, covering every possible position of the input. Increasing the stride results in fewer filter applications and a smaller output feature map, effectively reducing the spatial dimensions of the data. For example, a stride of 2 skips every other pixel, allowing the network to downsample the input while preserving computation efficiency. The choice of stride impacts the level of detail captured in the feature maps, with smaller strides retaining more spatial information and larger strides focusing on broader patterns. Figure 1.4 illustrates stride operations with stride values of 1 and 2.

1.3.2.2 Padding

Padding is a technique used in convolutional operations to preserve the spatial dimensions of the input data. By adding extra pixels (often zeros) around the edges of the input, the filter can effectively cover border elements without reducing the size of the resulting feature map. This approach ensures that edge information is preserved, maintaining the structural integrity of the data across successive layers. Figure 1.5 shows padding operation.

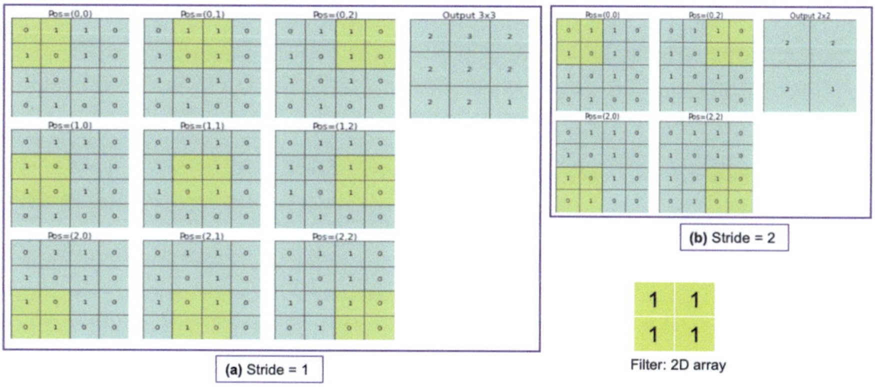

Fig. 1.4 Examples of stride operations with stride values of 1 (**a**) and 2 (**b**)

Fig. 1.5 Example of padding operation

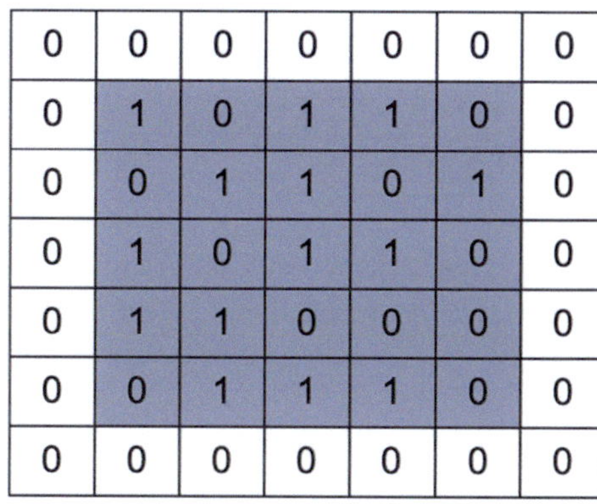

1.3.3 Pooling Layer

Pooling operations reduce the spatial dimensions of a feature map while retaining its most important information. This is useful in minimizing computation cost and complexity while obtaining the essential information. Max-pooling and average pooling operations are two commonly-used approaches for retaining the most important information within a local region.

The max-pooling operation selects the maximum value from the defined sliding window size. In max pooling, the maximum value within a defined region (e.g., an $R \times C$ window) is selected, effectively capturing the most salient feature in that region:

$$u(i, j) = \max_{0 \le r < R,\, 0 \le c < C} x(i + r, j + c) \tag{1.4}$$

where $u(i, j)$ is the output value at position (i, j) after max pooling. $x(i + r, j + c)$ is the input value at position $(i + r, j + c)$ within the pooling window. R is the height of the pooling window. C is the width of the pooling window. max is the maximum operation, which selects the largest value from the pooling window.

In average pooling, the output is the mean value of the inputs within the region, leading to a more smoothed representation:

$$u(i, j) = \frac{1}{R \times C} \sum_{r=0}^{R-1} \sum_{c=0}^{C-1} x(i + r, j + c) \tag{1.5}$$

where $u(i, j)$ is the output value at position (i, j) after average pooling. $x(i + r, j + c)$ is the input value at position $(i + r, j + c)$ within the pooling window. R is the height of the pooling window. C is the width of the pooling window. These pooling operations help in reducing computation, controlling overfitting, and providing a degree of translation invariance to the network. Figure 1.6 shows an example of max-pooling operation.

1.3.4 Feature Map Size

After performing a convolution with a kernel of size K, a padding of P, and a stride of S, the dimensions of the resulting feature map can be determined using a standard formula. For an input of size N, the output dimension M is given by:

$$M = \frac{N - K + 2P}{S} + 1 \tag{1.6}$$

This formulation ensures that by appropriately setting padding, stride, and kernel size, one can control how much spatial information is retained and how the feature maps evolve through successive layers. Floor or ceiling operations are used to ensure that divisions resulting in non-integer values are rounded appropriately.

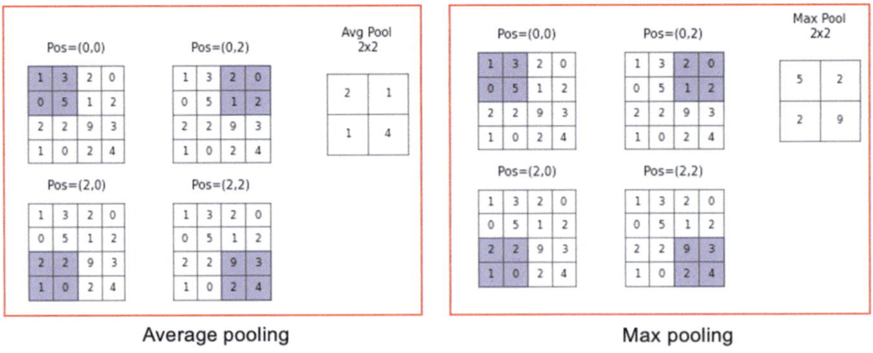

Fig. 1.6 Example of max pooling and average pooling with a stride of 2 and a kernel size of 2

Figure 1.7 illustrates the relationship among N, K, and M for the case where $S = 1$ and $P = 0$. In the example shown in Fig. 1.4a, with $N = 4$, $K = 2$, $S = 1$, and $P = 0$, the output size M is 3. In contrast, for the example in Fig. 1.4b, with $N = 4$, $K = 2$, $S = 2$, and $P = 0$, the output size M is 2.

1.3.5 Fully Connected Layer

Fully connected layers, often denoted as FC layers, are employed to perform downstream tasks following the extraction of features through CNNs. These layers are responsible for conducting classification or regression tasks based on the learned features. Neurons within the fully connected layer are connected to all neurons in both the preceding and following layers, resulting in dense interconnections. FC layers play a crucial role in modeling complex data relationships and are instrumental in task-specific operations, such as generating final classification probabilities or regressing the size of objects. Figure 1.8 illustrates an example of the fully connected layer.

1.3.6 Activation Functions

Activation functions play a crucial role in neural networks by introducing nonlinearity into the model, enabling it to learn and approximate complex relationships within the data. Without activation functions, a neural network would behave as a linear

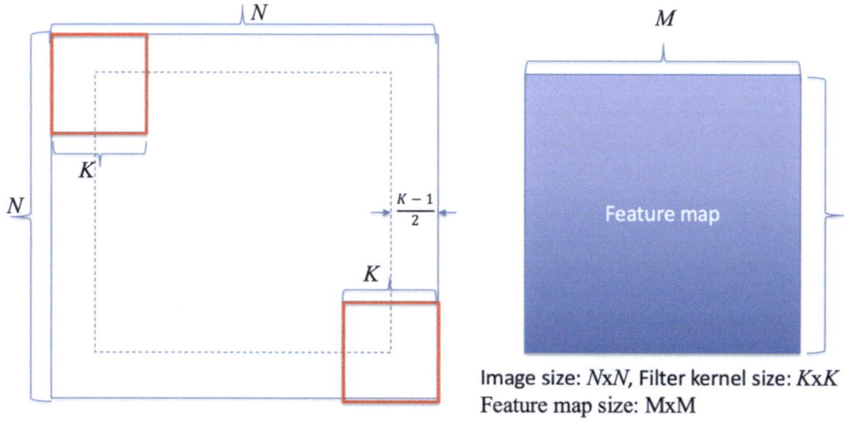

Fig. 1.7 The relationship among N, K, and M for the case where $S = 1$ and $P = 0$

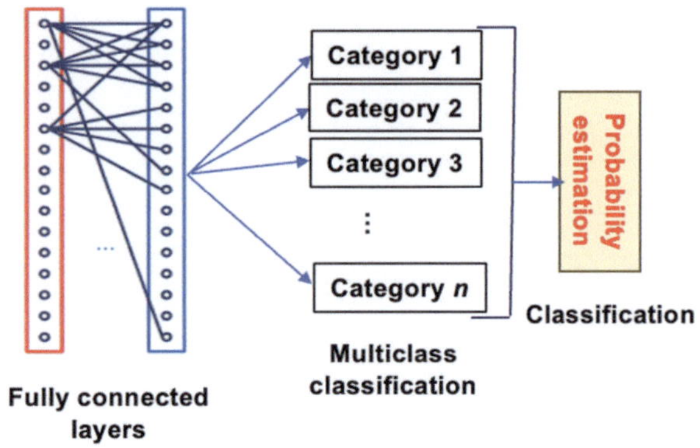

Fig. 1.8 Example of fully connected layer for classification

model, limiting its capacity to solve real-world problems that often involve nonlinear data patterns. These functions are normally applied after convolutional layers and do not include any learnable parameters or weights. These functions determine and adjust the output of the convolutional layer for a given input. With the DL frameworks, ReLU, Tanh, and Sigmoid activation functions are commonly used. Common activation functions include.

1.3.6.1 Rectified Linear Unit (ReLU)

ReLU is one of the most popular activation functions due to its simplicity and effectiveness. It outputs zero for any negative input and returns the input directly for any positive value. The formula for ReLU is:

$$z = f(u) = \mathrm{ReLU}(u) = \max(0, u) \tag{1.7}$$

where u is the input to the activation function, typically the output of the previous layer. z is the output of the activation function. This allows the model to introduce nonlinearity without complicating the computation, which makes it effective for training deep networks.

1.3.6.2 Sigmoid Function

The sigmoid function maps input values to a range between 0 and 1. It is particularly useful for binary classification tasks. However, the sigmoid function can lead to vanishing gradient problems, especially when the input is far from zero, as the gradients become very small. The formula for the sigmoid function is:

$$z = f(u) = \sigma(u) = \frac{1}{1 + e^{-u}} \tag{1.8}$$

The Sigmoid function, denoted as $\sigma(u)$, maps any real-valued input into the range (0, 1). e is Euler's number, a mathematical constant that serves as the base of natural logarithms. It is commonly used as an activation function in neural networks to introduce non-linearity and is particularly useful for binary classification tasks because its output can be interpreted as a probability.

1.3.6.3 Hyperbolic Tangent (Tanh)

The Tanh function is similar to the sigmoid function but outputs values in the range of -1 to 1, which makes it zero-centered and helps improve convergence during training. It is generally preferred over the sigmoid function when zero-centered data is desired. The formula for the tanh function is:

$$z = f(u) = \tanh(u) = \frac{\sinh(u)}{\cosh(u)} = \frac{e^u - e^{-u}}{e^u + e^{-u}} \tag{1.9}$$

The Tanh function maps any real-valued input u to the range $(-1, 1)$. It is a scaled and shifted version of the Sigmoid function, providing a symmetric output around zero. This symmetry makes it useful in neural networks to model outputs that can have both positive and negative values, aiding faster convergence during training.

1.3.6.4 Leaky ReLU

This is a variant of ReLU that allows a small, non-zero gradient for negative input values. This helps prevent issues where some neurons can become inactive and never recover, a problem sometimes called the "dying ReLU". The formula for Leaky ReLU is:

$$z = f(u) = \begin{cases} u, & \text{if } u \geq 0 \\ \alpha u, & \text{if } u < 0 \end{cases} \tag{1.10}$$

The Leaky ReLU function introduces a small slope α for negative input values instead of setting them to zero, as in standard ReLU. This allows the gradient to flow through the network even for negative inputs. The parameter α typically has a small positive value, such as 0.01, but it can be adjusted depending on the application. The function maps inputs to the range $(-\infty, \infty)$.

1.3.6.5 Gaussian Error Linear Unit (GeLU)

The GeLU is an activation function widely used in recent DL architectures, such as Transformers. It provides a smooth and differentiable non-linearity that allows for better gradient flow during training. GeLU is particularly effective because it combines the properties of both the ReLU (Rectified Linear Unit) and Sigmoid functions, balancing linear and non-linear regions for improved learning dynamics. The GeLU function can be expressed mathematically as:

$$z = f(u) = \text{GeLU}(u) = u \cdot \frac{1}{2}\left[1 + \text{erf}\left(\frac{u}{\sqrt{2}}\right)\right] \tag{1.11}$$

where erf(·) is the Gaussian error function, defined as, $\text{erf}(u) = \frac{2}{\sqrt{\pi}} \int_0^u e^{-t^2}\, \text{d}t$, which captures the cumulative distribution of the Gaussian distribution.

1.3.6.6 Swish Activation Function

The Swish activation function improves the performance of DL models by introducing a smooth and non-monotonic relationship between inputs and outputs. Its unique ability to retain small negative values (due to the sigmoid scaling) enhances gradient flow during backpropagation, reducing issues like vanishing gradients. The multiplication of the input with the sigmoid function allows Swish to act as a dynamic gate, modulating the input based on its value. This makes Swish effective in capturing complex patterns and improves training stability, especially in very deep neural networks. The Swish activation function is defined as:

$$z = f(u) = \text{Swish}(u) = u \cdot \sigma(u) \tag{1.12}$$

where $\sigma(u)$ is the sigmoid function, defined as $\sigma(u) = \frac{1}{1+e^{-u}}$, which scales the input u into a range between 0 and 1. Swish activation function can be generalized using a hyperparameter.

$$z = f(u) = \text{Swish}(u) = u \cdot \sigma(\beta u) \tag{1.13}$$

here, β is an optional parameter in the generalized Swish function ($u \cdot \sigma(\beta u)$) that controls the steepness of the sigmoid component. When $\beta = 1$, it reduces to the standard Swish function.

Figure 1.9 shows an comparison of six activation functions: (a) Sigmoid, (b) Tanh, (c) ReLU, (d) Leaky ReLU, (e) GeLU, and (f) Swish. The choice of activation function significantly impacts the performance and efficiency of DL models, with each function offering unique properties. ReLU is the most commonly used due to its simplicity and computational efficiency, making it suitable for general-purpose tasks like image classification and object detection. However, it can suffer from the "dying ReLU" problem, which Leaky ReLU addresses by allowing a small gradient for negative inputs, making it better for deeper networks with diverse features. Sigmoid and Tanh, while older, remain useful in specific scenarios; Sigmoid is ideal for binary classification tasks due to its probabilistic interpretation, while Tanh is preferred for zero-centered data and models like RNNs, where both positive and negative outputs are meaningful. More advanced functions like GeLU and Swish excel in modern architectures. GeLU, widely used in Transformer-based models, provides a

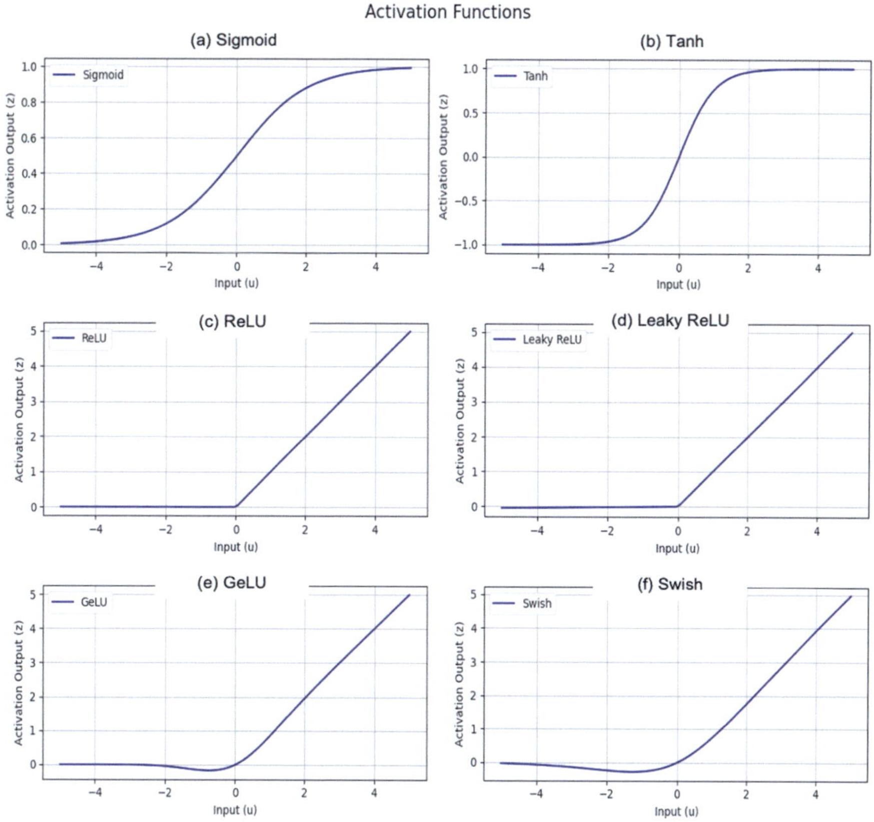

Fig. 1.9 Comparison of six activation functions: **a** Sigmoid, **b** Tanh, **c** ReLU, **d** Leaky ReLU, **e** GeLU, and **f** Swish. Each subplot demonstrates the relationship between the input (u) and the activation output (z), highlighting the unique characteristics and behaviors of each function

smooth and probabilistic gating mechanism that enhances gradient flow and feature learning. Similarly, Swish, with its smooth and non-monotonic nature, improves gradient flow and model convergence, making it a popular choice in architectures like EfficientNet. While ReLU and its variants are computationally efficient and sufficient for many tasks, GeLU and Swish offer superior performance for cutting-edge models requiring more nuanced gradient behavior and enhanced learning dynamics. The choice of activation function depends on the task, model architecture, and the need for computational efficiency versus advanced learning capabilities.

1.3.7 Softmax

The softmax function is commonly used in the output layer of classification models to convert logits into probabilities, assigning a value between 0 and 1 for each class such that the sum of all probabilities equals 1. The formula for the softmax function is:

$$\hat{y}_i = \sigma(z_i) = \frac{e^{z_i}}{\sum_{j=1}^{N} e^{z_j}}, \quad \text{for } i = 1, \ldots, N \tag{1.14}$$

The Softmax function converts the logits (z_i) into probabilities that sum to 1. It is widely used in the output layer of neural networks for multi-class classification tasks.

1.4 Loss Function

Loss functions (cost function or loss objective) quantify the difference between the predicted values (outputs of the model) and the actual ground truth values. In DL, the task of a loss function is to measure the performance of a neural network on a specific task to update weights and other parameters. The optimization algorithm calculates the gradient with respect to the loss to update the parameters of the model to adjust its parameters (weights and biases in neural networks) using optimization techniques [8]. The model iteratively updates these parameters with the aim of minimizing loss. This iterative process continues until the model achieves a satisfactory level of performance. The choice of particular loss function depends on the nature of specific task, such as classification, regression or more specialized task like object detection. We use cross-entropy loss for classification, while Mean Squared Error (MSE) is commonly used for regression tasks. Following are some commonly used loss functions for DL frameworks.

1.4.1 Binary Cross-Entropy (BCE)

BCE is mainly used for binary classification tasks. Basically, it may involve classifying the entire area or image or any input data into two categories. Additionally, such two-class classification can involve classifying individual pixels into two categories, normally back-ground or foreground for segmentation tasks.

$$BCE(y, \hat{y}) = -\left[y \cdot \log(\hat{y}) + (1 - y) \cdot \log(1 - \hat{y})\right] \tag{1.15}$$

where $BCE(\cdot)$ is the Binary Cross-Entropy Loss, y is the ground truth binary label (0 or 1), $\hat{y}$ is the predicted labels of the given input tensor. The ground truth is calculated by passing the output logits through a Sigmoid or Softmax function. This formula quantifies the difference between the predicted labels and the ground truth binary labels. The magnitude of the loss value determines the strength of optimization procedure to minimize this loss during training by fine-tuning model's parameters.

1.4.2 *Cross-Entropy*

The cross-entropy loss is widely used for multi-class classification tasks. This loss quantifies the difference between the predicted output and the ground truth labels in a multi-class setting. This applies to a range of tasks, such as classifying pixel values (image segmentation), entire images (image classification), or patches/regions contained within bounding boxes (object detection).

$$CE(y, \hat{y}) = -\sum_{i=1}^{c} y_i \log(\hat{y}_i) \tag{1.16}$$

where $CE(\cdot)$ is the cross-entropy loss, y_i is ground truth class label, y_i is equal to 1 if the input data belongs to class i and 0 otherwise. $\hat{y}_i$ denotes the predicted label of class i. The sum is taken over all classes in scenario of the multi-class problem.

1.4.3 *Mean Squared Error (MSE)*

The Mean Squared Error (MSE) loss, also known as L_2 loss, is primarily used in regression tasks. It quantifies the distance between predicted values and ground truth values by calculating the squared differences. MSE loss is effective in tasks involving continuous predictions, such as regression, image enhancement, and super-resolution. For object detection tasks, MSE can measure errors in predicting bounding box coordinates, such as the center point (x, y), width, and height. However, MSE is sensitive to outliers as it disproportionately penalizes larger errors.

$$\text{MSE} = \frac{1}{n}\|\hat{\mathbf{y}} - \mathbf{y}\|^2 \tag{1.17}$$

here, n is the number of data points, $\mathbf{y}$ is the ground truth, and $\hat{\mathbf{y}}$ is the predicted result.

1.4.4 Mean Absolute Error (MAE)

Mean Absolute Error (MAE or L_1 norm) loss measures the average absolute difference between predicted values and ground truth values. MAE provides a uniform penalty for errors and is less sensitive to outliers.

$$\text{MAE} = \frac{1}{n}\|\hat{\mathbf{y}} - \mathbf{y}\|_1 \tag{1.18}$$

here, n is the number of data points, $\mathbf{y}$ is the ground truth, and $\hat{\mathbf{y}}$ is the predicted result.

1.5 Optimization

Optimization is a crucial process in training neural networks, as it aims to minimize the loss function by adjusting the model's parameters (e.g., weights and biases) to achieve the best performance for a given task. In DL, optimization algorithms are used to efficiently navigate the complex and high-dimensional loss landscapes, ensuring that the model converges to an optimal or near-optimal solution. Figure 1.10 shows the optimization process. There are several key components and techniques involved in the optimization process:

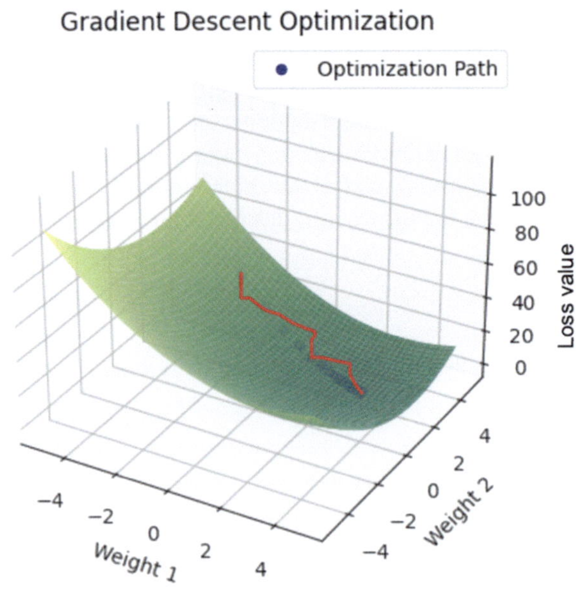

Fig. 1.10 Optimization process

1. **Gradient-Based Methods**: Gradient Descent is the backbone of most optimization techniques used in DL. It leverages the gradient of the loss function with respect to model parameters to iteratively update the parameters in the direction that minimizes the loss. Variations of Gradient Descent include Batch Gradient Descent, Stochastic Gradient Descent (SGD), and Mini-batch Gradient Descent, each offering different trade-offs between stability and computational efficiency.
2. **Learning Rate**: The learning rate is a hyperparameter that determines the step size for parameter updates. Selecting an appropriate learning rate is essential, for example, if it is too high, the training process may diverge. While if it is too low, convergence may be slow and may result in getting stuck in local minima.
3. **Momentum**: Momentum is an enhancement to gradient-based methods that helps accelerate convergence by adding a fraction of the previous update to the current one. This mechanism helps smooth out the optimization trajectory and mitigates oscillations, allowing the optimizer to move more quickly towards the minimum.
4. **Adaptive Optimization Algorithms**: Adaptive methods like AdaGrad [9], RMSprop [10], and Adam [11] dynamically adjust the learning rate for each parameter based on past gradients. These optimizers are particularly useful for dealing with sparse data and complex loss surfaces, as they adaptively scale learning rates to maintain balance between convergence speed and stability.
5. **Regularization**: Regularization techniques such as L_1 regularization, L_2 regularization, and dropout are often employed during the optimization process to prevent overfitting. These techniques add penalties to the loss function or drop connections during training, thereby helping the model generalize better to unseen data.
6. **Weight Initialization**: Proper initialization of the model's parameters plays a significant role in effective optimization. Techniques like Xavier [12] and He [13] initialization are used to ensure that weights are set in a way that prevents gradients from vanishing or exploding, allowing for smoother and more efficient convergence.

The choice of optimization algorithm, learning rate schedule, and regularization methods significantly influences the effectiveness of training DL models. Proper optimization ensures not only that the model converges to a solution but also that the solution generalizes well to unseen data.

1.5.1 Gradient Decent

In DL, Gradient Descent is a fundamental optimization algorithm used to minimize a loss function $L(\boldsymbol{\theta})$, where $\boldsymbol{\theta}$ represents the model's parameters (e.g., weights and biases). The algorithm iteratively updates the parameters in the direction of steepest descent, which is the opposite direction of the gradient of the loss function at the current point. The general update rule for Gradient Descent can be expressed as:

$$\boldsymbol{\theta}^{(k+1)} = \boldsymbol{\theta}^{(k)} - \alpha \nabla L(\boldsymbol{\theta}^{(k)}) \tag{1.19}$$

where

- $\boldsymbol{\theta}^{(k)}$ represents the parameters at iteration k,
- α is the learning rate, a hyperparameter that controls the step size taken in the direction of the negative gradient
- $\nabla L(\boldsymbol{\theta}^{(k)})$ is the gradient of the loss function L evaluated at $\boldsymbol{\theta}^{(k)}$. This gradient indicates the direction of steepest ascent of the loss function. Therefore, by moving in the opposite direction (i.e., subtracting the gradient), we aim to reduce the loss function.

Implementation Steps

1. **Initialize Parameters**: Start with a set of initial parameter values $\boldsymbol{\theta}^{(0)}$.
2. **Compute Gradient**: Calculate the gradient $\nabla L(\boldsymbol{\theta}^{(k=0)})$ at the current parameter values $\boldsymbol{\theta}^{(k)}$.
3. **Update Parameters**: Update the parameters using the update rule:

$$\boldsymbol{\theta}^{(k+1)} = \boldsymbol{\theta}^{(k)} - \alpha \nabla L(\boldsymbol{\theta}^{(k)}) \tag{1.20}$$

4. **Repeat**: Repeat steps 2 and 3 until convergence (i.e., until the loss function stops decreasing significantly or a predefined number of iterations is reached).

1.5.2 Variations of Gradient Descent

Gradient Descent, in its standard form, calculates the average gradient of the cost function for the entire dataset at each iteration. While effective, this approach can be computationally expensive and unsuitable for DL tasks involving large datasets. To address these limitations, several variations of Gradient Descent have been proposed:

- **Batch Gradient Descent**: Calculates the gradient based on the entire training set at each iteration. This provides accurate gradient estimates but can be computationally expensive for large datasets.
- **Stochastic Gradient Descent (SGD)**: Calculates the gradient based on randomly selected training example at each iteration. While this speeds up convergence, it introduces noise in the gradient estimation, leading to a less stable optimization trajectory.
- **Mini-batch Gradient Descent**: A compromise between Batch and Stochastic Gradient Descent, this method calculates the gradient based on a small subset (mini-batch) of the training set at each iteration. It balances the computational efficiency of SGD with the stability of Batch Gradient Descent.

1.6 Deep Learning Frameworks

There are several DL frameworks available, each providing unique features and benefits that cater to different use cases and user preferences. Some of the most popular ones are described below:

- **PyTorch**: PyTorch is a Python-based DL framework developed by Facebook's AI Research lab [14]. It provides tensor computation capabilities with strong GPU acceleration, along with deep neural networks built on a tape-based autograd system. PyTorch is known for its ease of use, flexibility, and dynamic computational graph, which makes it highly popular in both academia and industry for research and prototyping. The framework also supports efficient deployment with tools like TorchServe.
- **TensorFlow**: Developed by the Google Brain team, TensorFlow is one of the most widely used DL frameworks [15]. It uses dataflow graphs to represent computation, making it highly efficient for large-scale ML and DL tasks. TensorFlow supports multiple languages, including Python, C++, and R, and is known for its versatility. It offers high-level APIs, such as Keras, as well as low-level functionalities for customization. TensorFlow Extended (TFX) also provides tools for end-to-end model deployment, making it popular in production environments.
- **Keras**: Keras is an open-source neural network library that provides a high-level interface for building and training DL models [16]. Initially developed as a standalone library, Keras is now integrated into TensorFlow as its official high-level API. Keras is known for its user-friendly, modular, and extensible interface, which allows users to quickly prototype and experiment with models. It is designed to make DL accessible to beginners while also offering advanced capabilities for experienced developers.
- **Caffe**: Developed by the Berkeley Vision and Learning Center (BVLC), Caffe is a DL framework that focuses on speed, modularity, and expression [17]. It is particularly popular for image classification tasks and is known for its efficient implementation, which makes it suitable for real-time applications. Caffe's model zoo provides a repository of pre-trained models, making it easy for users to experiment and apply models to their own tasks. However, it lacks some of the flexibility of newer frameworks like PyTorch and TensorFlow.
- **MXNet**: Apache MXNet is a DL framework that supports both symbolic and imperative programming, giving developers flexibility in how they design models [18]. MXNet is highly scalable and can be used for training models across multiple GPUs and machines. It has been adopted by Amazon as the default DL framework for AWS, and it provides good integration with cloud services. MXNet's efficient memory usage and ability to scale make it suitable for both research and industrial use cases.

These frameworks offers distinct advantages, and the choice of which to use often depends on the specific requirements of the project, such as ease of use, scalability, community support, or deployment capabilities.

1.7 Summary

This chapter provides an introduction to the fundamentals of DL techniques. These components are the core of DL frameworks and are designed to learn and represent complex relationships in data. DL networks are mathematical functions designed to process and learn from input data to make predictions. These networks include Multi-Layer Perceptron (MLP), Convolutional Neural Network (CNN), Fully Connected Layers (FC), and some other architectures. These networks can have different layer structures and connectivity patterns, making them highly adaptable to address diverse tasks. In the next chapter, we talk about different feature extractor CNN networks. Additionally, we explore the design and various design choice advantages of the state-of-the-art DL frameworks for tasks like classification, detection and semantic segmentation.

References

1. Chen, Y. W., & Jain, L. C. (Eds.). (2020). *Deep learning in healthcare*. Springer.
2. LeCun, Y., Bengio, Y., & Hinton, G. (2015). Deep learning. *Nature, 521*(7553).
3. Chollet, F. (2017). *Deep Learning with Python*. Manning Publications.
4. Bishop, C. M. (2006). *Pattern recognition and machine learning*. Springer.
5. Glorot, X., Bordes, A., & Bengio, Y. (2011). Deep sparse rectifier neural networks. In *Proceedings of the Fourteenth International Conference on Artificial Intelligence and Statistics (AISTATS-11)*
6. Goodfellow, I., Bengio, Y., & Courville, A. (2016). *Deep learning*. MIT Press.
7. Ioffe, S., & Szegedy, C. (2015). Batch normalization: accelerating deep network training by reducing internal covariate shift. In *Proceedings of the 32nd International Conference on Machine Learning (ICML-15)*
8. Ruder, S. (2016). *An overview of gradient descent optimization algorithms*. arXiv:1609.04747
9. Duchi, J., Hazan, E., & Singer, Y. (2011). Adaptive subgradient methods for online learning and stochastic optimization. *Journal of Machine Learning Research, 12*
10. Hinton, G., Srivastava, N., & Swersky, K. (2012). *Lecture 6e RMSProp: divide the gradient by a running average of its recent magnitude*. Coursera: Neural Networks for Machine Learning.
11. Kingma, D. P., & Ba, J. (2015). Adam: a method for stochastic optimization. In *Proceedings of the 3rd International Conference on Learning Representations (ICLR)*
12. Glorot, X., & Bengio, Y. (2010) Understanding the difficulty of training deep feedforward neural networks. In *Proceedings of the 13th International Conference on Artificial Intelligence and Statistics*.
13. He, K., Zhang, X., Ren, S., & Sun, J. (2015). Delving deep into rectifiers: Surpassing human-level performance on ImageNet classification. In *Proceedings of the IEEE International Conference on Computer Vision (ICCV)*.
14. Paszke, A., et al. (2019). PyTorch: an imperative style, high-performance deep learning library. In *Advances in neural information processing systems (NeurIPS)*
15. Abadi, M., et al. (2016). TensorFlow: a system for large-scale machine learning. In *Proceedings of the 12th USENIX Symposium on Operating Systems Design and Implementation (OSDI)*
16. Chollet, F., et al. (2015). *Keras*. GitHub Repository. https://github.com/fchollet/keras
17. Jia, Y., et al. (2014). Caffe: convolutional architecture for fast feature embedding. In *Proceedings of the ACM International Conference on Multimedia (MM)*
18. Chen, T., et al. (2015). *MXNet: a flexible and efficient machine learning library for heterogeneous distributed systems*. arXiv.

Chapter 2
Deep CNNs for Image Classification, Object Detection, and Segmentation

2.1 Introduction

This chapter focuses on various computer vision tasks utilizing deep learning. A deep learning network is trained using input samples along with their corresponding task-specific labels. These labeled training samples guide the learning process, enabling the model to perform specific tasks such as classification, object detection, and segmentation by calculating an value (objective loss function) that measures the discrepancy between the model's output and the expected result (ground truth). During training, the network's weights and biases are adjusted or fine-tuned to learn the accurate prediction tasks. As discussed in the previous chapter, several components play a pivotal role in facilitating this process. Once trained, the model can be applied to unlabeled test inputs to predict results on unseen images. Deep learning networks can perform different tasks in an end-to-end manner, including:

1. **Image Classification**: This task involves categorizing the entire image based on detected features or patterns without specifying the location of objects within the image.
2. **Object Detection**: Detection frameworks not only identify multiple objects or patterns within an image but also precisely locate them by drawing bounding boxes around each detected object.
3. **Semantic Segmentation**: This task assigns each pixel in the image a label from a predefined set, effectively classifying every pixel into a specific category and segmenting the entire image. Pixel-wise classification is particularly useful in applications where understanding the context and structure of the entire image is essential, such as in medical imaging or autonomous vehicle navigation.

Figure 2.1 illustrates the tasks of image classification, object detection, and segmentation. Recently, deep learning techniques have been widely used for these tasks and achieved state-of-the arts results. As outlined in Chap. 1, a deep neural network typically comprises two primary components: a feature extractor (commonly referred to as an encoder) and a downstream task head. These components are trained

Y. Chen et al., *Recent Advances in Deep Learning for Medical Image Analysis*, Intelligent Systems Reference Library 278, https://doi.org/10.1007/978-3-031-94791-9_2

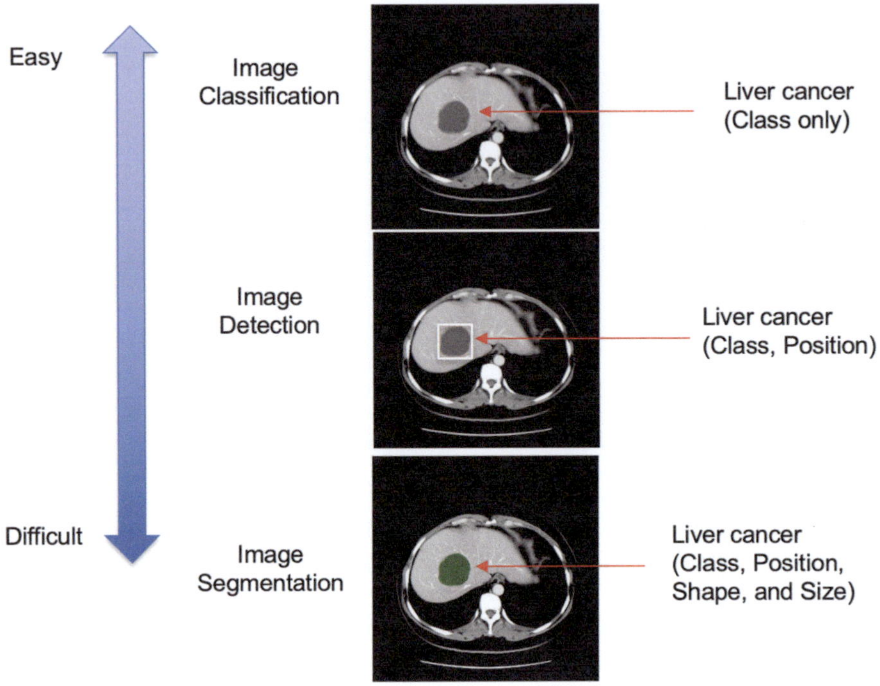

Fig. 2.1 Overview of computer vision tasks: image classification (categorizing entire images), object detection (locating and identifying objects with bounding boxes), and semantic segmentation (classifying each pixel to understand image context)

in an end-to-end fashion, enabling the optimization of filters for feature extraction tailored to specific downstream tasks, such as image classification, object detection, and semantic segmentation.

Among various feature extraction techniques, Convolutional Neural Networks (CNNs) have emerged as the most prominent and effective architecture. CNNs are composed of a hierarchical series of layers, including convolutional, pooling, and activation layers, that collectively learn to identify important spatial patterns within the input data. The primary advantage of CNNs lies in their ability to handle large datasets and automatically extract hierarchical feature representations without requiring manual feature engineering. This learning paradigm is commonly referred to as data-driven learning.

In CNNs, feature representations can be categorized into three general levels. Low-level features, learned in the early layers, typically capture basic elements such as edges, corners, and color gradients. Middle-level features, extracted in intermediate layers, represent more complex structures like textures and simple shapes. High-level features, obtained from the deeper layers of the network, encode abstract concepts such as object parts or entire object classes, enabling the model to perform high-level reasoning tasks. Figure 2.2 illustrates the progression from low- to high-level

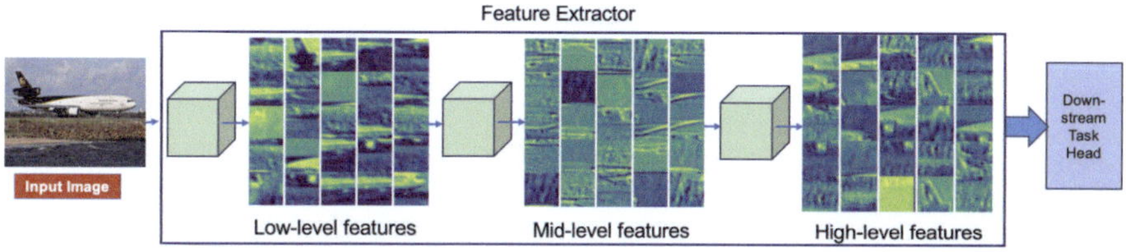

Fig. 2.2 Illustration of feature extraction levels in CNNs: low-level features (e.g., edges and colors), mid-level features (e.g., textures and shapes), and high-level features (e.g., abstract patterns and object representations

features in a CNN, showcasing how the model transforms raw input into meaningful representations.

Over the years, numerous CNN-based architectures have been proposed for automatically extracting feature patterns and recognizing structural details in images. These architectures serve as the foundation for a wide variety of computer vision tasks, including classification, detection, and segmentation. In the following subsections, we present an overview of several widely used CNN frameworks and describe how they are employed for different tasks in modern computer vision systems.

2.2 Deep CNNs for Image Classification

2.2.1 Image Classification Task

A representative architecture of a deep neural network for image classification is illustrated in Fig. 2.3. The feature extractor (encoder) is composed of a sequence of convolutional layers, where the learnable kernel weights function as filters for feature extraction. These weights are optimized during the training process, embodying a data-driven approach. The features extracted by the encoder are subsequently passed through a fully connected (FC) layer, which serves as the classifier (downstream head). The number of neurons in the FC output layer corresponds to the number of target classes, with each neuron outputting the probability of the input image belonging to the corresponding class. In this manner, the network maps input images to their respective classes, assigning probabilistic scores to predefined categories. Figure 2.3 illustrates an example of a deep neural network designed for digit classification. The input to the network is an image of a digit. Since the number of target classes is 10 (corresponding to the digits 0 through 9), the probability of the input image belonging to one of these classes. For instance, if the input image is the digit "7," its corresponding label vector can be expressed as $\mathbf{y} = [0, 0, 0, 0, 0, 0, 0, 1, 0, 0]^T$. The network, which comprises both convolutional layers and fully connected (FC) layers, is trained using a dataset of training samples (pairs of images and their corresponding labels) and optimized via a loss function,

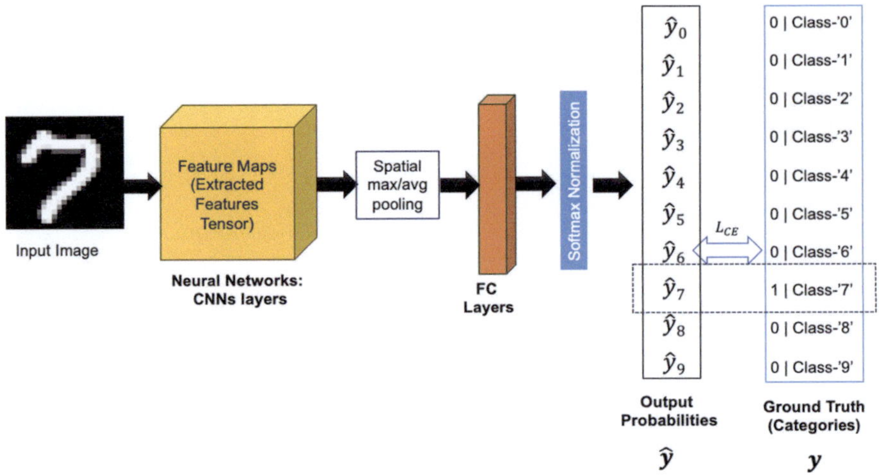

Fig. 2.3 Example of multiclass classification: calculating cross-entropy loss using Class-A as the ground truth, showcasing the relationship between model predictions and true labels

such as cross-entropy or mean squared error (MSE), as defined in Eqs. (1.16) and (1.17), respectively. During the inference stage, the input image is classified into the class associated with the highest output probability.

2.2.2 *Regression Task*

Similar network architectures can also be applied to regression tasks. Figure 2.4 demonstrates an example of a deep neural network designed for facial landmark detection. Assuming there are n landmarks, each landmark is represented by its coordinates (x, y). Consequently, the output layer consists of $2n$ neurons, which correspond to the coordinates of the landmarks in the form $[(x_1, y_1), (x_2, y_2), \ldots, (x_n, y_n)]$. For example, consider a case with three landmarks located at coordinates (50, 100), (150, 100), and (100, 50), respectively. The corresponding label vector can be expressed as $\mathbf{y} = [50, 100, 150, 100, 100, 50]^T$. The network is trained using a dataset of training samples, which consist of pairs of images and their corresponding landmark labels and is optimized via a loss function of mean squared error (MSE), as defined in Eq. (1.17). During the inference stage, the network outputs the predicted coordinates of the landmarks. In the following subsection, we provide an overview of the architectural structure and key design principles of some of the most widely adopted Convolutional Neural Network (CNN) frameworks. Specifically, we discuss AlexNet, VGG, and ResNet, highlighting their layer configurations, computational efficiency, and advancements in feature extraction.

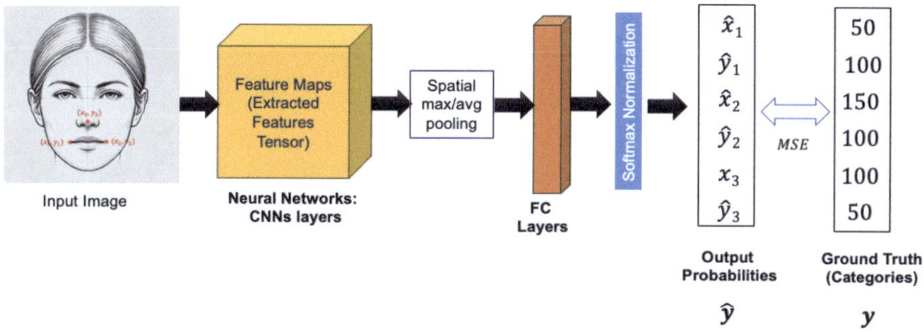

Fig. 2.4 Sketch of a human face used as a base for facial landmark annotation. This illustration shows as a reference for demonstrating regression-based facial landmark detection techniques

2.2.3 AlexNet

In 2012, AlexNet was proposed by Alex Krizhevsky et al. in their work titled "ImageNet Classification with Deep Convolutional Neural Networks" [1]. AlexNet significantly outperformed the state-of-the-methods at the time and won the ImageNet Large Scale Visual Recognition Challenge (ILSVRC) [2]. Their proposed network reduced the top-5 error rate from 26% to 15.3%, a significant improvement compared to previous machine learning models. AlexNet effectively demonstrated feature capturing capabilities of deep neural networks, especially in handling large volume of data. It encapsulates high-level feature representation using a deep architecture that consists of several CNN layers. AlexNet also introduced several key components into modern neural network architectures. It was one of the first networks to use the ReLU activation function instead of using the Sigmoid or Tanh functions. AlexNet implemented dropout and spatial pooling methods to control over-fitting and reduce network size. Additionally, its design utilized the parallel computational power of GPUs, a relatively new approach at the time. This achievement significantly drew attention of researcher towards deep learning, establishing it as a leading method in the field of AI. Many advanced networks such as VGG [3], GoogLeNet [4], and ResNet [5] are inspired by AlexNet, improving and building on the new concepts to use CNNs more effectively.

The architecture of the AlexNet network is shown in Fig. 2.5. AlexNet is composed of 8 layers, in which the first 5 layers are convolutional layers followed by 3 fully connected layers. This architecture is significantly deeper and larger than LeNet-5 [6]. AlexNet uses parallel computing on GPUs, dividing the network parameters across two GPUs. AlexNet operates with connected layers across separate GPUs at layers 2, 4, and 5, while layer 3 is connected to the two GPUs. Corresponding convolutional layers on different GPUs use the same size kernel.

The first convolutional layer convolves the input image with 96 kernels of size 11×11, and subsequent convolutional layers employ smaller filters of sizes 5×5 and 3×3 while increasing in depth (number of channels). The number of channels for

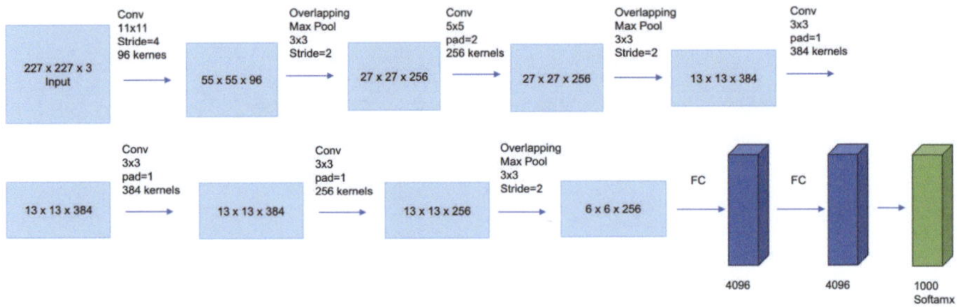

Fig. 2.5 AlexNet architecture: a deep CNN with 8 layers (5 convolutional layers followed by 3 fully connected layers), demonstrating feature extraction and classification capabilities

the other 4 layers are [256, 384, 384, 256]. Max pooling operations are used after some convolutional layers (i.e., layers 1, 2, and 5) to reduce the spatial size of the feature representation and the number of parameters. It reduces computations in the network while maintaining depth. The size of the pooling window is 3×3 with a stride of 2 between adjacent windows. For generating the same output size, an overlapping max pooling layer with a window size of 2×2 is employed at some places. Overlapping pooling layers function similarly to max pooling layers, with the difference that the windows of adjacent pooling elements overlap. The generated output from layer 5 is fed into the fully connected layers. There are two subsequent fully connected layers in the architecture. Each of the fully connected layers has 4096 neurons. The output of the second fully connected layer is passed to a final fully connected layer, which provides logits for a 1000-class classification. The output of the final fully connected layer is fed into a 1000-class softmax function, which generates a probability distribution over the 1000 class labels. AlexNet demonstrated outstanding performance on the challenging ImageNet dataset, marking the beginning of deep learning networks in the field of computer vision.

2.2.4 VGG: Very Deep Convolutional Networks

The VGG network is introduced by Simonyan et al. in their paper "Very Deep Convolutional Networks for Large-Scale Image Recognition" [3]. This network achieved the second position in the ILSVRC (ImageNet Large Scale Visual Recognition Challenge) in 2014 [2]. VGG also has different depth of the network in terms of the number of layers with learnable weights. VGG-16 and VGG-19 are the widely used networks.

The design architecture of VGG is simple. VGG structures a sequence of convolutional layers followed by max pooling operation. It uses 3×3 convolutional filters throughout the entire network with a stride of 1. The max pooling operation is performed over a 2×2 window with a stride of 2. After a series of convolutional and max pooling layers, 3 fully connected layers are used. The first two FC layers have

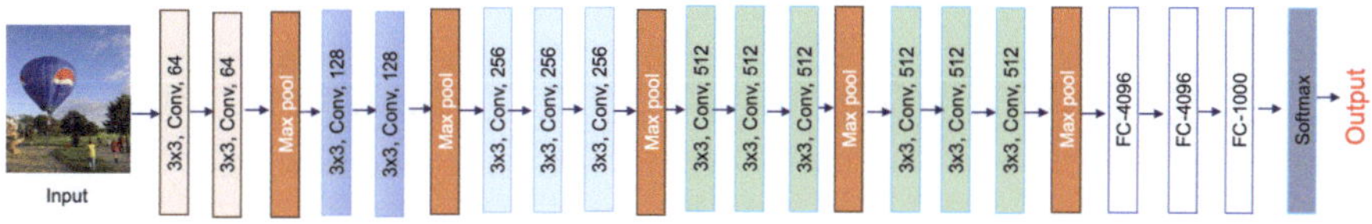

Fig. 2.6 An overview of VGG16 structure

4096 neurons each, and the third performs a 1000-class classification. For classification, the final operation involves a softmax layer for probability estimation. Along with the classification task, VGG has also been applied to the detection and segmentation task, necessitating adjustments in the final output layer. Figure 2.6 shows a simple overview of VGG16 (16 layers) structure.

2.2.5 ResNet: Residual Network

A residual network, commonly known as ResNet, is a CNN architecture that incorporates residual blocks to address the degradation problem in deep models. This innovative architecture was first introduced by Kaiming He et al. in their paper [5]. ResNet won the ImageNet Large Scale Visual Recognition Challenge in 2015 and is famous for its deep architecture capable of having hundreds or thousands of deep layers.

The depth or size of a deep learning network is a critical factor in extracting semantic, contextual, and structural features of objects. During training, a trade-off exists between the depth of the network and its efficiency, with several other factors influencing performance. Shallow networks often fail to meet the performance requirements for object recognition, particularly when dealing with large and complex datasets. However, as the depth of the network increases, challenges such as vanishing gradients can arise, where parameter updates during training become less effective. This issue affects the learning capability of convolutional neural networks (CNNs) and hinders their performance. Additionally, deeper networks may experience a degradation problem, where increasing depth leads to diminished returns in accuracy or even a decline in performance as the network struggles to converge.

To address the problem of vanishing gradients in deep networks, He et al. proposed a residual learning scheme in ResNet. Unlike the structure of conventional CNNs, where the output of one layer is fed as input to the next, ResNet introduces the concept of residual learning by implementing skip connections, also known as shortcut connections. For instance, within a residual block that contains two convolutional layers, residual learning employs a skip connection that adds the original input to the output of the second convolutional layer. Figure 2.7 illustrates the conventional approach and the residual skip connection.

In the figure, let $\mathbf{x}$ represent the input feature of the residual block. $F(\mathbf{x})$ denotes the function of the layer that generates an output feature maps. The input $\mathbf{x}$ is added to the

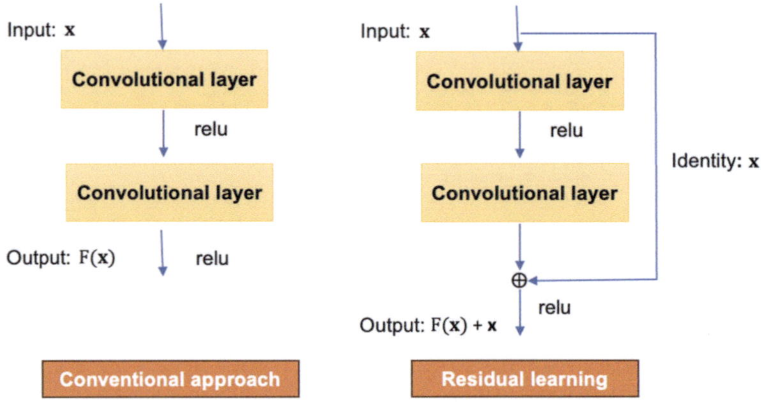

Fig. 2.7 Comparison between conventional CNN connections and ResNet's shortcut connections, showcasing how residual learning mitigates degradation in deep networks

output of the layer. Formally, if $F(\mathbf{x})$ represents the underlying mapping, the learned representation is $\mathbf{x}' = F(\mathbf{x}) + \mathbf{x}$. The final output is determined by incorporating the shortcut connection. This process enables the network to learn from the original input features, while shortcut connections help preserve feature representations across different layers. By facilitating gradient flow and retaining multi-level feature information, this approach effectively mitigates the problem of network degradation and enhances overall performance.

In deeper ResNet models (those with 50+ layers), residual blocks incorporate a 'bottleneck' design to manage channel dimensionality and computation costs. This design involves a stack of three layers: a 1×1 convolution (pointwise) that reduces the channel dimensions, a 3×3 convolution which outputs the same number of channels, and then another 1×1 convolution that expands the channel dimensions. The 1×1 kernel size, also known as linear transformation, which effectively reduces computational costs. The 3×3 kernel size layers extend coverage over spatial positions including more broad representations. Figure 2.8 shows a simple scheme of bottleneck structure. Each residual or bottleneck block has two or three convolutional layers (depending on the ResNet variant).

There are several variants of ResNet, such as ResNet-18, ResNet-34, ResNet-50, ResNet-101, and ResNet-152, with the number indicating the depth of the network in terms of layers. These network includes standard convolution, followed by a batch normalization layer, a ReLU activation, and a max pooling layer. Subsequently, a global average pooling layer reduces the spatial dimensions to a single value per feature map. Finally, a fully connected layer reduces the feature vector to the number of classes in the dataset. Then a softmax layer is employed to compute the classification probabilities. Figure 2.9 shows a simple overview of ResNet18 structure.

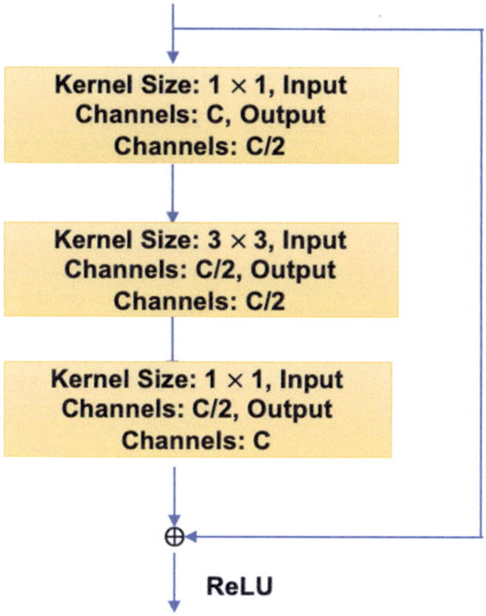

Fig. 2.8 ResNet bottleneck block structure: a three-layer configuration (1 × 1, 3 × 3, and 1 × 1 convolutions) designed to reduce computational cost while preserving feature richness

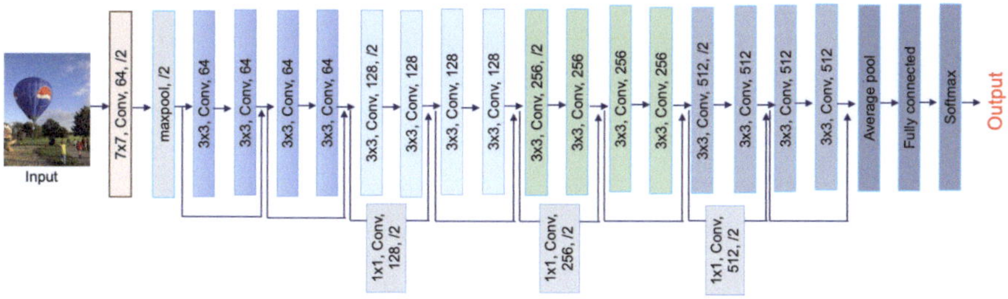

Fig. 2.9 An overview of ResNet18 structure

ResNet can be also utilized for feature extraction for segmentation and detection tasks, by connecting it with subsequent segmentation head or object detection components.

2.3 Deep CNNs for Object Detection

As aforementioned, image classification is the process of assigning a single label to an image from a fixed set of categories. This task can be extended to object detection, a computer vision task that enables the identification and localization of multiple

Fig. 2.10 Illustration of object detection using logo detection as an example, highlighting the identification and localization of objects within an image

objects from specific categories within images or videos. Object detection involves recognizing and pinpointing objects, which is achieved by drawing a rectangular boundary around each detected object. Object detection is widely useful in various applications, including crowd counting, person detection, logo detection, self-driving cars, video surveillance, face detection, anomaly detection, production inspection, and medical imaging task like tumor detection. Figure 2.10 shows an example of logo detection. An object detector comprises two main structures: a feature extractor network and a detection head. We will discuss both these components in detail in the following.

2.3.1 *Feature Extractor Backbone for Object Detection Tasks*

Feature extractor or backbone network transforms the input high-dimensional image into lower-dimensional feature maps for dense prediction. This process extracts the semantic and spatial information within image, such as patterns, shapes, edges, textures, and other semantic elements of different objects. This extracted feature maps or feature space is used for detecting objects and determining their sizes. We can use various deep neural networks such as ResNet [5] and VGG [3], which are described in Sect. 2.2, as the feature extractor backbone network, adjusting their final output for object detection. During training, data augmentation techniques, which include image translations, horizontal flipping, and cropping, are commonly employed to enhance the diversity of the training dataset in the detection tasks.

2.3.2 Detection Head

The detection head constitutes the second part of the deep learning object detection network, responsible for determining the size and location of objects while predicting their categories. The detection head analyzes feature maps, proposes candidate regions, draws bounding boxes around them and assigns an object (category) confidence level (probability score) to each proposed candidate region. Finally, some candidate regions with the highest probabilities are selected as detected objects. Detection layers in detection heads may comprise either 1D convolutional layers or fully connected layers, operating on the extracted feature maps or feature space to detect potential target objects. To predict object bounding boxes, detection head not only needs to identify object categories, but also perform a regression task to determine the exact bounding box coordinates. The entire detection framework (feature extraction backbone and head) can be trained using a multitask loss that combines both object classification loss and location regression loss, which have been discussed in the first chapter. Figure 2.11 shows an object detection pipeline. Some common functions of detection head are explained in the following subsection.

Object Localization: The detection head predicts bounding box locations by regressing the coordinates (x, y, w, h) using regression layers, where (x, y) are center coordinates, w and h are width and height of the bounding box. We can describe this step as a patch selection procedure with a high possibility of containing an object. Anchor boxes plays a key role in this process.

Candidate Region Classification: The detection head predicts class labels or object categories for each detected bounding box or selected patch with candidate proposals. For this task, multiclass classification with softmax activation is employed. This procedure is similar to image classification task (discussed in Chap. 1), where probabilities are assigned to each predicted candidate region.

Anchor Boxes: The detection head places anchor boxes, which are predefined bounding box priors of various sizes and aspect ratios. Anchor boxes are placed across the extracted feature space. The selected patch within an anchor box serves as a potential

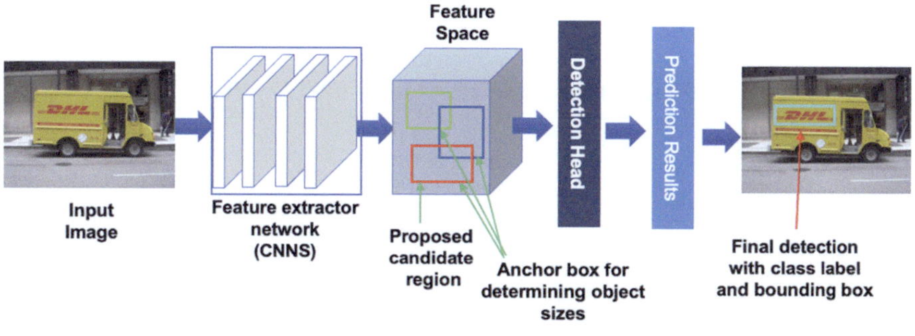

Fig. 2.11 Object detection pipeline: demonstrating feature extraction, region proposal, bounding box prediction, and classification processes

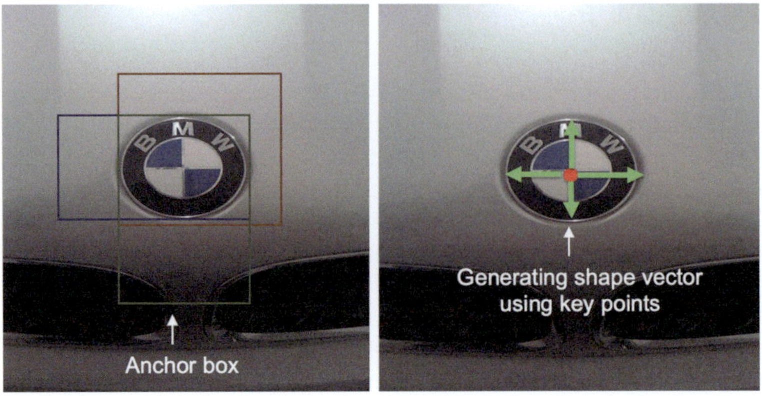

Fig. 2.12 Comparison of anchor-based (left) and anchorless (right) detection methods: anchor boxes provide predefined bounding box priors, while shape vectors dynamically adapt to object dimensions

object candidate proposal. We can classify the candidate proposals and determine a score indicating the likelihood of any target object being present in the selected patch. The detection layers are designed to learn how to select the most appropriate shape and size of an anchor box based on the object's characteristics. This also includes the selection of anchor boxes for the predicted regions to better align with the precise boundaries of detected objects. Recently, some anchorless detection methods have been introduced. Unlike conventional methods, these do not require a fixed set of anchor box priors. Instead, they adaptively learn a shape vector that corresponds to the possible candidate regions of an object. They identify a keypoint for the object within the feature maps and learn its size from the activated pixel values in that particular region or patch. Figure 2.12 shows examples of anchor box priors and anchorless design.

Non-maximum Suppression (NMS): This is a technique to eliminate redundant or overlapping bounding box predictions. This filtering process is based on a confidence threshold (estimated probability score). We can set a confidence threshold to filter out bounding boxes with low confidence scores. Detection frameworks filter out bounding boxes with low confidence scores to achieve precise detection.

In the next subsection we will discuss some widely used object detectors.

2.3.3 Faster R-CNN

For object detection using deep learning networks, Ross Girshick et al. introduced R-CNN (Regions with CNN features) in their work "Rich feature hierarchies for accurate object detection and semantic segmentation" [7]. In their work, for region proposals, they used a widely used algorithm known as selective search algorithm [8].

In this framework, they first select multiple image patches and then the selected image patches are processed through a CNN network to extract deep features, followed by using SVMs for classification. Subsequently, Ross Girshick presented Fast R-CNN in his paper "Fast R-CNN" [9]. The selective search segments the input image based on color, texture, and other low-level features to produce a diverse set of candidate regions. After generating region proposals, Fast R-CNN applies post-processing and filtering to refine the candidate regions. In Fast R-CNN, the entire image is processed through a CNN to generate feature maps just once. Once the region proposals are obtained and filtered, Region of Interest (RoI) pooling is then applied to these feature maps to extract features for each proposal. RoI (Region of Interest) pooling converts the variable-sized feature maps into fixed-sized feature vectors for each proposal. These candidate region proposals generated from the image using CNN feature extractor are used for further post processing and detection. The non-maximum suppression algorithm is used to select the final bounding box. This concept significantly improved speed and outperformed conventional object detection algorithms. Since then, many other deep learning detection frameworks for object detection have been introduced and applied to various applications. Ross Girshick and his team also proposed Faster R-CNN in their paper "Faster R-CNN: Towards Real-Time Object Detection with Region Proposal Networks" [10]. In the following subsection, we will discuss the functions and components of Faster R-CNN, which has been extensively using in various detection problems.

Faster R-CNN incorporates a feature extractor CNNs (backbone network) such as ResNet [5] or VGG [3], and a small CNN network known as the Region Proposal Network (RPN). The feature extractor CNNs generate a feature space by extracting semantic information from a given image. The RPN network then generates candidate object region (candidate proposals or patches), from generated feature space. RPN network eliminates the need for selective search or other similar algorithms. The RPN network identifies potential region proposals, determines their objectivity score and classifies them as containing any object (foreground) or being part of the background based on this score. The RPN classifies these candidate regions in a class-agnostic manner, considering all types of detected regions as a single object category.

To determine the candidate proposals, RPN network takes generated feature maps as input in a sliding window fashion. The candidate region proposals are defined in terms of "k" anchor boxes, each centered at the sliding window location. Anchor box priors are fixed-size boxes (patch) that serve as reference points for the model to identify potential objects and their shapes in an image. These anchor box priors are associated with specific scales and aspect ratios. If the objectness score exceeds some threshold, the rectangular image patch is considered as a potential candidate region. Within the RPN network, each sliding window (e.g., $n \times n$) is transformed into a low-dimensional feature vector. This vector is then processed through two separate fully connected (FC) layers. One FC layer is dedicated to bounding box regression, determining the coordinates of the bounding box. The function of the other FC layer is to classify the region as either foreground or background. For each simultaneously predicted "k" anchor box candidates, the first fully connected regression layer encodes the coordinates (x, y, w, h) of these "k" boxes (i.e., $4k$ scores of

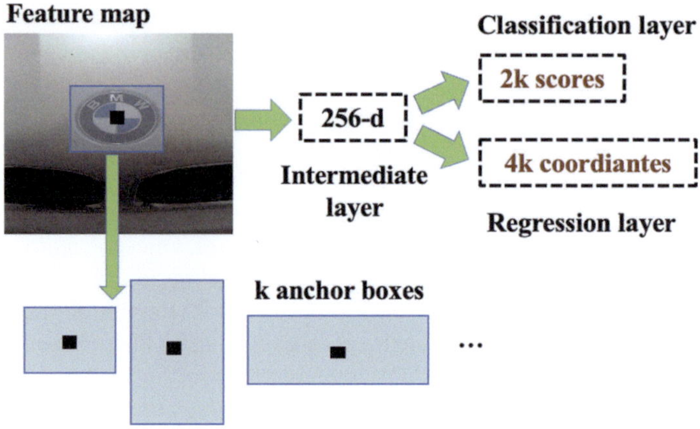

Fig. 2.13 Illustration of candidate region proposal generation in Faster R-CNN: sliding window approach identifies potential object regions based on feature maps

coordinates). Whereas the classification layer outputs $2k$ scores, indicating the probability of the presence or absence of an object for each proposal. This mechanism is demonstrated in the Fig. 2.13.

The Region of Interest (ROI) pooling procedure in Faster R-CNN is another essential step for object detection. After the RPN network generates region proposals, each proposed region (i.e., ROI) is projected onto the feature space of the image (i.e., generated feature map using main feature extractor). These ROIs can vary in size, but the subsequent fully connected layers in the network require inputs of a fixed size. To address this, ROI pooling converts these variable-sized ROIs into a fixed-size or uniformly sized feature maps. The ROI pooling layer divides each ROI into small and equal-sized bins. The most important feature is then selected from each bin using a max pooling operation. This process generates a new fixed-sized feature map for each ROI. The fixed-size feature maps are subsequently fed into separate fully connected layers to classify object category and determine precise object location. These separate fully connected layers are dedicated to performing classification and box regression tasks. This procedure is shown in the Fig. 2.14.

The entire Faster R-CNN framework can be trained in an end-to-end manner. The training includes fine-tuning the Region Proposal Network (RPN), Feature Extractor CNNs, and the fully connected (FC) layers. Faster R-CNN uses a combination of loss functions during training to effectively learn object classification and localization. These losses are given as follows:

Losses in the Region Proposal Network (RPN):

Objectness Classification Loss: This loss is a binary classification loss that determines whether each anchor box is an object ("objectivity") or background.

Bounding Box Regression Loss: This loss refines the coordinates of the anchor boxes proposed by RPN. It is designed to adjust and select appropriate boxes to

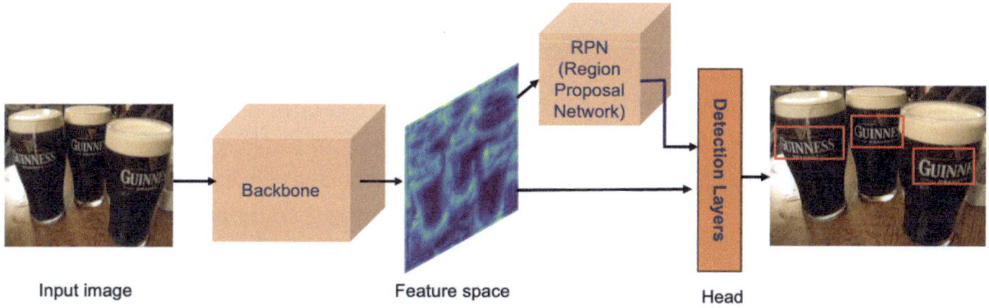

Fig. 2.14 Overview of faster R-CNN: showcasing region proposal network (RPN), ROI pooling, and object detection processes

better align them with the ground truth object location. Normally, smooth L_1 loss is used for bounding box regression.

Losses in the Final Fully Connected Layers:

Multiclass Classification Loss: After the ROI pooling layer, the proposed regions are classified into specific object categories or classified as background. This is a multiclass classification problem.

Bounding Box Regression Loss: Another bounding box refinement loss (L_1 loss) is applied using the final fully connected layers. This refinement is based on fixed-size feature maps generated by ROI pooling. While bounding box regression in the RPN network aligns the anchor box with the object, this step adjusts the box after ROI pooling to make it more accurate. Both stages of bounding box regression are useful. The first stage determines the rough location of objects, and the second stage fine-tunes these locations based on the more detailed information available after ROI pooling. The training object loss for Faster R-CNN framework can be given as:

$$L = L_{\text{cls\_rpn}} + \lambda_1 L_{\text{reg\_rpn}} + L_{\text{cls\_roi}} + \lambda_2 L_{\text{reg\_roi}} \tag{2.1}$$

$L_{\text{cls\_rpn}}$ is the Binary classification loss (objectness classification) in the Region Proposal Network (RPN), as defined in Eq. (1.15). It determines whether each anchor is an object or background. $L_{\text{reg\_rpn}}$ is the bounding box regression loss in the RPN (e.g., Smooth L_1 loss), as defined in Eq. (1.18). It refines the coordinates of anchor boxes proposed by RPN. $L_{\text{cls\_roi}}$ is the multiclass classification loss after ROI pooling, as defined in Eq. (1.16). It classifies proposed regions into specific object categories or as background. $L_{\text{reg\_roi}}$ is the Bounding box regression loss after ROI pooling (e.g., Smooth L_1 loss), as defined in Eq. (1.18). It fine-tunes bounding box coordinates for accurate alignment with ground truth objects. λ_1 and λ_2 are weights to balance the regression and classification losses at each stage. Typically, these weights are set to 1 but can be tuned based on the task. Faster R-CNN is an efficient object detector framework and has been used in many applications.

2.3.4 CenterNet: Anchorless Detection Approach

Previously discussed detection approaches used a fixed set of anchor boxes to determine and learn the height and width of object during training. These methods necessitate careful design in the selection, sizes, and number of anchor boxes. The performance and fine-tuning of these detector frameworks heavily depend on adjusting various hyperparameters including the size, number, and aspect ratio of densely placed anchor boxes. As a result, researchers have been motivated to design new detection schemes. In recent years, some anchorless frameworks have been proposed for the object detection task. These techniques identify objects in terms of keypoints such as center or corner points of the objects.

In anchorless detection frameworks, a set of category-wise heatmaps or keypoint maps are generated using the extracted feature maps. A specific object is activated in the respective category heatmap. By learning the height and width of the activated region as a shape vector we can determine the coordinates of the bounding box. After that, a distance loss function such as L_1 can be used to compute the difference between the predicted bounding box and the ground truth to learn localization. Among various anchorless detection methods, one notable technique we will discuss is CenterNet. Zhou et al. introduced an anchorless detection technique known as CenterNet in their paper "Object As Points" [11].

CenterNet locates objects by identifying their center points within bounding boxes and learns a shape vector to estimate each object's height and width. CenterNet does not require any additional grouping step for keypoints, as it generates only one keypoint to recognize the object. During training, the ground truth ROIs are first transformed into class-wise keypoint maps, each corresponding to a specific object category. In these maps, each object is marked at the coordinates of its center, creating a peak spot in the corresponding keypoint map, while other pixel values are set to zero. These keypoint maps are subsequently converted into Gaussian heatmaps based on the shape of the objects to activate the region of the object within the maps. After predicting the keypoints, CenterNet converts these keypoint maps into Gaussian heatmaps. Each keypoint corresponds to a heatmap, where the intensity of each pixel represents the likelihood of that pixel belonging to the object associated with the keypoint. The intensity typically follows a Gaussian distribution centered around the keypoint location, gradually decreasing as you move away from it. By converting the keypoint maps into Gaussian heatmaps, the entire region of the object is effectively activated within these heatmaps. This means that pixels within the heatmap corresponding to the object's location will have high intensity values, indicating a high confidence that those pixels belong to the object. Finally, these Gaussian keypoint maps serve as ground truth for both heatmaps classification and estimation of the shape vector of the objects. During training, CenterNet aims to generate class-wise heatmaps for object classification and region activation to learn shape vectors for shape estimation. A detection confidence score (threshold) is calculated using the predicted keypoint values from these heatmaps, functioning similarly to non-maximum suppression. The detection head of CenterNet comprises three sub-heads:

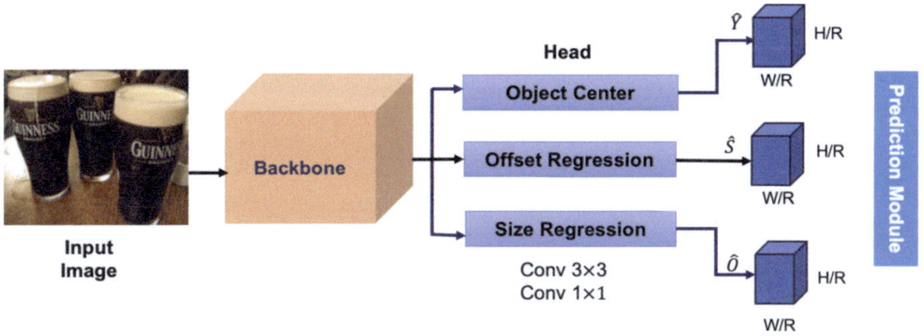

Fig. 2.15 CenterNet architecture: an anchorless object detection framework that uses center points and Gaussian heatmaps for localization and classification

- A heatmap head generates class-wise heatmaps for object categories, in which the classification loss is calculated against the ground truth heatmaps to learn the object category.
- A size head produces shape vectors to estimate the width and height of the object and regress the shape of object using ground truth coordinates.
- An offset head corrects discretization errors caused by stride (change in the spatial dimensions) operations during dimensionality reduction of the output feature map.

Figure 2.15 illustrates the architecture of CenterNet. As shown in the figure, the generated feature space is fed into different output heads, each responsible for a specific prediction task. Each head consists of two convolutional layers with kernel sizes 3×3 and 1×1, respectively. In the figure, the predicted outputs $\hat{Y}$, $\hat{S}$, and $\hat{O}$ represent the keypoint heatmaps (center points of objects) for c classes, the object size (height and width), and the positional offset, respectively. Here, W and H denote the width and height of the input image, while R represents the downsampling ratio of the feature maps with respect to the input image. Multiple objects, both within the same and different categories, can be detected by activating multiple regions in their respective keypoint maps. The training object loss for CenterNet can be expressed as:

$$L = L_{\text{heatmap}} + \lambda_1 L_{\text{offset}} + \lambda_2 L_{\text{size}} \tag{2.2}$$

where L_{heatmap} is the Focal loss for predicting the center points of objects in the heatmap. L_{offset} is the L_1 loss (Eq. 1.18) for refining the center point offsets relative to the nearest grid cell. L_{size} is L_1 loss (Eq. 1.18) for predicting the width and height of the bounding boxes. The L_1 loss can be given as:

$$\text{MAE} = \frac{1}{n}\|\hat{\mathbf{y}} - \mathbf{y}\|_1 \tag{2.3}$$

here, n is the number of data points, $\mathbf{y}$ is the ground truth, and $\hat{\mathbf{y}}$ is the predicted value.

The Focal loss is formulated as:

$$L_{\text{Focal}} = -\sum_{i=1}^{N} \left[\alpha\,(1-\hat{y}_i)^{\gamma}\, y_i \log(\hat{y}_i) + (1-\alpha)\,\hat{y}_i^{\gamma}\,(1-y_i)\log(1-\hat{y}_i)\right] \quad (2.4)$$

where $\hat{y}_i \in [0, 1]$ is the predicted probability at the i-th location in the heatmap, and $y_i \in \{0, 1\}$ is the corresponding ground truth label. The hyperparameter $\alpha \in [0, 1]$ is a weighting factor that balances the importance of positive and negative examples. The focusing parameter $\gamma \geq 0$ adjusts the rate at which easy examples are down-weighted, preventing the model from being overwhelmed by a large number of well-classified (easy) background samples.

2.3.5 YOLO: You Only Look Once

The first version of the YOLO (You Only Look Once) detection algorithm was introduced in 2016 by Joseph Redmon et al. [12]. A series of YOLO frameworks with improvements in terms of backbone, detection head, and data augmentation techniques have been proposed in various subsequent works. In this subsection, we will discuss the YOLOv3 detection head [13], which is used in most recent versions of the YOLO series detectors such as YOLOv5.

The YOLOv3 architecture uses a feature pyramid CNN network backbone to extract features from an input image. Feature pyramid CNN network generates feature spaces with different spatial dimension and depth at different scales. The YOLOv3 detection head performs multi-scale detection by operating on three different scales of the extracted feature space, enabling it to detect objects across various sizes. These scales are extracted from different layers of the feature pyramid CNN network. YOLOv3 utilizes different predefined anchor box priors with varying aspect ratios, suitable for the corresponding scale of the feature space.

The YOLOv3 architecture divides the input image into a grid. The grid size varies depending on the size of the feature map. Each cell in this grid is associated with multiple anchor boxes of different aspect ratios. The cell is responsible for detecting objects whose centers fall within the cell. YOLOv3 predicts the objectness confidence score and bounding box attributes such as center coordinates (x, y), width (w), and height (h) for each anchor box in a cell. It assigns an objectness confidence score to each anchor box, indicating the likelihood of absence/presence of any object. YOLOv3 employs a binary cross-entropy classification loss to calculate the object confidence score. Each ground truth bounding box is assigned an "objectness" score of 1, indicating the presence of an object, irrespective of its category. Additionally, each grid cell in YOLOv3 predicts the probabilities of different object classes for the contents within the bounding box. Finally, each bounding box can be represented by a set of parameters consisting of 5 values plus a class probability vector. These parameters include the center coordinates (x, y), width (w), height (h), objectness confidence score, and the "classes" vector. The "classes" vector has the same dimension as the number of classes (c), making each bounding box represented by $5 + c$ numbers, where "c" is the number of classes. These predictions, which include the bounding box parameters, objectness score and class probabilities, are initially relative to the cell position and the size of the anchor box. Subsequently, these parameters are transformed to absolute terms to project onto the original image dimensions.

To train the model, YOLOv3 encodes ground truth bounding boxes in a specific format for the detection head. During training, each ground truth bounding box's width and height are normalized with respect to the dimensions of a pre-defined set of anchor boxes. An anchor box is a predetermined box of a certain size and aspect ratio. It serves as a reference for the network to predict bounding box coordinates relative to these anchor boxes. The width and height of the bounding boxes are transformed using a logarithmic scale. Specifically, the ground truth width and height are first divided by the dimensions of the corresponding anchor box, and then the logarithm of this ratio is computed. This normalization process helps in making the bounding box size relative to the anchor box dimensions, making it easier for the network to predict adjustments to these dimensions during training. This transformation helps stabilize the learning process, particularly when there are large variations in object sizes. The training object loss for detection framework can be expressed as

$$L = L_{\text{box}} + L_{\text{obj}} + L_{\text{no\_obj}} + L_{\text{class}} \tag{2.5}$$

L_{box} is the Regression loss [as defined in Eq. (1.17)] for bounding box coordinates (e.g., center x, y, width w, and height h). It ensures accurate localization of detected objects. L_{obj} is the Objectness loss, a binary cross-entropy loss [as defined in Eq. (1.15)], indicating whether a grid cell contains an object. $L_{\text{no\_obj}}$ is the No-object loss [i.e., binary cross-entropy loss, Eq. (1.15)], penalizing false positives in grid cells that do not contain objects. L_{class} is the Classification loss, a cross-entropy loss [as defined in Eq. (1.16)] for predicting the correct class for the detected objects.

During training, the model learns to predict bounding boxes, objectness scores, and class probabilities for each anchor box at multiple scales on the feature maps. The multi-scale approach helps in the accurate detection of objects of various sizes, including small, medium, and large. After initial predictions, post-processing techniques such as non-maximum suppression are used to refine the bounding boxes, eliminating redundant boxes and keeping only those with the highest confidence scores. YOLOv3 employs a combination of loss function during training, which includes bounding box regression loss, objectness score prediction, and object category prediction. Figure 2.16 shows a simple architecture of YOLOv3 framework. YOLOv3 is designed for speed, allowing real-time object detection, which is crucial for applications like video analysis and autonomous driving.

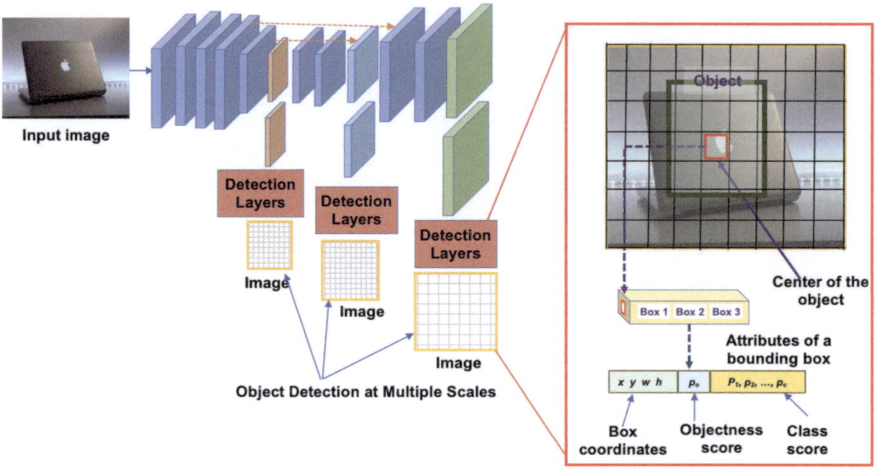

Fig. 2.16 YOLOv3 architecture: a real-time object detection framework using a multi-scale feature pyramid and anchor boxes for robust performance

2.4 Deep CNNs for Semantic Segmentation

Semantic segmentation is an important computer vision task that involves classifying each pixel in an image to a class label. Segmentation network generates output dense prediction maps or masks that partition the image into regions of semantic similarity. This task has wide range of applications across different fields such as navigation, autonomous driving and medical image diagnosis, including the segmentation of anatomical body structure. The introduction of deep learning networks has significantly advanced the segmentation process by enabling end-to-end learning frameworks, where feature extraction and pixel-wise classification are jointly learned from the input image and their corresponding ground truth mask. In the following subsection, we will introduce some commonly used segmentation networks and their design.

2.4.1 Fully Convolution Network (FCN)

Fully convolutional network was introduced by Jonathan Long et al. in their paper "Fully Convolutional Networks for Semantic Segmentation" [14]. This network has served as a foundation architecture for the segmentation tasks. It is one of the first networks to be comprised entirely of convolutional layers. This network does not include fully connected layers at the end. For the segmentation task, instead of using fully connected layers in the final stage of the framework, it adapts its design to

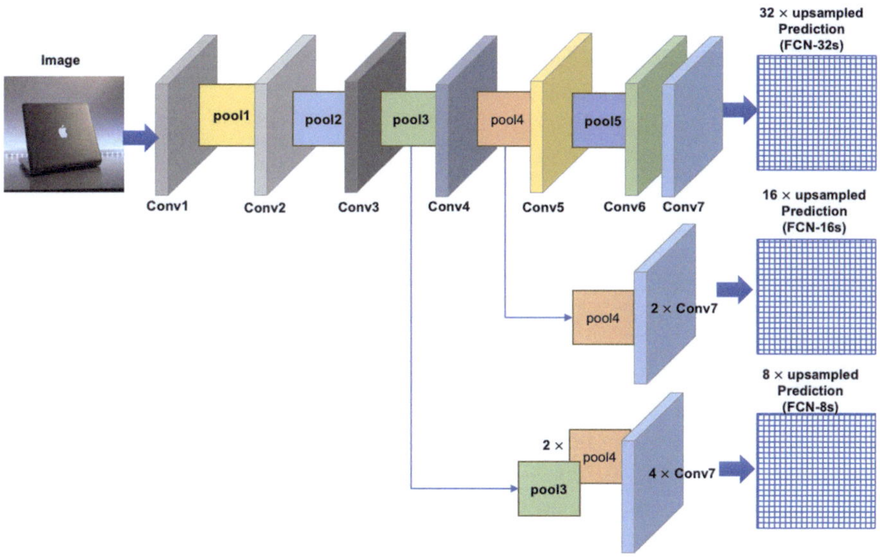

Fig. 2.17 Fully Convolutional Network (FCN): leveraging skip connections to combine low-level fine details with high-level coarse information for precise segmentation

generate dense outputs for making predictions. In other words, the spatial dimensions of the generated output are maintained to align with the size of the input image.

The FCN network can be trained in an end-to-end manner, using the entire image to produce an output of the same size. This network has surpassed existing segmentation methods. FCN networks are composed of the basic components of CNN network such as convolution layers, pooling layers, and activation functions. Additionally, alongside the primary output stream, skip and side connections are implemented to integrate feature responses from various scales into a unified output layer, as depicted in Fig. 2.17. The number of output channels from feature maps at different levels is aligned with the number of classes for segmentation through convolutional operations. Transposed convolution (or deconvolution) operators are employed to restore spatial resolution, increasing the spatial dimensions to match the size of the input image. The final prediction is then generated by aggregating these features. FCN applies skip connection concept more effectively and broadly. They leverage skip connections that combine semantic information from deep, coarse layers with appearance information from shallow, fine layers to achieve precise and detailed segmentation. The variability in the shape of generated feature maps also aids in recognizing objects of various sizes. The training object for the segmentation framework can be given as:

$$L_{\text{FCN}} = -\frac{1}{N}\sum_{i=1}^{N}\sum_{c=1}^{C} y_{i,c}\log(\hat{y}_{i,c}) \tag{2.6}$$

where N is the total number of pixels in the image. C is the total number of classes. $y_{i,c}$ is ground truth label (one-hot encoded) for pixel i and class c. $\hat{y}_{i,c}$ is the predicted probability for pixel i belonging to class c.

2.4.2 UNet

Olaf Ronneberger et al. introduced the UNet architecture based on Encoder-Decoder structure in their paper "U-Net: Convolutional Networks for Biomedical Image Segmentation" [15]. UNet has a structure similar to a "U" shape and is specifically designed for biomedical image segmentation. UNet performs effectively with limited training images while achieving highly precise segmentations. Inspired by the design of FCN [14], UNet has implemented several modifications and extensions. The architecture such as, UNet++ [16] and UNet3+ [17], adopts an Encoder-Decoder structure, where the encoder progressively reduces spatial dimensions, while the decoder gradually restores spatial dimensions. This feature representation learning process enables the network to effectively capture semantic and high-level context information and facilitates precise pattern recognition.

In encoder, convolution layers decrease the spatial dimension while increasing the number of feature channels (depth), the number of channels is [64, 128, 256, 512, 1024]. The decoder in UNet performs a series of upsampling operations to gradually increase the spatial dimensions same as the input size while reducing the number of channels. A key innovation in UNet is the introduction of skip connections that merge local information from the encoder with global information in the decoder during the upsampling process. To effectively utilize and fuse visual information with semantic information, skip connections are implemented from the encoder to the decoder, concatenating channels at the corresponding levels. For example, channels are concatenated, such as [512 (encoder features) + 256 (decoder features) = 768 (output)], to enhance the extracted information. Each convolutional stage consists of two 3×3 convolution layers, each followed by a rectified linear unit (ReLU) and a 2×2 max pooling operation with a stride of 2 for downsampling. With each downsampling step, the network doubles the number of feature channels. Decoder includes a 2×2 upsampling with a stride of 2, two 3×3 convolution layers, each followed by a ReLU. The final layer is a 1×1 convolution that maps the final output feature maps to the desired number of classes. Figure 2.18 illustrates the UNet structure using a 512×512 spatial dimension for input and output. UNet has been employed in CT and MRI image segmentation for tasks such as liver and lesion segmentation, multi-organ segmentation, and pancreas segmentation, among other medical applications. The U-Net training loss combines the Dice coefficient loss to handle class imbalance and the cross-entropy loss for precise pixel-wise classification.

$$L_{\text{U-Net}} = L_{\text{dice}} + L_{\text{cross-entropy}} \tag{2.7}$$

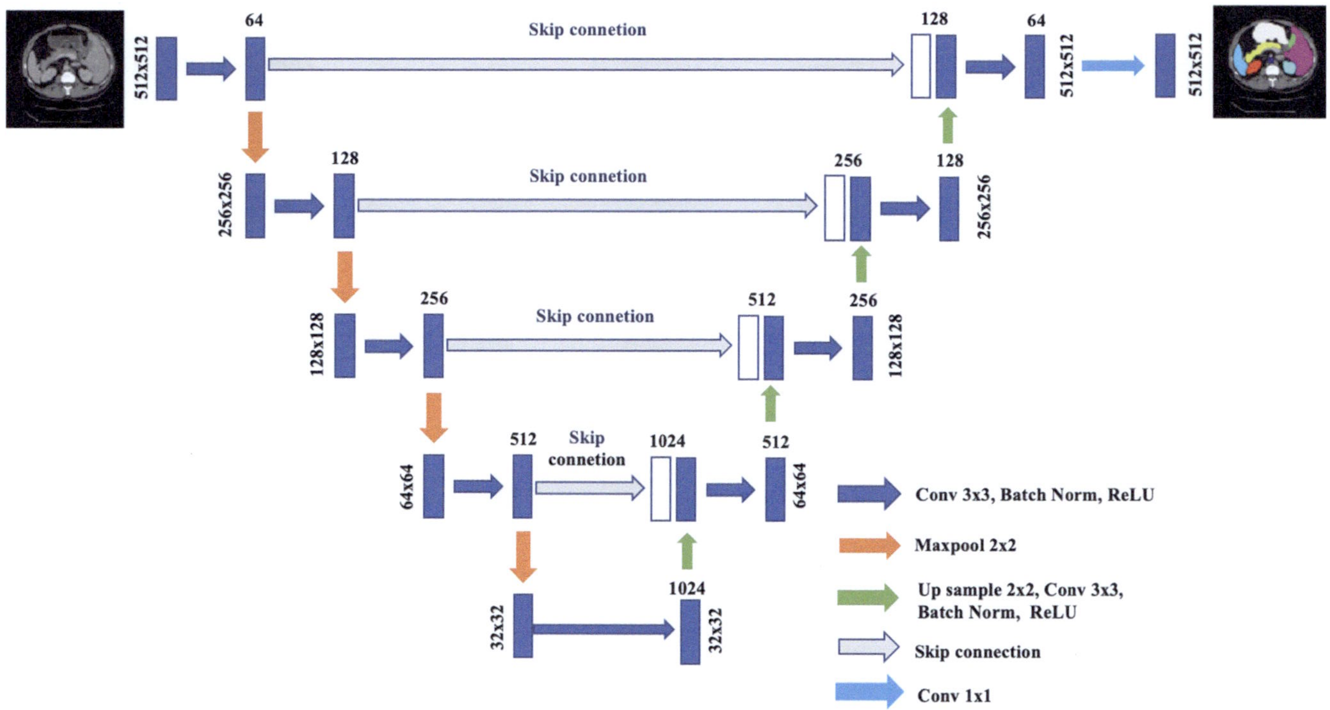

Fig. 2.18 UNet architecture: a U-shaped encoder-decoder network with skip connections, designed for detailed segmentation tasks like medical imaging

L_{dice} is Dice coefficient loss, which measures the overlap between the predicted segmentation ($\hat{y}$) and the ground truth segmentation (y):

$$L_{\text{Dice}} = 1 - \frac{2\sum_{i=1}^{N} \hat{y}_i y_i}{\sum_{i=1}^{N} \hat{y}_i + \sum_{i=1}^{N} y_i + \epsilon} \tag{2.8}$$

$\hat{y}_i$ is the predicted probability for pixel i. y_i is the ground truth label for pixel i. N is the total number of pixels. The constant ϵ is a small value (e.g., 10^{-6}) added to the denominator to prevent division by zero and ensure numerical stability during training.

Pixel-wise cross-entropy loss:

$$L_{\text{cross-entropy}} = -\frac{1}{N} \sum_{i=1}^{N} \sum_{c=1}^{C} y_{i,c} \log(\hat{y}_{i,c}) \tag{2.9}$$

C is the number of classes. $y_{i,c}$ is the ground truth label (one-hot encoded) for pixel i and class c. $\hat{y}_{i,c}$ is the predicted probability for pixel i belonging to class c.

2.5 Measuring Metrics

Measuring metrics are essential tools for assessing the effectiveness of deep learning algorithms and models. There are different evolution metrics for different tasks like classification, segmentation, object detection, image enhancement, and other regression tasks [18]. These metrics are helpful in establishing algorithms and models based on the quality of the prediction performance and suitability for specific tasks.

True Positive (TP), False Positive (FP), True Negative (TN), and False Negative (FN) are key components of evaluation metrics in various tasks. A True Positive (TP) happens when the model correctly predicts a positive class (e.g., 1) and the ground truth is also positive. A False Positive (FP) occurs when the model incorrectly predicts a positive class (e.g., 1) but the ground truth is negative (e.g., 0). A True Negative (TN) is considered when the model correctly predicts a negative class (e.g., 0) while the ground truth is also negative. A False Negative (FN) occurs when the model incorrectly predicts a negative class (e.g., 0) while the ground truth class is positive (e.g., 1). These components are essential for calculating various evaluation metrics such as Accuracy, Precision, Recall (Sensitivity), F1-Score (Dice Coefficient), Confusion Matrix and so on. A True Negative (TN) is considered when the model correctly predicts a negative class (e.g., 0) while the ground truth is also negative. A False Negative (FN) occurs when the model incorrectly predicts a negative class (e.g., 0) while the ground truth class is positive (e.g., 1). These components are essential for calculating various evaluation metrics such as Accuracy, Precision, Recall (Sensitivity or TPR—True Positive Rate), False Positive Rate (FPR), F1-Score, Confusion Matrix and so on. The confusion matrix is shown in Table 2.1.

Table 2.1 Confusion matrix representing classification results

	Predicted positive	Predicted negative
Actual positive	True positive (TP)	False negative (FN)
Actual negative	False positive (FP)	True negative (TN)

For classification, a common evaluation metric is the ratio of correctly classified inputs to the total inputs.

$$\text{Accuracy} = \frac{\text{True Positive} + \text{True Negative}}{\text{Total Predictions}} \tag{2.10}$$

Precision, for a specific category, represents the percentage of correct predictions among the total predicted positives.

$$\text{Precision} = \frac{\text{True Positive}}{\text{True Positive} + \text{False Positive}} \tag{2.11}$$

On the other hand, Recall (Sensitivity) assesses the proportion of true positives, defined as:

$$\text{Recall} = \frac{\text{True Positive}}{\text{True Positive} + \text{False Negative}} \tag{2.12}$$

The F1 Score is a metric that combines both precision and recall, offering a single value for assessing a classification model's performance.

$$\text{F1 Score} = \frac{2 \times (\text{Precision} \times \text{Recall})}{\text{Precision} + \text{Recall}} \tag{2.13}$$

The False Positive Rate (FPR) measures the proportion of actual negatives that are incorrectly classified as positive. It is calculated as:

$$\text{FPR} = \frac{\text{False Positive}}{\text{False Positive} + \text{True Negative}} \tag{2.14}$$

Specificity (also called True Negative Rate or Selectivity) measures how well a model correctly identifies negative cases. It is often used alongside Sensitivity (Recall/TPR) to evaluate classification models, especially in imbalanced datasets.

$$\text{Specificity} = 1 - \text{FPR} \tag{2.15}$$

Figure 2.19 illustrates a Venn diagram representing the relationships among accuracy, precision, recall, and F1-score. Figure 2.20 shows examples of the ROC and Sensitivity curves.

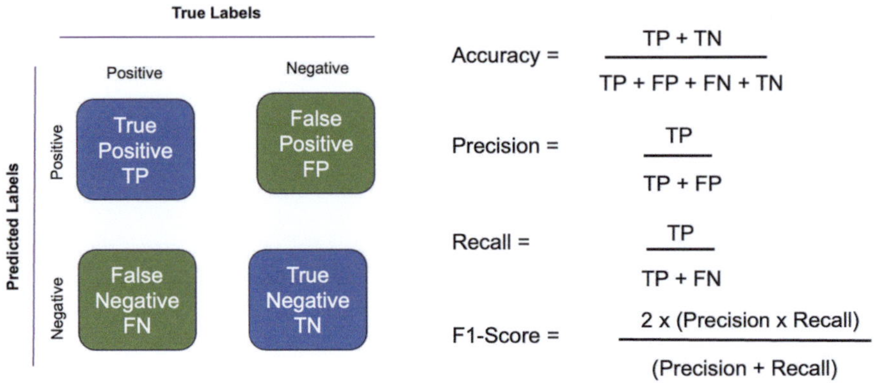

Fig. 2.19 The calculation of the accuracy, precision, recall, and F1-score

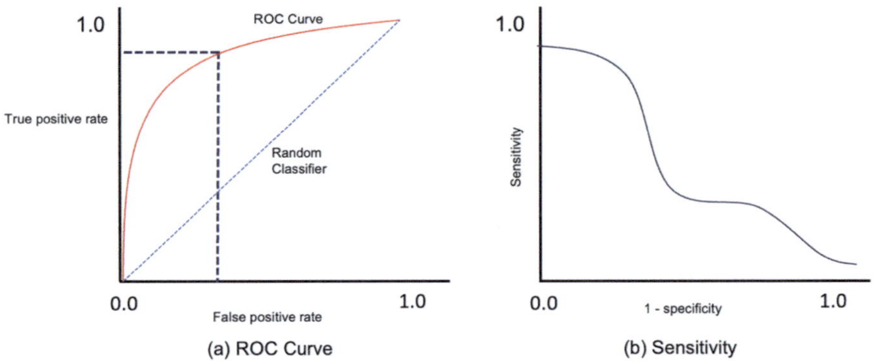

Fig. 2.20 Example of ROC and sensitivity curves

In segmentation tasks, we consider pixel-wise accuracy using segmentation ground truth mask and predicted segmentation mask. These masks are binarized output showing the activated region of interest. We calculate the overlapping region between the predicted segmentation mask and the ground truth mask by finding both the intersection and the union of these regions. In the context of image segmentation, Dice Coefficient calculates the overlap between the predicted segmentation mask and the ground truth mask.

$$\text{Dice Coefficient} = \frac{2 \times |\text{Predicted mask} \cap \text{Ground Truth Mask}|}{|\text{Predicted mask}| \cup |\text{Ground Truth Mask}|} \tag{2.16}$$

Based on the Dice Coefficient, the Dice loss value can be calculated as:

$$\text{Dice Loss} = 1 - \text{Dice Coefficient}$$

Intersection over Union (IoU) is another commonly used term for segmentation or detection. To determine the Intersection over Union (IoU), we divide the intersection by the union of these regions. If there are multiple classes in the segmentation task Mean Intersection over Union (mIoU) is used.

$$\text{IoU} = \frac{|\text{Predicted mask} \cap \text{Ground Truth Mask}|}{|\text{Predicted mask} \cup \text{Ground Truth Mask}|} \tag{2.17}$$

For evaluation in object detection tasks, the predicted candidate region is considered as a True Positive if the IoU of the predicted candidate region box and the ground truth (GT) bounding box surpasses a specific threshold. We usually consider a correct object region true for IoU 50% or IoU 30% overlapping. Average Precision (AP) is a widely used metric in object detection to evaluate the effectiveness of model for ranking the predicted bounding boxes based on their confidence scores (i.e., category). In general, the Average Precision (AP) for any given category is the area under the Precision-Recall curve (AUC-PR—Area Under the Curve- Precision-Recall curve). AP is the weighted sum of Precision values where the weight is the increase in Recall. For multiple object categories we can use Mean Average Precision (mAP) to measure the efficiency of detection algorithm.

The Area Under the Curve (AUC) of the Receiver Operating Characteristic (ROC) curve is used for classification and evaluates the model's capacity to distinguish between positive and negative classes. In the context of detection tasks, the model generates predictions, assigning category scores or probabilities to candidate regions (i.e., potential objects of interest) and negative samples (e.g., non-objects). By varying the classification threshold (changing the score that designates a sample as positive), we can plot points on the ROC curve. The AUC summarizes the model's performance at various threshold settings. This measuring metric is essential to understand the performance of model, especially in terms of its ability to correctly identify

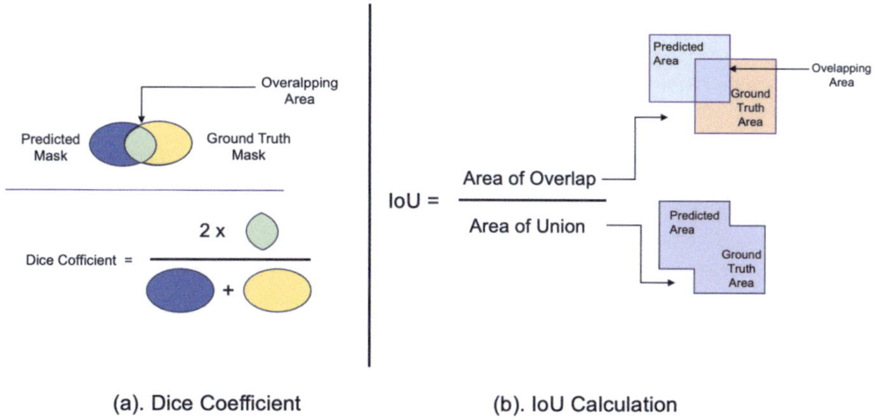

Fig. 2.21 Calculation of **a** dice coefficient and **b** intersection over union (IoU)

positive cases and avoid false positives and false negatives. Figure 2.21 illustrates the calculation of Dice Coefficient (Fig. 2.21a) and Intersection over Union (IoU) (Fig. 2.21b).

2.6 Summary

The chapter discusses deep learning-based feature extraction networks and their applications in various tasks such as image classification, object detection and segmentation. Image classification involves categorizing entire images based on detected features, while object detection frameworks not only identify objects but also precisely localize them using bounding boxes. Semantic segmentation further refines this process by assigning labels to each pixel in the image, facilitating detailed understanding of image context, crucial in fields like medical imaging and autonomous vehicle navigation. The chapter also highlights the training and testing scheme of deep learning models.

References

1. Krizhevsky, A., Sutskever, I., & Hinton, G. E. (2012). ImageNet classification with deep convolutional neural networks. In *Advances in neural information processing systems*.
2. https://image-net.org/challenges/LSVRC/2010/index.php . Accessed on November 23, 2023.
3. Simonyan, K., & Zisserman, A. (2014). *Very deep convolutional networks for large-scale image recognition*. arXiv:1409.1556 [cs].
4. Szegedy, C., Liu, W., Jia, Y., Sermanet, P., Reed, S., Anguelov, D., Erhan, D., Vanhoucke, V., & Rabinovich, A. (2014). *Going deeper with convolutions*. arXiv:1409.4842.
5. He, K., et al. (2016). Deep residual learning for image recognition. In *Proceedings of the IEEE Conference on Computer Vision and Pattern Recognition*.
6. Lecun, Y., et al. (1998). Gradient-based learning applied to document recognition. In *Proceedings of the IEEE*. https://doi.org/10.1109/5.726791
7. Girshick, R., Donahue, J., Darrell, T., & Malik, J. (2014). Rich feature hierarchies for accurate object detection and semantic segmentation. In *Proceedings of the IEEE Conference on Computer Vision and Pattern Recognition*.
8. Uijlings, J. R. R., van de Sande, K. E. A., Gevers, T., & Smeulders, A. W. M. (2013). Selective search for object recognition. *International Journal of Computer Vision, 104*(2).
9. Girshick, R. (2015). Fast R-CNN. In *Proceedings of the IEEE Conference on Computer Vision and Pattern Recognition*.
10. Ren, S., He, K., Girshick, R., & Sun, J. (2017). Faster R-CNN: towards real-time object detection with region proposal networks. *IEEE Transactions on Pattern Analysis and Machine Intelligence*
11. Zhou, X., Wang, D., & Kráhenbühl, P. (2019). *Objects as points*. arXiv:1904.07850
12. Redmon, J., et al. (2015). *You only look once: Unified, real-time object detection*. arXiv:1506.02640
13. Redmon, J., & Farhadi, A. (2018). *YOLOv3: An incremental improvement*. arXiv:1804.02767
14. Long, J., Shelhamer, E., & Darrell, T. (2015). Fully convolutional networks for semantic segmentation. In *Proceedings of the IEEE Conference on Computer Vision and Pattern Recognition*.

15. Ronneberger, O., Fischer, P., Brox, T. (2015). U-Net: convolutional networks for biomedical image segmentation. In *Medical image computing and computer-assisted intervention (MICCAI)*.
16. Zhou, Z., et al. (2018). UNet++: A nested U-Net architecture for medical image segmentation. In *Deep learning in medical image analysis and multimodal learning for clinical decision support. DLMIA ML-CDS 2018. Lecture notes in computer science* (Vol. 11045). Springer.
17. Huang, H., et al. (2020). UNet3+: A full-scale connected UNet for medical image segmentation. In *ICASSP 2020–2020 IEEE International Conference on Acoustics, Speech and Signal Processing (ICASSP)*. IEEE.
18. Powers, D. M. (2011). Evaluation: from precision, recall and F-measure to ROC, informedness, markedness & correlation. *Journal of Machine Learning Technologies*, *2*(1).

Chapter 3
Attention and Transformer Networks

3.1 Introduction

Attention is a technique utilized to enhance discriminative features (channel attention) and significant regions of interest (spatial attention) within deep learning networks. Attention mechanisms have found extensive applications in areas such as fine-grained image recognition and weakly-supervised learning, demonstrating their remarkable effectiveness. Transformer has revolutionized natural language processing (NLP) by introducing self-attention mechanisms that capture long-range dependencies and improve sequence modeling. Recently, Vision Transformer (ViT) has showcased the potential of Transformer models in computer vision (CV) tasks by effectively capturing global-wise spatial information within images. Numerous self-attention and Transformer-based methods have emerged, demonstrating their effectiveness in fundamental CV tasks and establishing Transformers as the next evolution in deep learning, succeeding Convolutional Neural Networks (CNNs). The following sections delve into self-attention mechanisms, exploring their functionalities and performance in detail. Additionally, an overview of Transformer principles and their applications in imaging is provided. Finally, Vision Transformers (ViTs) and their variants are introduced, highlighting their capabilities in computer vision tasks.

3.2 Attention and Transformer Networks

3.2.1 Self-attention in Transformer Networks

Attention is a technique that selectively enhances relevant parts of the input in relation to a query. When the query and input are the same, it is referred to as self-Attention, a mechanism that enables the model to weigh the importance of different parts of the input relative to each other. The flow of self-attention is illustrated in Fig. 3.1. In the

Y. Chen et al., *Recent Advances in Deep Learning for Medical Image Analysis*, Intelligent Systems Reference Library 278, https://doi.org/10.1007/978-3-031-94791-9_3

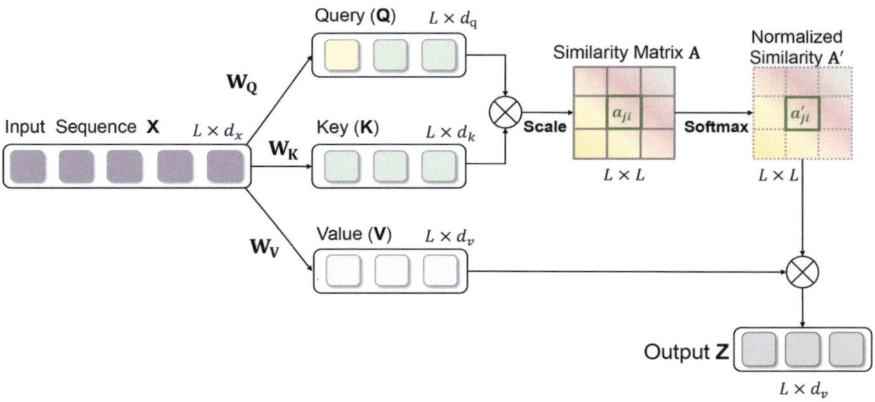

Fig. 3.1 The calculation process of self-attention, where ⊗ represents matrix multiplication. a_{ji}, a'_{ji} are elements of **A** and **A**′

self-attention mechanism, the attention module maps a query (**Q**) and a set of key (**K**)—value (**V**) pairs to an attention value (**Z**). Here, **Q**, **K**, and **V** are derived from linear projections of the input sequence (**X**) through learned weight matrices. The relationship between these elements is then computed using a scaled dot-product function, which calculates the attention score between **Q** and **K** to determine how much influence the corresponding value **V** will have on the final output. The final output is computed as a weighted sum of the values **V**, where the weights are determined by the attention scores **A**. This mechanism allows the model to focus on the most relevant parts of the input sequence, thereby capturing dependencies across different positions. The calculations for this process are detailed in Eqs. (3.1)–(3.4):

(1) **Projection**: The input sequence $\mathbf{X} \in \mathbb{R}^{L \times d_x}$ is mapped to the query (**Q**), key (**K**), and value (**V**) through three different linear projection matrices $\mathbf{W}_Q$, $\mathbf{W}_K$ and $\mathbf{W}_V$:

$$\mathbf{Q} = \mathbf{X}\mathbf{W}_Q \in \mathbb{R}^{L \times d_q}, \quad \mathbf{K} = \mathbf{X}\mathbf{W}_K \in \mathbb{R}^{L \times d_k}, \quad \mathbf{V} = \mathbf{X}\mathbf{W}_V \in \mathbb{R}^{L \times d_v} \tag{3.1}$$

where L is the sequence length, d_x is the input dimension, d_q, d_k and d_v are the mapped dimensions, $d_q = d_k$.

(2) **Dot Product and Scaling**: The dot product is computed between the query (**Q**) and the key (**K**), which indicates the relevance of the key to the query. The resulting raw similarity matrix (**A**) is then scaled through division by the square root of the dimension ($\sqrt{d_k}$) of the key (**K**):

$$\mathbf{A} = \frac{\mathbf{Q}\mathbf{K}^\top}{\sqrt{d_k}} \in \mathbb{R}^{L \times L}, \quad a_{ji} = \frac{q_j k_i}{\sqrt{d_k}} \tag{3.2}$$

where a_{ji}, q_j and k_i are elements of **A**, **Q** and **K**. The scaling step is crucial as it prevents excessively large dot-product values when the dimension of the key (d_k) is large, which could otherwise lead to very small gradients during training and hinder the learning process.

(3) **Softmax**: After scaling, the raw attention scores (**A**) are passed through a softmax function, which normalizes them into a probability distribution ($\mathbf{A}'$). Each attention value a'_{ji} in $\mathbf{A}'$ can be formulated as:

$$a'_{ji} = \mathrm{softmax}(a_{ji}) = \frac{\exp(a_{ji})}{\sum_{k=1}^{L} \exp(a_{jk})} \tag{3.3}$$

(4) **Weighted Sum**: Finally, the normalized attention weights ($\mathbf{A}'$) are used to compute a weighted sum of the value (**V**), resulting in the output (**Z**):

$$\mathbf{Z} = \mathbf{A}'\mathbf{V} \in \mathbb{R}^{L \times d_v}, \quad z_j = \sum_{i=1}^{L} a'_{ji} v_i \tag{3.4}$$

where z_j, a'_{ji} and v_i are elements of **Z**, $\mathbf{A}'$ and **V**, respectively. This weighted sum represents the output of the attention mechanism, emphasizing the parts of the input that the model should focus on most, as dictated by the attention scores.

3.2.2 *Transformer Networks*

Transformer was first introduced in 2017 [1], revolutionizing the field of NLP through the introduction of a novel mechanism known as self-attention. It allows deep learning models to focus on different parts of the input data when making predictions. The structure of transformer encoder is shown in Fig. 3.2. Based on self-attention mechanism, Transformer incorporates multiple-head self-attention to jointly attend to information from various representation subspaces. In addition to Multi-head Attention, the key components of a Transformer include positional encoding, add

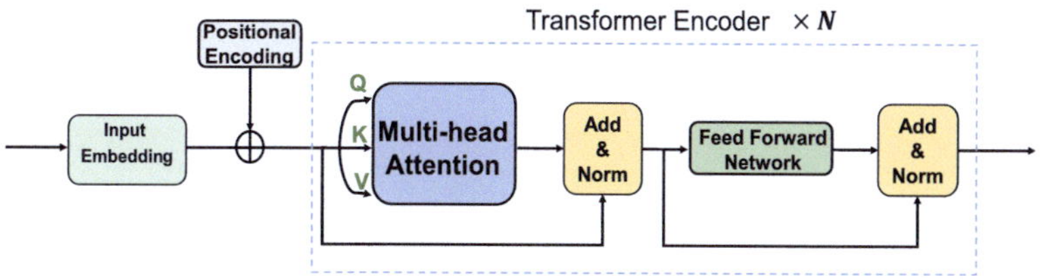

Fig. 3.2 The overview of transformer block, where ⊕ represents element-wise addition. *N* is the number of blocks

& norm, the feed-forward network, and input embedding, which are described as follows:

Multi-head Attention. It is a fundamental component of Transformer-based models, enabling the model to capture long-range dependencies within the input sequence. Unlike traditional attention mechanisms, which consider the relationship between a query and a single key, multi-head attention allows the model to focus on different parts of the input sequence simultaneously. This is achieved by dividing the input into multiple parallel subspaces, called heads, and applying attention independently within each head. The output of each attention head is computed following Eqs. (3.1)–(3.4). Rather than performing a single attention function with d_{model}-dimensional keys, values, and queries, experimental results demonstrate that it is advantageous to linearly project the queries, keys, and values h times using distinct learned linear projections into d_k, d_k, and d_v dimensions, respectively. For each of these projected versions of queries, keys, and values, the attention function is then computed in parallel. The final output of the multi-head attention mechanism is produced by concatenating (i.e., Concat($\cdot$)) the outputs from all heads and passing the result through a linear layer, which can be expressed as:

$$\begin{aligned} \text{MultiHead}(\mathbf{X}) &= \text{Concat}(\text{head}_1, ..., \text{head}_h)\mathbf{W}_O \\ \text{head}_i &= \text{Attention}(\mathbf{QW}^i_Q, \mathbf{KW}^i_K, \mathbf{VW}^i_V) \end{aligned} \tag{3.5}$$

where head_i is the attention value of the i-th head, h is the number of heads. The projections are parameter matrices $\mathbf{W}^i_Q \in \mathbb{R}^{d_{\text{model}} \times d_k}$, $\mathbf{W}^i_K \in \mathbb{R}^{d_{\text{model}} \times d_k}$, $\mathbf{W}^i_V \in \mathbb{R}^{d_{\text{model}} \times d_v}$, and $\mathbf{W}_O \in \mathbb{R}^{hd_v \times d_{\text{model}}}$. By using multiple heads, the model can effectively capture different patterns within the input sequence including both long-term and short-term dependency, thus improving its representation power.

Positional Encoding. Since the Transformer lacks an inherent sense of order, unlike RNNs [2] or LSTMs [3], positional encodings are added to the input embeddings to convey information about the position of tokens in a sequence. The formula of Positional Encoding is shown in Eq. (3.6):

$$\begin{aligned} PE(pos, 2i) &= \sin\left(\frac{pos}{10000^{\frac{2i}{d_{\text{model}}}}}\right) \\ PE(pos, 2i+1) &= \cos\left(\frac{pos}{10000^{\frac{2i}{d_{\text{model}}}}}\right) \end{aligned} \tag{3.6}$$

where pos represents the position. Given a sentence length of L, $pos = 0, 1, \ldots, L-1$. Each dimension of the positional encoding is represented by a sinusoidal function, with wavelengths following a geometric progression from 2π to $10000 \cdot 2\pi$. The dimension index i corresponds to the embedding dimension: the sine function is applied when i is even, and the cosine function is applied when i is odd.

This design is chosen as it allows the model to effectively learn relative positional relationships. Specifically, for any fixed offset k, PE_{pos+k} can be expressed as a linear function of PE_{pos}.

Add and Norm. Residual connections (Add) help mitigate the issue of vanishing or exploding gradients during training. Meanwhile, Layer Normalization (Norm) in Transformers stabilizes and accelerates the training process by standardizing the inputs to each layer.

Feed-forward Network. In addition to the attention layers, the Transformer architecture incorporates a feed-forward network (FFN) to facilitate the capture of complex patterns within the data. This consists of two linear transformations with a ReLU activation in between, which can be formulated as:

$$\mathrm{FFN}(\mathbf{X}) = \max\left(0, \mathbf{X}\mathbf{W}_1 + \mathbf{b}_1\right)\mathbf{W}_2 + \mathbf{b}_2 \tag{3.7}$$

where $\mathbf{W}_1$, $\mathbf{b}_1$, $\mathbf{W}_2$ and $\mathbf{b}_2$ are the parameters of the two linear projection layers.

Input Embedding. Transformer is first used for NLP and then is also used for time sequence data analysis. Its expanded version (i.e., Vision Transformer) is now widely used for vision tasks. In NLP, each word in a sentence is converted to a numerical representation called a token. The token is then embedded to a vector used as input of the transformer. During the training process, the embedding matrix is treated as a set of parameters of the model and it is collaboratively optimized with the transformer. In time sequence data analysis, each time signal is treated as a token. For computer vision tasks, the image is divided to several patches and each patch is treated as a token. Then, we use learned embeddings to convert the input tokens to vectors of dimension d_{model}. The details about ViT and its variants will be described in Sect. 3.3.

3.3 Vision Transformers: An Exploration of Transformers for Image Understanding

The ViT [4] is a pioneering model that has revolutionized the landscape of computer vision. Unlike traditional CNNs, ViT embraces the Transformer architecture initially designed for NLP tasks and adapts it for image data. It achieves this by dividing images into non-overlapping patches and formulating them as sequences of tokens. Then a neural network called Transformer is employed to understand them better. By relying on self-attention, ViT can consider relationships between all patches simultaneously, which can be advantageous for recognizing complex patterns and contexts in images, enabling it to capture long-range dependencies in images effectively. This unique approach to image processing has led to remarkable achievements in various computer vision tasks, particularly image classification.

ViT's remarkable achievements have led to the development of various specialized variants, including DeiT [5], Pyramid Vision Transformer(PVT) [6], Swin Transformer [7], and CoAtNet [8]. These frameworks are carefully designed to address specific challenges in computer vision. They offer solutions for enhancing data efficiency, handling multi-scale features, improving scalability, and providing versatility in feature extraction. This diversity of models has significantly expanded the capabilities of ViT, inspiring further research and fostering practical applications. The following section provides a detailed introduction to ViT and its variants, with a particular focus on their design principles and how these designs address specific challenges.

3.3.1 Vision Transformer

Self-attention based deep learning models, in particular Transformers [1], are now widely used for NLP tasks. Inspired by the success of Transformers in NLP, researchers investigated the potential of applying Transformer-based designs and approaches to images, making necessary design modifications to accommodate the unique characteristics of visual data. This led to the development of the ViT model for image classification tasks. The design of ViT framework is depicted in Fig. 3.3.

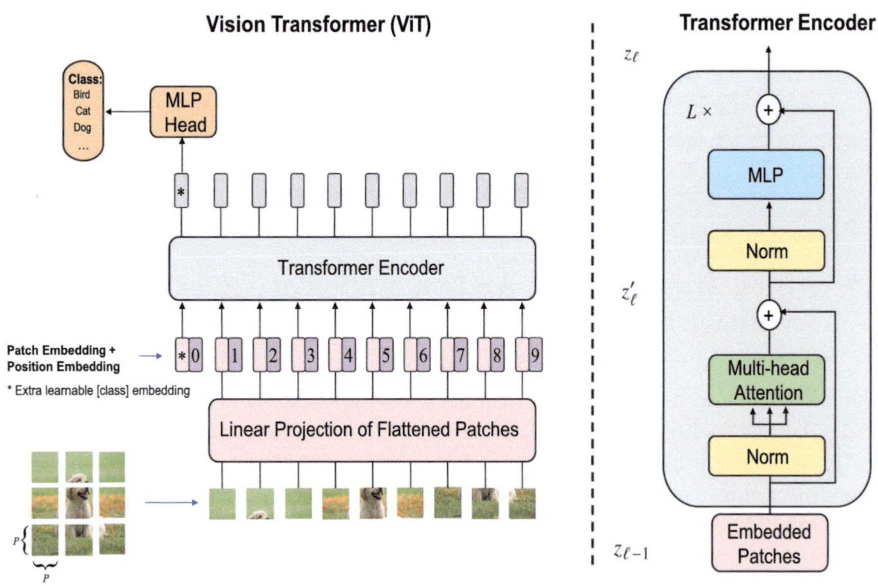

Fig. 3.3 ViT overview: ViT splits the image into fixed-size patches. It adds linear embeddings to each patch, includes position embeddings, and feeds the resulting sequence of vectors into a conventional transformer encoder. For classification, ViT also includes an additional learnable classification token into the sequence

Within the context of the standard Transformer model, which conventionally processes 1D sequences of token embeddings as input, handling 2D images requires splitting the input image $\mathbf{X} \in \mathbb{R}^{H \times W \times C}$ into a sequence of 2D flattened vectors, forming non-overlapping patches denoted as $\mathbf{X}_p \in \mathbb{R}^{N \times (P^2 \cdot C)}$. Here, (H, W) is the size of input image, C is the number of channels, (P, P) denotes the size of each patch, and $N = HW/P^2$ indicates the resulting number of patches. This N also serves as the functional input sequence length for the Transformer. Considering that the Transformer maintains a consistent latent vector size D across all layers, the ViT adopts a distinctive strategy: it flattens the patches and maps them to D dimensions through a trainable linear projection, as illustrated in Eq. (3.8). This projected outcome is referred to as the patch embedding.

$$z_0 = [\mathbf{X}_{class};\ \mathbf{X}_p^1\mathbf{E};\ \mathbf{X}_p^2\mathbf{E};\ ...;\ \mathbf{X}_p^N\mathbf{E}] + \mathbf{E}_{pos}$$
$$\mathbf{E} \in \mathbb{R}^{(P^2 \cdot C) \times D},\ \mathbf{E}_{pos} \in \mathbb{R}^{(N+1) \times D} \tag{3.8}$$

Similar to the class token in Bidirectional Encoder Representations from Transformers (BERT) [9], ViT introduces a learnable embedding ($z_0^0 = \mathbf{X}_{class}$) to learn object classes (in the case of image classification) at the start of the sequence of embedded patches. To retain positional information, ViT incorporates standard learnable 1D position embeddings into the new sequence of embeddings with a length of $N + 1$. Following this, ViT passes these generated embeddings into the Transformer encoder. In the subsequent layers, a multi-head self-attention mechanism is utilized, where the class embedding interacts with the semantics of each patch, extracting crucial information and global dependencies. Consequently, the state of the class embedding at the output of the Transformer encoder (z_L^0) can serve as the image representation y (as defined in Eq. 3.9, where LN is the Layer Normalization, LN could stabilize the training process of neural networks, speed up convergence, and improve the model's generalization ability).

$$y = LN(z_L^0) \tag{3.9}$$

The Transformer encoder in the ViT model, as described in [1], is organized into a series of alternating layers. Each layer is composed of Multi-head Self-Attention (MSA) and Multi Layer Perceptron (MLP) blocks, as detailed in Eq. (3.10) and Eq. (3.11), respectively. LN is applied before each block, and residual connections are established after each block, following established practices [10, 11]. The MLP comprises two linear layers and employs a Gaussian Error Linear Units (GELU) non-linearity [12].

$$z'_\ell = MSA(LN(z_{\ell-1})) + z_{\ell-1},\ \ \ell = 1...L \tag{3.10}$$

$$z_\ell = MLP(LN(z'_\ell)) + z'_\ell,\ \ \ell = 1...L \tag{3.11}$$

After obtaining the image representation (y), ViT utilizes an MLP layer to perform the image classification task. This demonstrates ViT's successful adaptation of the Transformer architecture—originally developed for NLP—to the domain of computer vision.

3.3.2 DeiT: Data-Efficient Image Transformers

The Transformer architecture, initially successful in NLP, has been extended to computer vision with the ViT model. However, ViT still lags behind state-of-the-art CNNs, particularly in scenarios with limited training data. This issue stems from the lack of certain beneficial inductive biases in vanilla Transformer layers, which are inherent in CNNs. As a result, ViT requires substantial data and computational resources to mitigate this limitation, posing challenges in resource-constrained environments. To address this issue, DeiT [5] introduces a teacher-student strategy designed for Transformers. It employs a distillation token to facilitate learning from the teacher through attention mechanisms. This token-based distillation, particularly when a ConvNet serves as the teacher, delivers competitive results in image classification tasks with limited training data, demonstrating a more data-efficient approach compared to the original ViT model.

Consider a proficient image classifier as the teacher model, which could be a CNN or a combination of classifiers. Typically, soft distillation [13, 14] is employed to minimize the Kullback-Leibler divergence between the softmax probabilities generated by the teacher and student models. Let $\mathbf{Z}_t$ denote the logits produced by the teacher model and $\mathbf{Z}_s$ represent the logits from the student model. The parameter τ is introduced as the temperature for distillation. Here, we denote KL as the Kullback-Leibler divergence, L_{CE} as the cross-entropy loss with respect to ground truth labels y, ψ as the softmax function, and τ as the distillation temperature, a parameter used in the process of knowledge distillation. The role of τ is to control the smoothness of the softmax function, thereby affecting the calculation of the distillation loss. The $L_{teacher}$ is expressed in Eq. (3.12), and soft distillation objective is expressed in Eq. (3.13):

$$L_{teacher} = \tau^2 KL(\psi(\mathbf{Z}_s/\tau), \psi(\mathbf{Z}_t/\tau)) \tag{3.12}$$

$$L_{global} = (1-\lambda)L_{CE}(\psi(\mathbf{Z}_s), y) + \lambda\tau^2 KL(\psi(\mathbf{Z}_s/\tau), \psi(\mathbf{Z}_t/\tau)) \tag{3.13}$$

DeiT introduced a variation of the distillation method in which it adopts the teacher's hard decision as the true label. Let $y_t = argmax_c \mathbf{Z}_t(c)$ represent the teacher's hard decision. The objective associated with this distillation approach using hard labels is shown in Eq. (3.14):

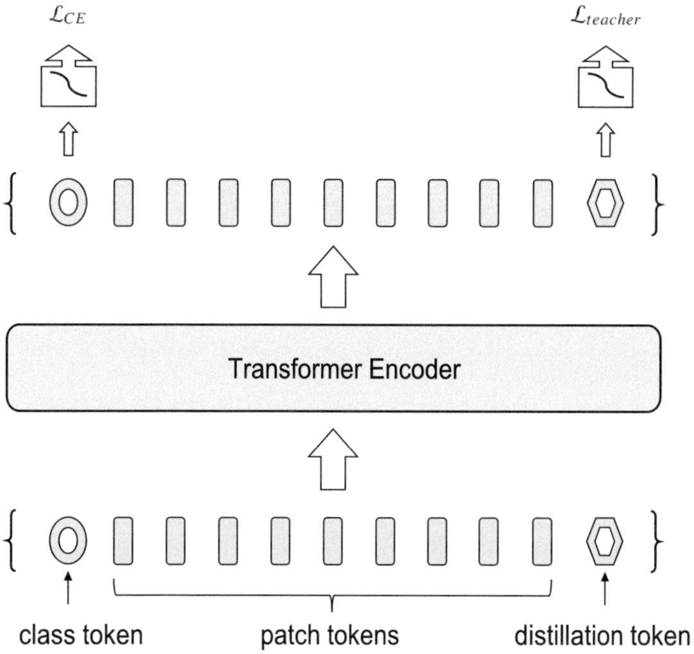

Fig. 3.4 DeiT introduces a novel distillation token that interacts with both class and patch tokens through self-attention layers. Similar to the class token, the distillation token is employed during the network's output phase, aiming to replicate the (hard) label predicted by the teacher rather than the true label. The learning of both class and distillation tokens in the transformer is achieved through back-propagation

$$L_{global}^{hardDistill} = \frac{1}{2} L_{CE}(\psi(\mathbf{Z}_s), y) + \frac{1}{2} L_{CE}(\psi(\mathbf{Z}_s), y_t) \tag{3.14}$$

As depicted in Fig. 3.4, DeiT integrates the distillation token into the initial embeddings, which encompass patches and the class token. Similar to the class token, the distillation token interacts with other embeddings via self-attention, with its output generated by the network after passing through the final layer. The primary objective of the distillation token is defined by the distillation component of the loss function in Eq. (3.14). By incorporating the distillation embedding, this model can gain insights from the teacher's output, aligning with standard distillation practices while complementing the role of the class embedding.

In summary, DeiT improves ViT's data efficiency through knowledge distillation, particularly beneficial for limited data scenarios.

3.3.3 PVT: Pyramid Vision Transformer

In contrast to ViT, which is tailored specifically for image classification, the PVT [6] introduces a novel pyramid structure into the Transformer architecture, aiming to create a CNN-free universal backbone. This innovation enables PVT to tackle a range of downstream dense prediction tasks, including detection and segmentation. PVT offers several distinct advantages.

- PVT facilitates dense partitioning of images for high-resolution output, unlike the low-resolution output of ViT. Additionally, it employs a progressively shrinking pyramid structure to reduce computational complexity and memory usage when processing large feature maps.
- PVT combines the strengths of both CNNs and Transformers, making it a versatile, convolution-free backbone that can directly replace CNN-based backbones.

While ViT segments images into fixed-size patches, the PVT takes a different approach by utilizing a pyramid structure to process images at multiple scales. This innovative architecture enables PVT to capture a diverse range of features, spanning from fine details to broader contextual information. Figure 3.5 compares the designs of CNNs, ViT, and PVT.

The overall network structure of PVT is shown in Fig. 3.6. In PVT, an image is processed through a series of stages each producing feature maps at different resolutions. At each stage, the feature map is partitioned into non-overlapping patches, similar to ViT, but with varying resolutions. These patches are then flattened and linearly embedded. As we progress deeper into the network, the number of patches decreases, and their resolution decreases, accordingly, forming a pyramid structure.

The PVT consists of four stages. Each stage produces feature maps of different scales and follows a similar structure, consisting of a patch embedding layer and Transformer encoder layers. In the first stage, the input image size is $H \times W \times 3$, where channel dimension C is 3. It is initially split into $\frac{H \times W}{4 \times 4}$ patches, each patch size is 4×4. These patches are then flattened and linearly projected to obtain embedded patches of size $\frac{H \times W \times C_1}{4 \times 4}$, thus the spatial size of the input is reduced using patching

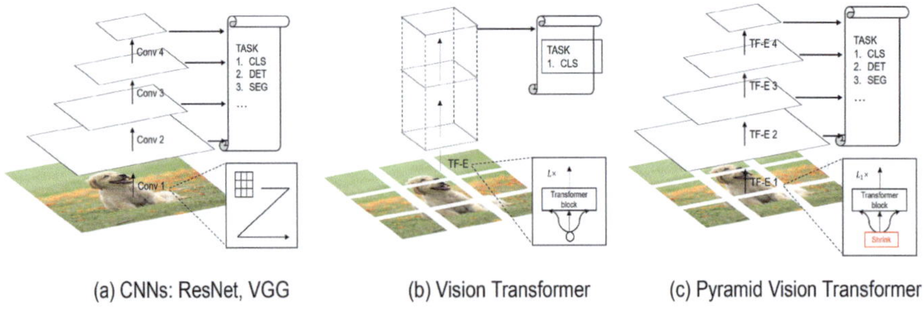

Fig. 3.5 Comparisons of CNNs, ViT and PVT architectures

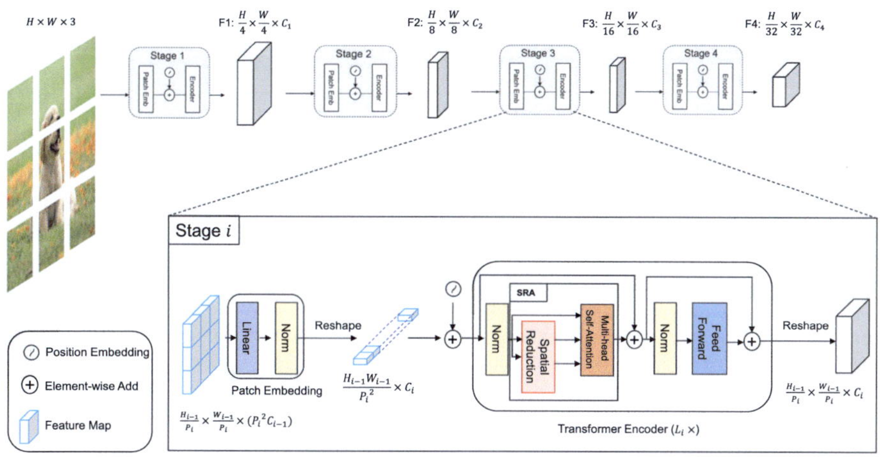

Fig. 3.6 The overall network structure of PVT

embedding, and the channel dimension is remapped to C_1. Subsequently, the embedded patches, along with position embeddings, are fed into an L_1-layer Transformer encoder, and then we reshape the resulted feature back to the spatial space to produce $F1$. In a manner similar to the first stage, the outputs $F2$ to $F4$ for stages 2 to 4 are obtained, with respective output strides of 8, 16, and 32, resulting in a feature pyramid $(F1, F2, F3, F4)$.

The encoder in PVT is similar to the standard Transformer, consisting of MSA and MLP blocks, with layer normalization (LN) applied before each block and residual connections after each block. However, PVT introduces a novel Spatial Reduction Attention (SRA) layer, which replaces the conventional MSA layer. The proposed SRA also accepts **Q**, **K**, **V** as inputs, but differs in that SRA reduces the width and height of **K** and **V** by a factor of R_i, as shown in Fig. 3.7.

As shown in Eq. (3.15), let us denote the **K** or **V** in stage i as $x \in \mathbb{R}^{h_i \times w_i \times C_i}$. The operation $Reshape(x, R_i)$ reshapes x into $\frac{h_i \times w_i}{R_i^2} \times (R_i^2 C_i)$. Additionally, $\mathbf{W}^s \in \mathbb{R}^{R_i^2 C_i \times C_i}$ represents a linear projection, which reduces the dimension of the input to C_i.

$$SR(x) = Norm(Reshape(x, R_i)\mathbf{W}^s) \tag{3.15}$$

In summary, the pyramid structure of PVT offers increased flexibility and efficiency compared to ViT, particularly for tasks necessitating multi-scale feature extraction. This attribute renders PVT a potent tool suitable for a wide range of computer vision applications.

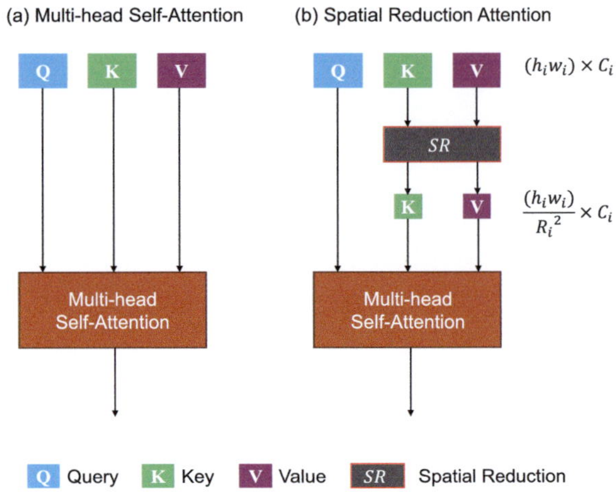

Fig. 3.7 The multi-head self-attention (MSA) layer and the spatial reduction attention (SRA) layer

3.3.4 Swin Transformer

The emergence of the Swin Transformer [7] stems from the necessity to overcome specific limitations posed by the ViT in handling large-scale image data. While ViT introduced a groundbreaking approach by considering images as sequences of fixed-size patches, this method is computationally intensive and less effective in complex scenes with varying scales. The Swin Transformer aims to enhance flexibility and efficiency, especially in capturing multi-scale features within images. To achieve this goal, the Swin Transformer introduces different key changes in design to overcome these challenges:

- Hierarchical Representation: Unlike ViT's uniform patch size, Swin Transformer adopts a hierarchical structure that allows for variable-sized patch representations (as shown in Fig. 3.8). This method effectively captures features at multiple scales.
- Shifted Window Self-Attention (as shown in Fig. 3.9): Swin Transformer reduces computational complexity by implementing self-attention within local windows. These windows are shifted across layers, facilitating cross-window connections and preserving long-range interactions.
- Cross-Window Connections: To ensure information flow across windows, the Swin Transformer employs a novel mechanism enabling cross-window interaction. This feature enhances the model's capacity to capture broader contextual information.
- Scalable Architecture: The design of Swin Transformer is inherently scalable, making it adaptable to various image sizes and complexities. This scalability makes it applicable across a wide range of computer vision tasks.

The basic concept of the architecture of Swin Transformer lies in the shifting window self-attention mechanism. Initially, the feature map is divided into non-overlapping windows, and self-attention is computed within each of these windows. This local self-attention approach significantly reduces computational complexity

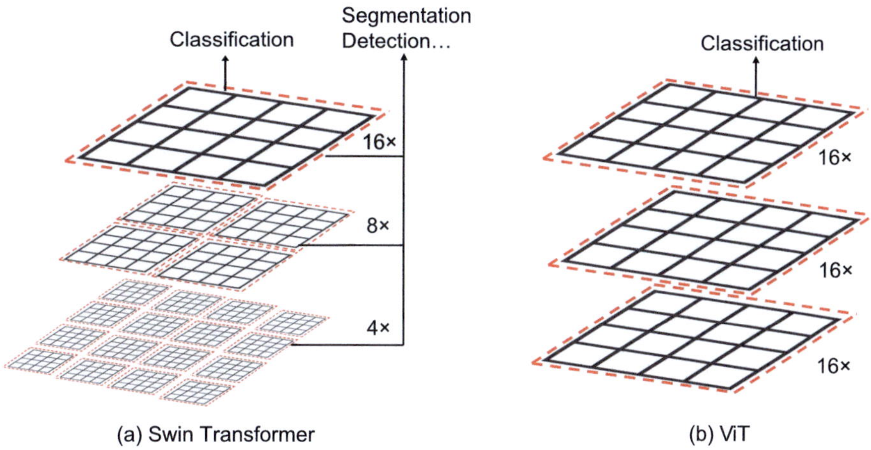

Fig. 3.8 Comparison between swin transformer and ViT: **a** The swin transformer constructs hierarchical feature maps by aggregating image patches (depicted in black frames) in deeper layers. Its linear computational complexity with respect to the input image size arises from the self-attention computation limited to each local window (depicted in the red dashed frame). This characteristic enables it to serve as a versatile backbone for various tasks, including image classification and dense recognition. **b** In contrast, previously proposed vision Transformers generate feature maps with a singular low resolution, resulting in quadratic computational complexity concerning input image size

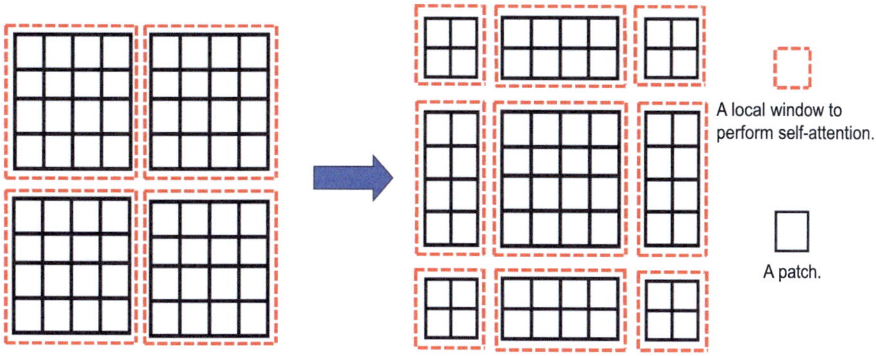

Fig. 3.9 The shifted window mechanism operates as follows: In the initial layer (left), a conventional window partitioning scheme is employed, where self-attention is computed within each window. As we move to the subsequent layer $(l + 1)$ (right), the window partitioning undergoes a shift, creating new windows. Importantly, the self-attention computation within these new windows extends beyond the boundaries of the previous windows in layer (l), establishing connections among them

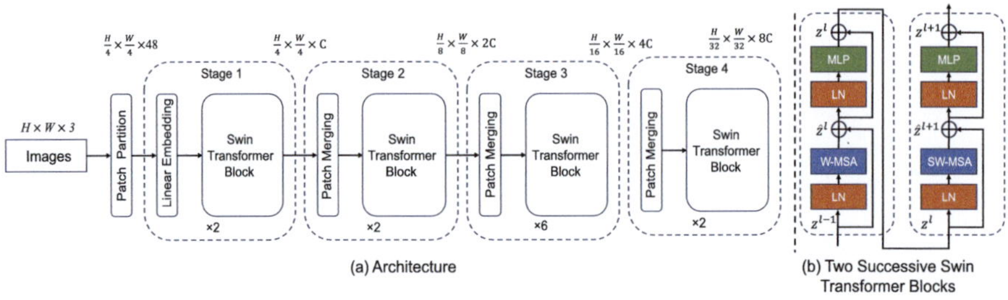

Fig. 3.10 Swin transformer overview: **a** The structure of a swin transformer (denoted as Swin-T); **b** a pair of consecutive swin transformer blocks

compared to ViT's global self-attention, as it limits the number of pairwise comparisons. To capture global context and enable interaction between adjacent windows, a shifting operation is introduced in subsequent layers. This shift in window partitioning between layers enables the model to capture relationships across the entire image, preserving its ability to understand long-range dependencies.

The overview of the Swin Transformer is depicted in Fig. 3.10. The Swin Transformer initiates by partitioning the input image into small patches (e.g., each patch is of size 4 × 4 pixels). These patches are subsequently flattened and linearly embedded. The patch embedding process can be represented as follows:

$$\mathbf{X}_0 = [\mathbf{X}_p^1\mathbf{E}; \mathbf{X}_p^2\mathbf{E}; ...; \mathbf{X}_p^N\mathbf{E}] + \mathbf{E}_{pos} \tag{3.16}$$

where $\mathbf{X}_p^i$ is the flattened vector of the ith patch, $\mathbf{E}$ is the embedding matrix, and $\mathbf{E}_{pos}$ is the positional embedding.

In order to generate a hierarchical representation, the network reduces the number of tokens through patch merging layers as it progresses in depth. Suppose the input patch dimension is C, the patch merging layer combines the features of each group of neighboring 2 × 2 patches and employs a linear layer on the concatenated features, resulting in a dimension of $4C$. This process reduces the number of tokens by a factor of $2 \times 2 = 4$ (equivalent to a 2× downsampling of resolution), and the output dimension is adjusted to $2C$. At each subsequent stage, the Swin Transformer merges adjacent patches to form new patches, effectively reducing the resolution and increasing the receptive field. This merging strategy can be represented as:

$$\mathbf{X}^{l+1} = PatchMerging\left(\mathbf{X}^l\right) \tag{3.17}$$

where $\mathbf{X}^l$ is the output of the lth layer.

Swin Transformer employs shifted window self-attention to balance computational efficiency and the ability to capture long-range dependencies:

- Window Partitioning: The feature map is divided into non-overlapping windows. Within each window, self-attention is computed.

- Shifting Operation: To facilitate cross-window connection and prevent the isolation of windows, the partitioning windows are shifted for alternate layers.

The shifted window mechanism can be represented as:

$$\hat{z}^l = W\text{-}MSA\left(LN\left(z^{l-1}\right)\right) + z^{l-1} \tag{3.18}$$

$$z^l = MLP\left(LN\left(\hat{z}^l\right)\right) + \hat{z}^l \tag{3.19}$$

$$\hat{z}^{l+1} = SW\text{-}MSA\left(LN\left(z^l\right)\right) + z^l \tag{3.20}$$

$$z^{l+1} = MLP\left(LN\left(\hat{z}^{l+1}\right)\right) + \hat{z}^{l+1} \tag{3.21}$$

where $\hat{z}^l$ and z^l denote the output features of the $(S)W\text{-}MSA$ module and the MLP module for block l, respectively. $W\text{-}MSA$ is the window-based multi-head self-attention, and $SW\text{-}MSA$ is the shifted window multi-head self-attention.

The self-attention mechanism within each window is expressed as:

$$SA\left(\mathbf{Q},\ \mathbf{K},\ \mathbf{V}\right) = softmax\left(\frac{\mathbf{QK}^T}{\sqrt{d_k}}\right)\mathbf{V} \tag{3.22}$$

where $\mathbf{Q}$, $\mathbf{K}$, and $\mathbf{V}$ represent the queries, keys, and values respectively, and d_k is a scaling factor. The output of the final layer of Swin Transformer is used as the representation for downstream tasks, such as classification, detection, or segmentation.

In Summary, the Swin Transformer overcomes the limitations of traditional ViT in processing large-scale image data through its innovative design. It incorporates a hierarchical structure that adapts to varying scales of image features and employs a computationally efficient shifted window self-attention mechanism.

3.3.5 CoAtNet: Marrying Convolution and Attention

Owing to their distinct structural designs, CNN and ViT each present unique advantages. CoAtNet [8] is introduced to combine these two strengths: the generalization ability of convolution in CNNs and the long-range modeling capacity of self-attention in Transformers. A systematic understanding of the respective roles of these two components is achieved by combining convolution and self-attention. These are the two key insights behind this argument:

- It is observed that the commonly employed depthwise convolution can be seamlessly integrated into attention layers using a straightforward relative positional encoding mechanism.

- The simple stacking of convolutional and attention layers, when appropriately arranged, proves remarkably effective in enhancing both generalization performance and model capacity.

Regarding the first insight, it is observed that both depthwise convolution and self-attention can be represented as a weighted sum of values across dimensions within a predefined receptive field. For convolution, this involves using a fixed kernel to extract information from a localized receptive field.

$$y_i = \sum_{j \in L(i)} w_{i-j} \odot x_j \tag{3.23}$$

where x_i, y_i are the input and output at position i respectively, and $L(i)$ denotes a local neighborhood of i, such as a 3×3 window centered at i in image processing. $\odot$ denotes the Hadamard product, i.e., element-wise multiplication between corresponding elements of the tensors w_{i-j} and x_j. These weights are determined during training and remain constant during inference, irrespective of input values. The weight in the summation depends only on the relative positions of pixel i and its neighbors due to the dependency on w_{i-j}. This reliance on relative positions in w_{i-j} for convolution produces the "translation equivalence" property, which has been found to improve generalization on datasets of limited size [15].

On the other hand, self-attention extends the receptive field to encompass the entire spatial domain and calculates weights by normalizing pairwise similarities between elements (x_i, x_j) in the sequence.

$$\mathbf{A}_{i,j} = \frac{\exp\left(x_i^\top x_j\right)}{\sum_{k \in \mathcal{G}} \exp\left(x_i^\top x_k\right)} \tag{3.24}$$

$$y_i = \sum_{j \in \mathcal{G}} \mathbf{A}_{i,j} x_j \tag{3.25}$$

where $\mathcal{G}$ refers to the global spatial space and the attention weights are denoted as $\mathbf{A}_{i,j}$. Unlike convolutional operations, these attention weights dynamically depend on the input representation, making self-attention more adept at capturing intricate relational interactions among various spatial positions. This characteristic is particularly desirable when handling high-level concepts. The extent of the receptive field stands out as a critical distinction between self-attention and convolution. In broad terms, a larger receptive field yields greater contextual information, potentially resulting in increased model capacity. Consequently, the pursuit of a global receptive field has been a primary incentive for incorporating self-attention in vision tasks.

Considering these properties, a combination of the two components (Eq. 3.26) is presented, which corresponds to a particular variant of relative self-attention block [16, 17]. This proposed self-attention block can incorporate the three advantageous characteristics mentioned in Table 3.1 to produce feature y_i: Translation Equivariance in w_{i-j} and Input-adaptive Weighting in $x_i^\top x_j$ and Global Receptive Field in $j \in \mathcal{G}$.

Table 3.1 Advantages present in convolution and self-attention mechanisms

Properties	Convolution	Self-attention
Translation equivariance	✓	
Input-adaptive weighting		✓
Global receptive field		✓

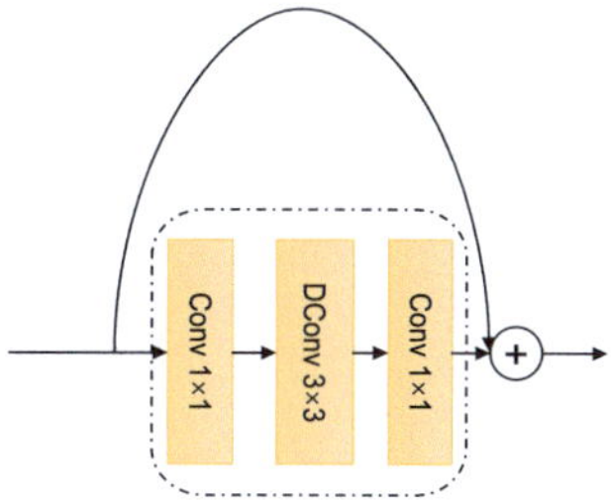

Fig. 3.11 The MBConv block comprises three main operations: (1) The first operation is 1×1 convolution. (2) then, 3×3 depthwise convolution. (3) Finally, 1×1 convolution, followed by a residual connection to complete the process

$$y_i = \sum_{j \in \mathcal{G}} \frac{\exp\left(x_i^\top x_j + w_{i-j}\right)}{\sum_{k \in \mathcal{G}} \exp\left(x_i^\top x_k + w_{i-j}\right)} x_j \tag{3.26}$$

In terms of the second insight, which involves vertically stacking different types of computational blocks (convolution and attention) to form a complete network, the approach is as follows: For the convolution operation, the MBConv block [18] (MobileNetV2 Inverted Bottleneck block, depicted in Fig. 3.11) is selected. This choice is inspired by a shared design feature between Transformers and the feed-forward neural network (FFN) module in MBConv, called the "inverted bottleneck". This design involves initially enlarging the channel size of the input by a factor of 4 and subsequently projecting the 4×-expanded hidden state back to the original channel size, facilitating the establishment of a residual connection. For the attention operation, the relative attention mechanism introduced in Eq. (3.26) is selected. The combination of these two components yields the CoAtNet architecture.

Let's denote C as Convolution and T as relative attention. They compare five model designs: $C-C-C-C$, $C-C-C-T$, $C-C-T-T$, $C-T-T-T$, and VIT_{REL}, where VIT_{REL} signifies replacing the self-attention in the original ViT with relative attention. These five designs are evaluated based on generalization and model capacity to develop CoAtNet. For generalization comparison, they focus on the gap between the training loss and the evaluation accuracy. When two models exhibit identical training losses, the one with superior evaluation accuracy demonstrates enhanced generalization capabilities, indicating better performance on unseen evaluation datasets. Regarding model capacity comparison, they evaluate the ability to handle extensive training datasets. In scenarios where sufficient training data is available and overfitting is not a concern, the model with greater capacity achieves superior final performance with a reasonable number of training steps. Through

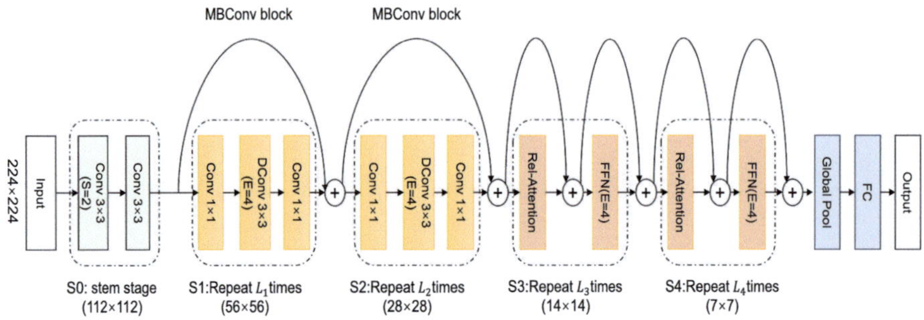

Fig. 3.12 CoAtNet overview: following the initial stage, CoAtNet employs two MBConv blocks to process low-level image information. Here, $E = 4$ indicates the squeeze-and-excitation network (SENet [19]) to perform channel attention after the depthwise convolution (DConv). Subsequently, after the two MBConv blocks, CoAtNet utilizes two relative attention layers to capture high-level interactions. Finally, classification is performed using a pooling layer and a fully connected (FC) layer

Table 3.2 The CoAtNet model family comprises blocks (L), with D representing the hidden dimension (number of channels)

Stages	Size	CoAtNet-0	CoAtNet-1	CoAtNet-2	CoAtNet-3	CoAtNet-4
S0-Conv	1/2	L=2, D=64	L=2, D=64	L=2, D=128	L=2, D=192	L=2, D=192
S1-MBConv	1/4	L=2, D=96	L=2, D=96	L=2, D=128	L=2, D=192	L=2, D=192
S2-MBConv	1/8	L=3, D=192	L=6, D=192	L=6, D=256	L=6, D=384	L=12, D=384
S3-TFMReal	1/16	L=5, D=384	L=14, D=384	L=14, D=512	L=14, D=768	L=28, D=768
S4-TFMReal	1/32	L=2, D=768	L=2, D=768	L=2, D=1024	L=2, D=1536	L=2, D=1536

Throughout all convolution and MBConv blocks, a kernel size of 3 is consistently used. For transformer blocks, the size of each attention head is configured to be 32. The inverted bottleneck's expansion rate is consistently set to 4, and the expansion (shrink) rate for the squeeze-and-excitation (SE) operation remains consistently at 0.25

experiments, they discover that the $C - C - T - T$ architecture yields promising results in terms of both generalization and model capacity. Consequently, they propose the CoAtNet design, depicted in Fig. 3.12. The details of the CoAtNet model family are outlined in Table 3.2.

In summary, CoAtNet boosts ViT's feature extraction versatility by unifying depthwise convolution and self-attention through relative attention, stacking hybrid convolutional-transformer stages (C-C-T-T), and implementing gradual downsampling for multi-scale hierarchical feature learning.

3.4 Summary

The Transformer model, leveraging self-attention, has revolutionized NLP by capturing long-range dependencies and improving sequence modeling. Its extension to computer vision through the ViT demonstrates similar success in capturing spatial information in images. The chapter provides a detailed discussion of attention mechanisms, Transformer principles, and their applications in imaging, concluding with an introduction to ViT and its variants, emphasizing their capabilities in computer vision tasks. As shown in Table 3.3, DeiT focuses on enhancing data efficiency by utilizing a distillation process that transfers knowledge from a teacher model to a

Table 3.3 Comparative analysis of vision transformer and its variants

Model	Core innovation	Features and advantages	Typical applications
ViT [4]	Transformer adaptation for CV	(1) Global self-attention mechanism (2) Long-range dependency modeling (3) Large-scale pretraining required	Image classification
DeiT [5]	Knowledge distillation	(1) Data-efficient training (2) No massive pretraining (3) Maintains ViT benefits	Image classification
PVT [6]	Pyramidal feature hierarchy	(1) Spatial reduction attention (2) Multi-scale feature extraction (3) Computational efficiency gains	Image classification dense prediction
Swin [7]	Hierarchical shifted window attention	(1) Linear computational complexity (2) Local-global fusion (3) High-resolution input handling	Image classification dense prediction
CoAtNet [8]	Convolution-attention hybrid architecture	(1) Translation equivariance (2) Input-adaptive weighting (3) Global receptive field	Image classification dense prediction

student model. PVT introduces a pyramid structure that effectively handles multi-scale features, allowing the model to capture information at various resolutions. Swin Transformer excels in improving scalability through its hierarchical architecture, which is well-suited for processing high-resolution images. CoAtNet provides versatility in feature extraction by combining the strengths of convolutional and attention mechanisms. These design philosophies have significantly broadened the capabilities of ViT and have encouraged further research and practical applications.

References

1. Vaswani, A., Shazeer, N., Parmar, N., Uszkoreit, J., Jones, L., Gomez, A. N., Kaiser, ł. & Polosukhin, I. (2017). Attention is all you need. *Advances in neural information processing systems*, (p. 30).
2. Zaremba, W., Sutskever, I., & Vinyals, O. (2014). *Recurrent neural network regularization*. arXiv preprint arXiv:1409.2329.
3. Hochreiter, S., et al. (1997). Long short-term memory. *Neural Comput, 9*(8), 1735–1780.
4. Dosovitskiy, A., Beyer, L., Kolesnikov, A., et al. (2020). *An image is worth* 16 *words: Transformers for image recognition at scale*. arXiv preprint arXiv:2010.11929.
5. Touvron, H., Cord, M., Douze, M., Massa, F., Sablayrolles, A., & Jégou, H., (2021). Training data-efficient image transformers and distillation through attention. *PMLR* (pp. 10347–10357).
6. Wang, W., Xie, E., Li, X., Fan, D. P., Song, K., Liang, D., Lu, T., Luo, P., & Shao, L. (2021). Pyramid vision transformer: A versatile backbone for dense prediction without convolutions. *Proceedings of the IEEE/CVF international conference on computer vision* (pp. 568–578).
7. Liu, Z., Lin, Y., Cao, Y., Hu, H., Wei, Y., Zhang, Z., Lin, S., & Guo, B., (2021). Swin transformer: Hierarchical vision transformer using shifted windows. *Proceedings of the IEEE/CVF international conference on computer vision* (pp. 10012–10022).
8. Dai, Z., Liu, H., Le, Q. V., & Tan, M. (2021). Coatnet: Marrying convolution and attention for all data sizes. *Advances in neural information processing systems* (pp. 3965–3977).
9. Devlin, J., Chang, M. W., Lee, K., & Toutanova, K. (2018). *Bert: Pre-training of deep bidirectional transformers for language understanding*. arXiv preprint arXiv:1810.04805.
10. Baevski, A., & Auli, M. (2019). *Adaptive input representations for neural language modeling*. ICLR.
11. Wang, Q., Li, B., Xiao, T., Zhu, J., Li, C., Wong, D. F., & Chao, L. S. (2019). *Learning deep transformer models for machine translation*. ACL.
12. Hendrycks, D., & Gimpel, K. (2016). Gaussian error linear units (gelus) [J] . arXiv preprint arXiv:1606.08415.
13. Hinton, G. E., Vinyals, O., & Dean, J. (2015). *Distilling the knowledge in a neural network*. arXiv preprint arXiv:1503.02531.
14. Wei, L., Xiao, A., Xie, L., Zhang, X., Chen, X., & Tian, Q., (2020). Circumventing outliers of autoaugment with knowledge distillation. *European conference on computer vision*.
15. Mohamed, M., Cesa, G., Cohen, T. S., & Welling, M., (2020). *A data and compute efficient design for limited-resources deep learning*. arXiv preprint arXiv:2004.09691.
16. Raffel, C., Shazeer, N., Roberts, A., Lee, K., Narang, S., Matena, M., Zhou, Y., Li, W., & Liu, P.J., (2020). Exploring the limits of transfer learning with a unified text-to-text transformer. *The Journal of Machine Learning Research, 21*(1), 5485–5551.
17. Shaw, P., Uszkoreit, J., & Vaswani, A. (2018). *Self-attention with relative position representations*. arXiv preprint arXiv:1803.02155.
18. Sandler, M., Howard, A., Zhu, M., Zhmoginov, A., & Chen, L. C., (2018). Mobilenetv2: Inverted residuals and linear bottlenecks. In: *Proceedings of the IEEE conference on computer vision and pattern recognition* (pp. 4510–4520).

19. Hu, J., Shen, L., & Sun, G. (2018). Squeeze-and-excitation networks. *Proceedings of the IEEE conference on computer vision and pattern recognition* (pp. 7132–7141).
20. Brown, T., Mann, B., Ryder, N., Subbiah, M., Kaplan, J. D., Dhariwal, P., et al. (2020). Language models are few-shot learners. *Advances in Neural Information Processing Systems, 33*, 1877–1901.
21. Choromanski, K., Likhosherstov, V., Dohan, et al., Song, X., Gane, A., Sarlos, T., Hawkins, P., Davis, J., Mohiuddin, A., Kaiser, L., & Belanger, D., (2020). *Rethinking attention with performers*. arXiv preprint arXiv:2009.14794.
22. Lin, Z., Feng, M., Santos, C. N. D., Yu, M., Xiang, B., Zhou, B., & Bengio, Y. (2017). *A structured self-attentive sentence embedding*. arXiv preprint arXiv:1703.03130.
23. Misra, D., Nalamada, T., Arasanipalai, A. U., & Hou, Q. (2021). Rotate to attend: Convolutional triplet attention module. *Proceedings of the IEEE/CVF winter conference on applications of computer vision* (pp. 3139–3148).
24. Qin, Z., Zhang, P., Wu, F., & Li, X., (2021). Fcanet: Frequency channel attention networks. *Proceedings of the IEEE/CVF international conference on computer vision* (pp. 783–792).
25. Radford, A., Narasimhan, K., Salimans, T., & Sutskever, I. (2018). *Improving language understanding by generative pre-training*.
26. Radford, A., Wu, J., Child, R., Luan, D., Amodei, D., et al. (2019). Language models are unsupervised multitask learners. *OpenAI Blog, 1*(8), 9.
27. Raffel, C., Shazeer, N., Roberts, A., Lee, K., Narang, S., Matena, M., et al. (2020). Exploring the limits of transfer learning with a unified text-to-text transformer. *The Journal of Machine Learning Research, 21*(1), 5485–5551.
28. Wang, X., Girshick, R., Gupta, A., & He, K. (2018). Non-local neural networks. *Proceedings of the IEEE conference on computer vision and pattern recognition* (pp. 7794–7803).
29. Yang, L., Zhang, R. Y., Li, L., & Xie, X. (2021). Simam: A simple, parameter-free attention module for convolutional neural networks. *International conference on machine learning* (pp. 11863–118740).
30. Woo, S., Park, J., Lee, J. Y., & Kweon, I. S. (2018). Cbam: Convolutional block attention module. *Proceedings of the European conference on computer vision (ECCV)* (pp. 3–19).

Part II
Advanced Deep Learning for Medical Image Analysis

Chapter 4
Transformer-Based Approaches for Medical Image Analysis

4.1 Introduction

Over the past decade, CNNs have significantly advanced the field of medical image segmentation, demonstrating remarkable success in various applications. However, CNNs face inherent limitations in capturing explicit long-range dependencies beyond local regions, as the receptive field of convolutional kernels is typically constrained.

In recent years, the Transformer architecture, originally developed for sequence-to-sequence tasks, has been adapted for visual applications. Transformers have shown great promise in modeling global context by capturing long-range dependencies effectively. Despite their potential, directly applying Transformers to medical image analysis presents significant challenges due to the unique characteristics of medical images, such as large spatial dimensions, complex anatomical structures, intricate boundary details, and highly heterogeneous textures.

While previous CNN-based methods also face these challenges, many effective solutions have been developed to address them. Consequently, recent research has focused on integrating the strengths of CNNs and Transformers. More specifically, CNNs excel at extracting local features and can maintain relatively consistent computational efficiency even with large spatial dimensions, whereas Transformers are highly effective at modeling global dependencies that can be effective for identifying the complete target organs. This complementary nature has driven efforts to develop hybrid architectures that leverage the advantages of both approaches to enhance performance in medical image analysis. Efforts have been directed toward designing hybrid backbone architectures that leverage both approaches effectively while prioritizing computational efficiency and reducing model complexity. From the perspective of model architecture design, these methods can be mainly categorized into the following three types:

(a) **Pure Transformer Structure**. Pure Transformer architectures are composed entirely of Transformer modules, closely mirroring the original design from the natural language processing (NLP) domain. However, unlike NLP tasks,

Y. Chen et al., *Recent Advances in Deep Learning for Medical Image Analysis*, Intelligent Systems Reference Library 278, https://doi.org/10.1007/978-3-031-94791-9_4

which primarily rely on sequential and contextual relationships, many visual tasks depend heavily on local features for optimal performance. This reliance on local information presents a potential limitation for pure Transformer architectures, as they treat local information in the same way as global context, lacking the spatial inductive bias that is inherent in CNNs. This can hinder the model's ability to more effectively capture fine-grained local features, which are especially crucial for tasks such as image segmentation and object detection.

(b) **Hybrid Transformer Structure**. Hybrid Transformer architectures combine Transformer modules with other feature extraction or decoding mechanisms to enhance efficiency. In these designs, Transformers are typically employed to capture long-range dependencies within high-level feature maps, while low-level features are extracted using computationally less expensive operators. By leveraging the strengths of both approaches, a well-designed hybrid structure can effectively balance performance and computational cost, achieving strong results with greater efficiency.

(c) **Advanced Transformer Structures**. The rapid advancement of vision Transformers has led to the development of numerous modified Transformer modules tailored for diverse applications. For instance, various self-attention mechanism variants have been proposed, aiming to improve computational efficiency or enhance the model's ability to capture local features. Furthermore, state-of-the-art approaches often incorporate specialized pipelines alongside these advanced modules to address specific limitations of existing methods, enabling more customized and effective training outcomes. Overall, the ongoing innovations in this domain suggest a promising future for Transformer-based methodologies in visual tasks.

Based on these categories, the following sections introduce several notable Transformer-based methods for medical image segmentation. For pure Transformer architectures, we primarily discuss the classic Swin-UNet [1] in Sect. 4.2 as a representative example. Hybrid Transformer structures are exemplified by TransUNet [2] in Sect. 4.3 and ScaleFormer [3] in Sect. 4.4. For architectures incorporating advanced Transformer modules, we present MT-UNet [4] and ClassFormer [5] in Sects. 4.5 and 4.6, respectively, to highlight different design strategies inspired by distinct motivations. Finally, we list the experimental results of all the above methods on the corresponding tasks to verify their effectiveness.

4.2 Pure Transformer Structure: Swin-UNet

Motivated by the Swin Transformer [6] originally designed for the natural image domain, Cao et al. proposed Swin-UNet in their 2021 paper titled "Swin-UNet: Unet-like Pure Transformer for Medical Image Segmentation" [1]. Swin-UNet leverages the pretrained Swin Transformer to construct its overall architecture, effectively balancing local and global dependencies.

4.2.1 Overall Architecture

As illustrated in Fig. 4.1, Swin-UNet is a hierarchical, UNet-like pure Transformer designed specifically for medical image segmentation. Its architecture consists of four key components: an encoder, a bottleneck, a decoder, and skip connections.

In the encoder phase, medical images are divided into non-overlapping patches of size 4×4, where each patch is represented by a feature dimension of $4 \times 4 \times 3$. A linear embedding layer is then applied to project these features into a configurable dimension, denoted as C. The resulting patch tokens are processed through multiple Swin Transformer blocks and patch-merging layers to build hierarchical feature representations. The patch-merging layers downsample the spatial resolution while increasing the feature dimension, whereas the Swin Transformer blocks are responsible for learning rich feature representations.

In the decoder phase, a Transformer-based symmetric structure is used to reconstruct high-resolution feature maps. The decoder integrates Swin Transformer blocks with patch-expanding layers, which are designed specifically for up-sampling. Contextual features from the encoder are fused with multiscale features via skip connections, mitigating the spatial detail loss caused by down-sampling. Unlike patch-merging layers, patch-expanding layers reorganize feature maps into higher-resolution representations, achieving a $2\times$ up-scaling at each stage. At the final stage, a patch-expanding layer performs $4\times$ up-sampling, fully restoring the feature maps to the original input resolution $H \times W$.

4.2.2 Detailed Introduction to the Components

The core components of Swin-UNet's architecture include Swin Transformer blocks [6], Patch Merging layers, and Patch Expanding layers, each of which will be described in detail in the following sections.

Swin Transformer. Unlike traditional Multi-head Self Attention (MSA) modules, the Swin Transformer block employs shifted windows, significantly reducing computational load while effectively establishing both local and global dependencies within a single block. As shown in Fig. 4.2, each block comprises a LayerNorm (LN) layer, a Window-based Multi-head Self Attention (W-MSA) module, a Shifted Window-based Multi-head Self Attention (SW-MSA) module, and a Multi-Layer Perceptron (MLP). These components are used in two consecutive transformer blocks, respectively. The process can be formulated as:

$$\begin{aligned} \hat{\mathbf{z}}^l &= \text{W-MSA}(\text{LN}(\mathbf{z}^{l-1})) + \mathbf{z}^{l-1} \\ \mathbf{z}^l &= \text{MLP}(\text{LN}(\hat{\mathbf{z}}^l)) + \hat{\mathbf{z}}^l \\ \hat{\mathbf{z}}^{l+1} &= \text{SW-MSA}(\text{LN}(\mathbf{z}^l)) + \mathbf{z}^l \\ \mathbf{z}^{l+1} &= \text{MLP}(\text{LN}(\hat{\mathbf{z}}^{l+1})) + \hat{\mathbf{z}}^{l+1} \end{aligned} \tag{4.1}$$

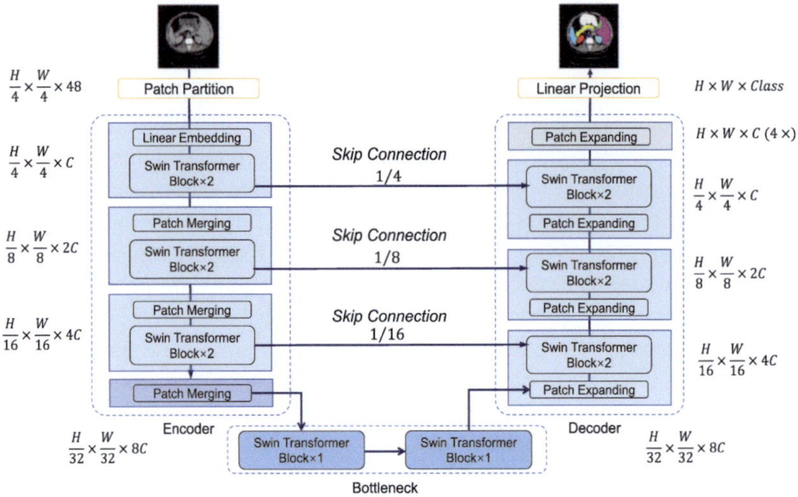

Fig. 4.1 The architecture of Swin-UNet. Encoder, bottleneck and decoder are all constructed based on swin transformer blocks. H, W, and C denote the height, width, and number of channels of the image, respectively

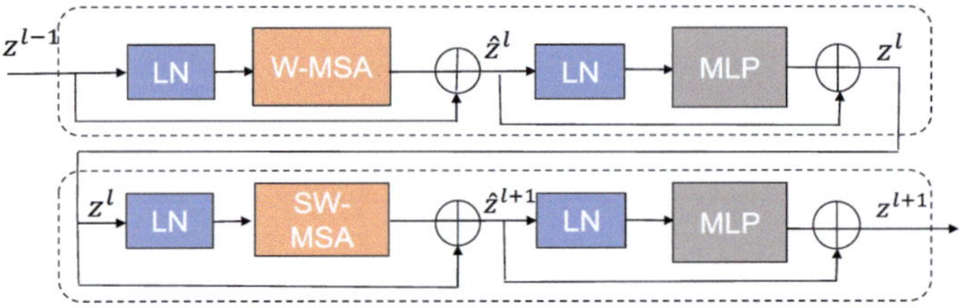

Fig. 4.2 The architecture of swin transformer block

where $\hat{\mathbf{z}}^l$ and $\hat{\mathbf{z}}^{l+1}$ represent the outputs of the W-MSA module and the SW-MSA module, respectively. The window-based multi-head self-attention is computed as follows:

$$\text{Attention}(\mathbf{Q}, \mathbf{K}, \mathbf{V}) = \text{Softmax}(\frac{\mathbf{QK}^{\mathrm{T}}}{\sqrt{d}} + \mathbf{B})\mathbf{V} \tag{4.2}$$

where $\mathbf{Q} \in \mathbb{R}^{M^2 \times d_q}$, $\mathbf{K} \in \mathbb{R}^{M^2 \times d_k}$, $\mathbf{V} \in \mathbb{R}^{M^2 \times d_v}$ denote the query, key and value matrices. M^2, d_q, d_k and d_v represent the number of patches in a window and the dimension of the query, key and value, respectively. $\mathbf{B} \in \mathbb{R}^{M^2 \times M^2}$ is a relative position bias to each head in computing similarity. Since each axis lies in the range $[-M+1, M-1]$, we parameterize a smaller-sized bias matrix $\hat{\mathbf{B}} \in \mathbb{R}^{(2M-1)\times(2M-1)}$, and values in $\mathbf{B}$ are taken from $\hat{\mathbf{B}}$.

Patch Merging Layer. The patch merging layer reduces the spatial resolution while increasing the feature dimension for enriched representation. Input patches are divided into four parts and concatenated, resulting in a spatial down-sampling by a factor of 2. This process increases the feature dimension by a factor of 4. A linear layer is then applied to the concatenated features to normalize the feature dimension, reducing it to twice its original size.

Patch Expanding Layer. The patch expanding layer restores spatial resolution through up-sampling, which is essential for generating high-resolution outputs. For example, in the first patch expanding layer depicted in Fig. 4.1, input features of size $\frac{H}{32} \times \frac{W}{32} \times 8C$ are processed through a linear layer that doubles the feature dimension to $\frac{H}{32} \times \frac{W}{32} \times 16C$. Subsequently, a rearrange operation expands the spatial resolution by a factor of 2 and reduces the feature dimension to one-quarter of its input size, resulting in $\frac{H}{32} \times \frac{W}{32} \times 16C \rightarrow \frac{H}{16} \times \frac{W}{16} \times 4C$.

In summary, the success of Swin-UNet can be primarily attributed to the innovative design of the Swin Transformer's window-based attention mechanism. On one hand, this mechanism enables the network to simultaneously capture both local and global dependencies, which is especially advantageous for visual tasks. By focusing on fine-grained local details, it ensures precise boundary delineation, while its ability to capture larger structural variations accommodates differences in organ shapes, sizes, and spatial relationships. On the other hand, the window-based design significantly reduces the number of learnable parameters, making the model lightweight and computationally efficient. This efficiency is particularly valuable in medical imaging, where datasets are often small due to the challenges of data collection and annotation.

4.3 Hybrid Transformer Structure: TransUNet

In addition to using the shifted window mechanism in Swin Transformers to maintain local information, directly leveraging CNNs to capture local features is also a feasible option. Based on this concept, Chen et al. presented their work "TransUNet: Transformers Make Strong Encoders for Medical Image Segmentation" in 2021 [2]. TransUNet becomes the first study to explore the potential of combining Transformers with CNNs for medical image segmentation. This hybrid approach enables the model to take advantage of the global context captured by Transformers while preserving the detailed local features typically extracted by CNNs, leading to improved segmentation accuracy.

4.3.1 *Overall Architecture*

As illustrated in Fig. 4.3, the overall structure of TransUNet comprises two primary components: the Hybrid Encoder and the Cascaded Decoder, which will be detailed in the following sections.

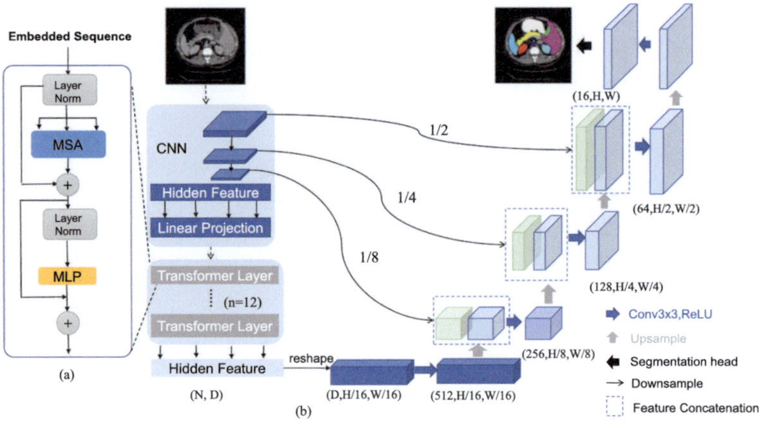

Fig. 4.3 Overview of the framework. **a** schematic of the Transformer layer; **b** architecture of the proposed TransUNet

4.3.2 Detailed Introduction to the Encoder and Decoder

Hybrid Encoder. The CNN and Transformer layers are arranged in the shallow and deep layers, respectively. For an input image $\mathbf{x} \in \mathbb{R}^{H \times W \times C}$, H and W represent the spatial resolution, and C denotes the number of channels. CNN (e.g., ResNet-50 [7]) is first used as a feature extractor to generate feature maps. These feature maps are then reshaped into a sequence of 2D patches. The vectorized patches, $\mathbf{x}_p$, are mapped into a latent D-dimensional embedding space through a trainable linear projection. To preserve the spatial information of the patches, position embeddings are learned and added to the patch embeddings, ensuring positional awareness in the representation as follows:

$$\mathbf{z}_0 = [\mathbf{x}_p^1\mathbf{E}; \mathbf{x}_p^2\mathbf{E}; \cdots; \mathbf{x}_p^N\mathbf{E}] + \mathbf{E}_{\text{pos}} \tag{4.3}$$

where $\mathbf{E} \in \mathbb{R}^{(P^2 \cdot C) \times D}$ is the patch embedding projection, and $\mathbf{E}_{\text{pos}} \in \mathbb{R}^{N \times D}$ denotes the position embedding. P is the patch size and $N = H \times W/P^2$.

The following Transformer encoder (e.g. pretrained Vision Transformer (ViT) [8]) consists of n layers of Multihead Self-Attention (MSA) and Multi-Layer Perceptron (MLP) blocks. n is set to 12 in the specific implementation. The output of the ℓ-th layer can be written as follows:

$$\begin{aligned} \mathbf{z}'_\ell &= \text{MSA}(\text{LN}(\mathbf{z}_{\ell-1})) + \mathbf{z}_{\ell-1} \\ \mathbf{z}_\ell &= \text{MLP}(\text{LN}(\mathbf{z}'_\ell)) + \mathbf{z}'_\ell \end{aligned} \tag{4.4}$$

where $\text{LN}(\cdot)$ denotes the layer normalization operator and $\mathbf{z}_\ell$ is the encoded image representation. Benefiting from the leverage the strengths of CNN and Transformer, both the local and global dependencies can be captured in a more powerful feature representation, leading to more accurate medical image segmentation.

Cascaded Decoder. The decoder includes multiple upsampling steps to decode the hidden features and produce the final segmentation mask. Each block consists of a 2× upsampling operator, a 3 × 3 convolution layer, and a ReLU layer successively. To preserve detail that may be lost at deeper levels with the Transformer, high-resolution CNN feature maps from the encoder are incorporated into the decoder path through skip connections.

In summary, TransUNet [2] is the first public study to employ pretrained Transformers for medical image segmentation, adopting a deep hybrid architecture that combines the strengths of CNNs and Transformers. While this approach has demonstrated notable success, it relies heavily on pretrained Transformers, which may restrict its adaptability to smaller medical image segmentation datasets. Additionally, the use of large pretrained models significantly increases computational complexity and introduces a substantial number of trainable parameters.

4.4 Hybrid Transformer Structure: ScaleFormer

Different from TransUNet, which simply adopts a Vision Transformer (ViT) structure within the model pipeline, Huang et al. introduced a more sophisticated approach in their 2022 work titled "ScaleFormer: Revisiting the Transformer-Based Backbones from a Scale-Wise Perspective for Medical Image Segmentation" [3] in 2022. ScaleFormer aims to leverage the Transformer's global context awareness to model correlations across different scales of feature maps.

Specifically, ScaleFormer addresses two primary challenges within existing segmentation network architectures from a scale perspective:

(1) **Intra-scale problem**: Current methods often struggle to capture both local and global information effectively within each scale. This limitation can hinder the accurate segmentation of small objects, as the network may fail to propagate signals effectively across different spatial extents.
(2) **Inter-scale problem**: Existing approaches often do not efficiently utilize distinctive information across multiple scales. This deficiency limits the network's ability to learn robust representations for objects that vary significantly in size, shape, and spatial location.

An intuitive illustration of these problems is presented in Fig. 4.4. As shown, the CNN-based ResUNet (depicted in (b) and (f)) struggles to capture long-range dependencies, resulting in missed detections of objects outside nearby areas—such as the liver, gallbladder, and pancreas not being identified. In contrast, Swin-UNet [1], a pure Transformer-based model (shown in (c) and (g)), tends to over-activate background regions, making it challenging to distinguish objects from their surroundings. Similarly, TransUNet [2] (depicted in (d) and (h)) struggles with accurately localizing small objects, such as the gallbladder and pancreas, which appear blurred due to

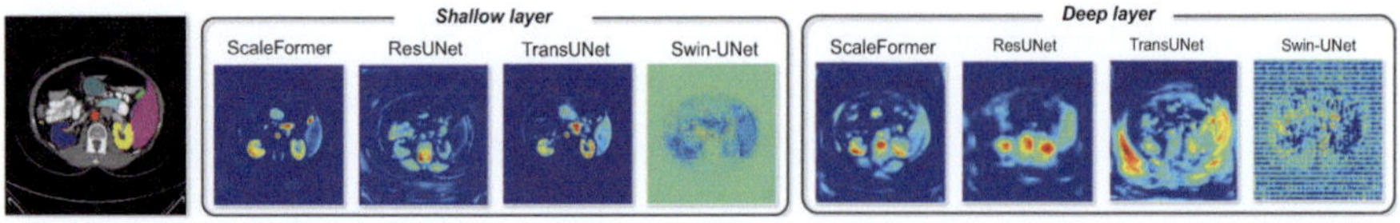

Fig. 4.4 Visualization of feature maps from different layers of four methods, including ScaleFormer [3], ResUNet, TransUNet [2], and Swin-UNet [1]

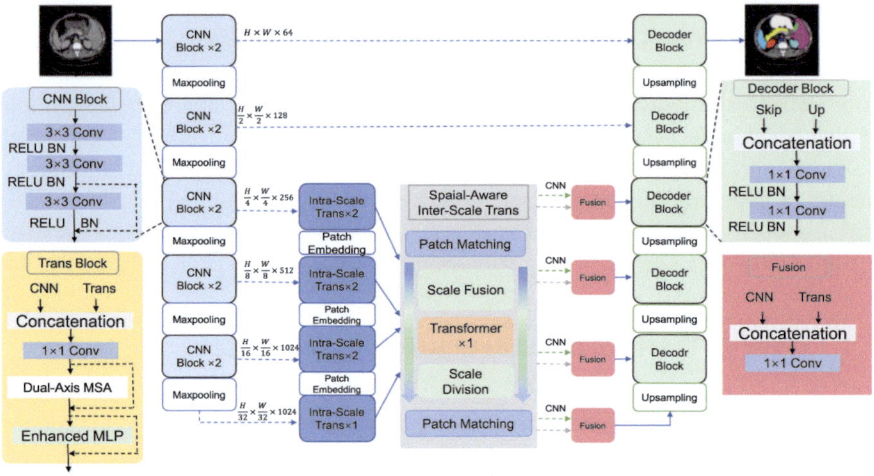

Fig. 4.5 Overview of ScaleFormer. CNN blocks extract local-level details from the input image. These details are forwarded to intra-scale transformers to capture global-level cues. Subsequently, an inter-scale transformer integrates enhanced representations across different scales, modeling mutual information between objects

their small size. Finally, ScaleFormer [3] (illustrated in (a) and (e)) addresses these limitations by enhancing Transformer-based backbones with a scale-aware design. This approach enables the network to recognize objects across varying sizes and distant contexts, effectively identifying both large structures, like the liver, and smaller ones, like the pancreas, even in challenging long-range scenarios.

4.4.1 Overall Architecture

As illustrated in Fig. 4.5, ScaleFormer incorporates a ResUNet backbone to hierarchically capture local features. To complement this, it introduces a scale-wise intra-scale Transformer, which bridges the gap between CNN-based local feature extraction and Transformer-based global information modeling within each scale. This integration is achieved through a lightweight Dual-Axis MSA mechanism, which efficiently captures row-wise and column-wise global dependencies.

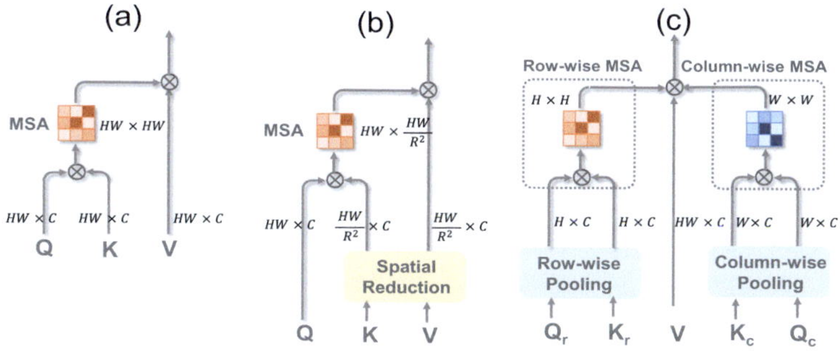

Fig. 4.6 Comparison of previous MSAs **a** and **b** with proposed dual-axis MSA **c**

In addition to addressing intra-scale challenges, ScaleFormer employs an efficient spatial-aware inter-scale Transformer to enable effective interactions among corresponding regions across multiple scales. This mechanism highlights cross-scale dependencies, thereby addressing complex scale variations in medical image segmentation.

In the following sections, we will delve into the detailed designs of the scale-wise intra-scale Transformer and the spatial-aware inter-scale Transformer, highlighting their specific contributions to ScaleFormer's overall architecture.

4.4.2 Scale-Wise Intra-Scale Transformer Block

In the original MSA setup (illustrated in Fig. 4.6a), the computational complexity is $O(H^2W^2C)$, where H, W, and C represent the height, width, and channel dimensions, respectively. To mitigate this computational expense, recent approaches have introduced a spatial reduction technique that scales down the dimensions of the key and value matrices (as shown in Fig. 4.6b). This reduces the complexity to $O(H^2W^2C/R^2)$, where R denotes the reduction ratio. However, while this approach lowers computational cost, using a large reduction ratio R on high-resolution feature maps often compromises critical horizontal and vertical information.

To address these limitations, ScaleFormer introduces a lightweight and adaptable Dual-Axis MSA mechanism, which redefines the traditional full-image relation ($HW \times HW$) by decomposing it into separate row-wise ($H \times H$) and column-wise ($W \times W$) relations, as illustrated in Fig. 4.6c. This innovative design reduces computational overhead while preserving essential spatial information.

The implementation of Dual-Axis MSA involves four distinct depth-wise convolutions applied to the query ($\mathbf{Q}$) and key ($\mathbf{K}$) matrices. These convolutions enhance row-wise ($\mathbf{Q}_r$, $\mathbf{K}_r$) and column-wise ($\mathbf{Q}_c$, $\mathbf{K}_c$) features. Subsequently, average pooling operations in both directions generate condensed global representations

$(\mathbf{Q}_{rp}, \mathbf{K}_{rp}, \mathbf{Q}_{cp}, \mathbf{K}_{cp})$ across the entire feature map. This process can be expressed mathematically as:

$$\mathbf{A}(\mathbf{Q}, \mathbf{K}, \mathbf{V}) = \underbrace{(\text{softmax}(\frac{\mathbf{Q}_{rp} K_{rp}^T}{\sqrt{d_k}}))}_{Row-wise \quad MSA} \mathbf{V} \underbrace{(\text{softmax}(\frac{\mathbf{Q}_{cp} K_{cp}^T}{\sqrt{d_k}}))}_{Column-wise \quad MSA} \tag{4.5}$$

In the application of row-wise MSA, the value matrix $\mathbf{V}$ is reshaped to a form of $H \times WC$, enabling efficient computation along the horizontal axis. Conversely, during column-wise MSA, $\mathbf{V}$ is reshaped into $HC \times W$, facilitating vertical attention. By combining these two approaches, Dual-Axis MSA achieves a computational complexity of $O(H^2WC + HW^2C)$.

This approach stands apart from traditional Axial Attention, which requires two consecutive blocks to fully capture long-range dependencies in both dimensions. In contrast, Dual-Axis MSA enables the simultaneous learning of global dependencies along both axes within a single block, offering a more computationally efficient and conceptually streamlined method for global feature understanding.

4.4.3 Spatial-Aware Inter-Scale Transformer Block

Multi-scale information has been proved to be highly effective in addressing the complex scale variations critical for medical image segmentation. To harness this potential, ScaleFormer introduces a Transformer-based block explicitly designed to enable seamless interaction across different scales within its hierarchical structure.

The process begins by rearranging the input sequence of enhanced intra-scale tokens into a 2D lattice representation. Unlike conventional methods that compute dependencies across a long concatenated sequence of multi-scale tokens (flattened from 2D feature maps at each scale), ScaleFormer adopts a more efficient approach by leveraging spatial correspondences among patches across different scales. This is achieved using a novel downsampling operation termed **Patch Matching**.

As illustrated in Fig. 4.7, Patch Matching identifies spatially corresponding patches across consecutive feature maps, such as from the i-th to $(i+2)$-th scale. To ensure consistency, the feature maps are first standardized to the same channel dimension C. Let $\mathbf{p_j^i}$ represent the j-th patch in the i-th scale. Corresponding downsampled patches that align spatially share the same bounding box color (e.g., blue), maintaining spatial coherence and minimizing redundancy.

Finally, ScaleFormer employs a **Scale Fusion** mechanism to concatenate spatially corresponding inter-scale patches, enabling the model to effectively capture multi-scale dependencies while preserving spatial relationships across the hierarchical structure. This process can be formulated as:

$$\text{flatten}(\mathbf{p}_j^1), \ldots, \text{flatten}(\mathbf{p}_j^n) \longrightarrow \mathbf{p}_j^{cat}, \tag{4.6}$$

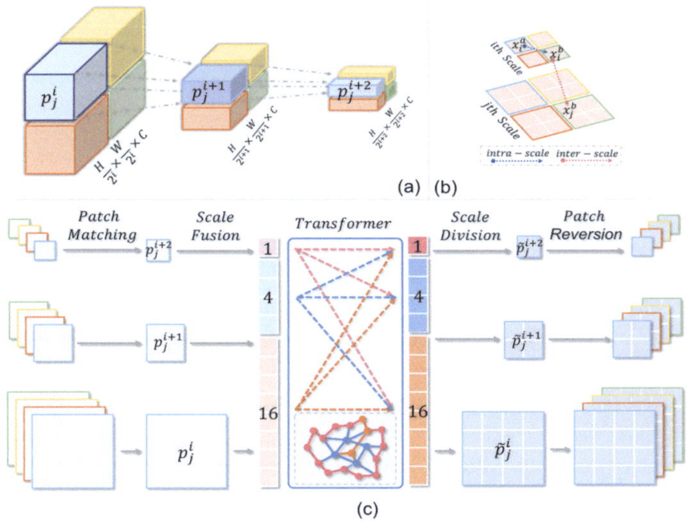

Fig. 4.7 An illustration of spatial-aware inter-scale transformer

where $flatten(\cdot)$ operation rearrange the patch $\mathbf{p_j^i}$ into the shape of $HW/2^{2i} \times C$ and $[\cdot]$ represents the concatenation operation. The spatial-aware inter-scale dependencies among patches can be obtained as follows:

$$\begin{aligned} \hat{\mathbf{P}}_j^{cat} &= \text{MSA}(\text{LN}(\mathbf{P}_j^{cat})) + \mathbf{P}_j^{cat} \\ \tilde{\mathbf{P}}_j^{cat} &= \text{MLP}(\text{LN}(\hat{\mathbf{P}}_j^{cat})) + \hat{\mathbf{P}}_j^{cat} \end{aligned} \tag{4.7}$$

where $LN(\cdot)$ denotes the layer normalization operator. Then we utilize **Scale Division** to reverse the enhanced sequence back to patches according to the order of concatenation:

$$\text{Reshape}(\text{Split}(\tilde{\mathbf{P}}^{cat})) \longrightarrow \tilde{\mathbf{P}}_j^i, ...\tilde{\mathbf{P}}_j^N \tag{4.8}$$

where the $Split(\cdot)$ is an inverse operation of previous concatenation operation $[\cdot]$. In the reverse procedure of patch matching, **Patch Reversion** assembles all patches within the same scale into a unified feature map. Following this, we concatenate the global-level Transformer features with the local-level CNN features via a skip connection, which are subsequently forwarded to the decoder stage.

In summary, from a scale-wise perspective, ScaleFormer [3] integrates CNNs and Transformers at each scale and incorporates an inter-scale Transformer to seamlessly fuse information across scales. This comprehensive architectural approach enables the model to capture both global and local features simultaneously, meanwhile fully leveraging the multi-scale information, making ScaleFormer outperforms prior methods. However, this advancement comes with a trade-off: the substantial increase in learnable parameters leads to greater computational complexity, which must be considered in practical applications.

4.5 Advanced Transformer Structures: MT-UNet

The introduction of Transformers has significantly advanced the accuracy of medical image segmentation tasks. However, these models often require extensive pre-training on large datasets and entail high computational complexity as they lack structure prior knowledge for visual tasks. To overcome these challenges, Wang et al. introduced an innovative solution in their 2022 work titled "Mixed Transformer U-Net for Medical Image Segmentation" [4]. MT-UNet adopts a hybrid network architecture that combines convolutional stems with Transformer blocks, introducing a strong prior structure knowledge for the pipeline.

4.5.1 Overall Architecture

The overall architecture of MT-UNet is illustrated in Fig. 4.8, which follows an encoder-decoder structure with skip connections to retain low-level features during decoding. Specifically, convolutional layers are employed in the shallow stages to serve as feature extractors. These layers focus on local relationships, preserve high-resolution details, and introduce structural priors—key advantages for medical imaging tasks with relatively small datasets. In the deeper layers, Transformer blocks are utilized to model global relationships, leveraging the reduced spatial resolution to mitigate computational demands.

To simultaneously learn intra- and inter-affinities, MT-UNet introduces a novel Mixed Transformer Module, which replaces conventional Transformer blocks. This module first efficiently computes self-affinities using a specially designed Local-Global Gaussian-Weighted Self-Attention mechanism. Subsequently, it explores connections across different data samples through an External Attention mechanism.

4.5.2 Local-Global Gaussian-Weighted Self-Attention (LGG-SA)

Traditional self-attention aims to capture the interconnections between all entities in the input sequence, introducing three matrices: key (**K**), query (**Q**), and value (**V**). However, correlations between nearby regions are often more significant than those farther apart, making it inefficient to allocate equal computational resources to distant areas. To address this, as shown in Fig. 4.9, self-attention is divided into local

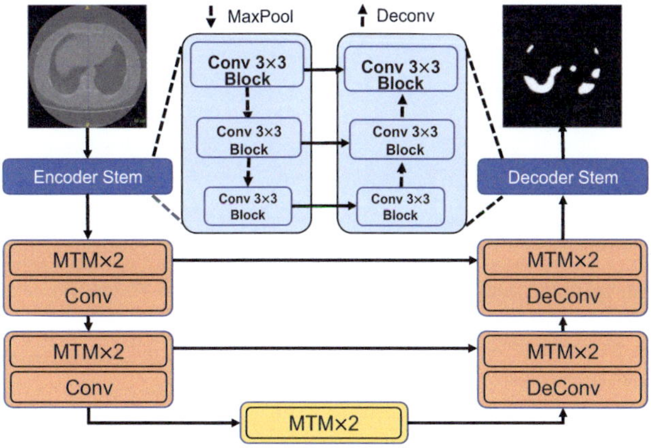

Fig. 4.8 The overall architecture of mixed transformer U-Net

self-attention (LSA) and global self-attention (GSA). LSA calculates self-affinities within each window, after which the tokens within each window are aggregated into a global token, representing the main information of the window. Mathematically, for an input feature map $\mathbf{X} \in \mathbb{R}^{H \times W \times C}$, if we set the window size to p, the overall process can be formulated as:

$$\mathbf{z}_{local} = \text{LSA}(\mathbf{X}), \tag{4.9}$$

$$\mathbf{z}_{global} = \text{GSA}(\text{Aggragate}(\mathbf{z}_{local})) \tag{4.10}$$

$$\mathbf{z} = \text{Concat}(\mathbf{z}_{local}, \text{Upsample}(\mathbf{z}_{global})) \tag{4.11}$$

where $\mathbf{z}$ refers to the module output, LSA represents Local Self-Attention, and GSA denotes the corresponding global operation.

As shown in Fig. 4.10, the key operation of LSA is Gaussian-Weighted Axial Attention (GWAA), which enhances each query's perception of nearby tokens through a learnable Gaussian matrix, and meanwhile has a lower time complexity due to Axial Attention [9]. Moreover, this matrix can be also viewed as relative positional encoding, emphasising the unique position information for each query. The final output at position (i, j) can be formulated as:

$$\mathbf{z}_{i,j} = \text{softmax}(-\frac{1}{2\sigma^2}\mathbf{D}_{i,j}^2 + S(\mathbf{q}_{i,j}, \mathbf{K}_{i,j}))\mathbf{V}_{i,j} \tag{4.12}$$

where $\mathbf{D}_{i,j}$ is defined as the Euclidean distances matrix between $\mathbf{q}_{i,j}$ and its corresponding $\mathbf{K}_{i,j}$ and $\mathbf{V}_{i,j}$, where $\mathbf{K}_{i,j}$ and $\mathbf{V}_{i,j}$ are vectors composed of key and value

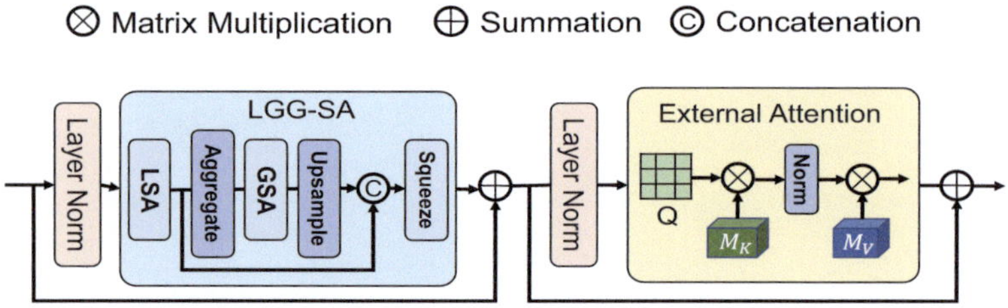

Fig. 4.9 Overview of mixed transformer module architecture: local-global Gaussian-weighted self-attention (LGG-SA) and external attention

tokens on ith row and jth column. Let the similarity between $\mathbf{q}$ and $\mathbf{K}$ being $S(\mathbf{q}, \mathbf{K})$ and Gaussian weight being $e^{-\frac{D_{i,j}^2}{2\sigma^2}}$. It improves the model performance for explicitly providing relative relation, which the ordinary absolute positional embedding can not.

4.5.3 *External Attention*

Different from LGG-SA that only build long-distance dependencies in-sample, External Attention (EA) is proposed to solve the problem that SA cannot exploit relations between different samples. Unlike Self-Attention using each sample's own linear transformations to calculate the attention score, in EA, all the samples share two memory units $\mathbf{M}_K$ and $\mathbf{M}_V$ (as is shown in Fig. 4.9), depicting the most essential information of the entire dataset. Moreover, an additional linear mapping is used for Q to enlarge its channel, improving the representation learning ability of this module.

4.5.4 *Complexity Analysis*

On the whole, for a given image with n voxels, the time complexity of LSA is $O(n)$ when p is fixed. In contrast, time complexity for GSA is $O(n\sqrt{n})$ due to Axial Attention. Therefore, the overall complexity of LGG-SA is just $O(n\sqrt{n})$.

In summary, MT-UNet introduces a hybrid architectural design with advanced Transformer modules that set it apart from prior models. Unlike TransUNet, which incorporates a hybrid structure solely within the encoder, MT-UNet extends this design to both the encoder and decoder. Simultaneously, it ensures the preservation of fine-grained details throughout the segmentation process, which is critical for achieving high accuracy in medical imaging tasks. Additionally, the model also introduces an advanced Transformer module, featuring a local-global self-attention mechanism, which effectively captures internal relationships within individual sam-

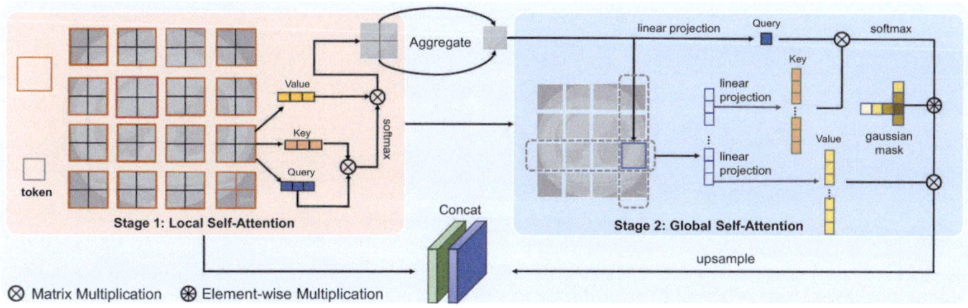

Fig. 4.10 Detailed components of LGG-SA: local self-attention (LSA) and Gaussian-weighted axial attention in global self-attention (GSA)

ples. These architectural advancements eliminate the reliance on pretrained weights, making MT-UNet particularly advantageous for medical image segmentation tasks, where access to large-scale labeled datasets is often a significant constraint.

4.6 Advanced Transformer Structures: ClassFormer

Different kinds of Vision Transformers have demonstrated impressive performance in medical image segmentation. However, two main concerns persist when considering a class level perspective, as illustrated in Fig. 4.11. **(1) Deficiency in Class-Specific Correspondences**: There is a lack of specific pixel-level correspondences for different classes, which can lead to inadequate coverage of objects and inaccurate boundary predictions. **(2) Neglect of Explicit Category Dependencies**: Current methods often overlook the explicit dependencies between different object categories, which results in inaccuracies in object localization. In Fig. 4.11, TransUNet, Swin-UNet, and ResUNet exhibit issues with inter-class confusion (e.g., misclassifying stomach as spleen) and intra-class incompactness (e.g., incomplete representation of stomach) as indicated by pink boxes. This can be observed in yellow boxes in panels (f)-(h), where the color is darker for stomach-to-stomach and brighter for stomach-to-spleen. Motivated by these challenges, Huang et al. made significant advancements in their 2023 work titled "ClassFormer: Exploring Class-Aware Dependency with Transformer for Medical Image Segmentation" [5]. ClassFormer improves upon Transformers by addressing these class-level concerns, leading to more accurate localization and improved compactness in segmentation tasks.

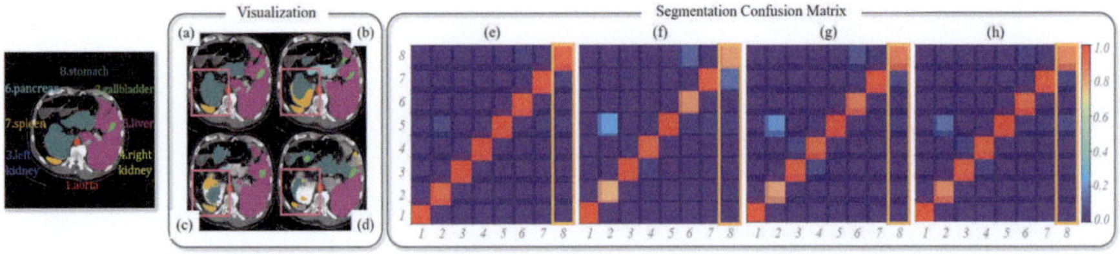

Fig. 4.11 Visualization of eight-organ segmentation and confusion matrix with column normalization: (**a**, **e**) ClassFormer [5], (**b**, **f**) TransUNet [2], (**c**, **g**) Swin-UNet [1], (**d**, **h**) ResUNet. Typically, in a confusion matrix, the diagonal represents correct classifications (intra-class compactness), while off-diagonal elements represent misclassifications (inter-class mis-segmentation)

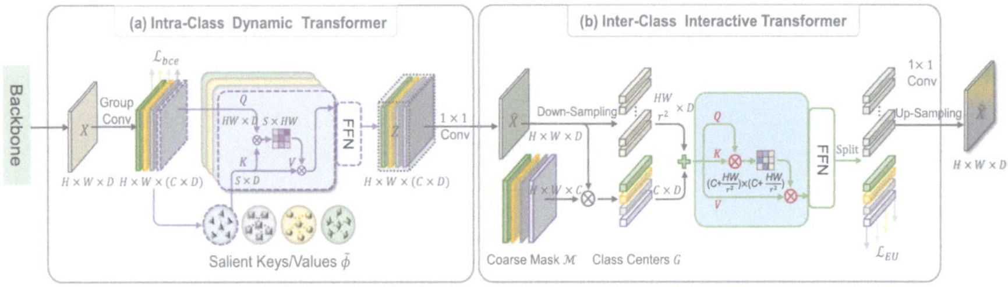

Fig. 4.12 Overview of ClassFormer: **a** intra-class dynamic transformer and **b** inter-class interactive transformer

4.6.1 Overall Architecture

As depicted in Fig. 4.12, to address these challenges, ClassFormer introduces two key components: intra-class dynamic Transformer and inter-class interactive Transformer. These components aim to enhance compactness and address discrepancies comprehensively. The intra-class dynamic Transformer initially separates representations of distinct categories using an adaptive selection mechanism for focused learning. This method optimizes the highlighting of informative features, emphasizing salient keys and values across multiple scales. Subsequently, the inter-class interactive Transformer is designed to capture category dependencies among diverse objects. It utilizes class tokens as representative centers to facilitate global semantic reasoning. This approach ensures feature consistency by penalizing intra-class variations while reinforcing discriminative features between different categories through inter-class constraints.

Given an input image $\mathbf{I}$, a backbone network (such as ResUNet) first processes it to obtain a pixel representation $\mathbf{X} \in \mathbb{R}^{H \times W \times D}$. Next, the class-agnostic feature map $\mathbf{X}$ is replicated C times to form $\mathbf{X} \in \mathbb{R}^{H \times W \times (C \times D)}$. This concatenated feature map is then input into a group convolution with C groups, resulting in a set of decoupled features $\hat{\mathbf{X}} = [\mathbf{X}_1, \mathbf{X}_2, \cdots, \mathbf{X}_c]$, where $\mathbf{X}_c \in \mathbb{R}^{H \times W \times D}$ represents the features specific to the c-th category. Here, $[\cdot]$ denotes the concatenation operation. To explicitly incorporate class-aware information into each decoupled feature $\mathbf{X}_c$, intermediate supervision is

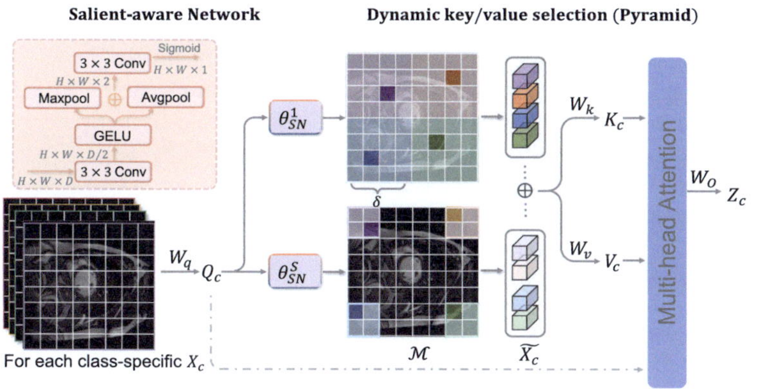

Fig. 4.13 An illustration of intra-class dynamic transformer

applied. This step ensures that each $\mathbf{X}_c$ emphasizes regions relevant to the object category it represents while suppressing irrelevant information.

Specifically, for each decoupled representation $\mathbf{X}_c$, a binary mask is predicted to determine whether each pixel belongs to the target class or not. This mask corresponds to its respective category. Subsequently, a binary cross-entropy loss L_{bce} is computed based on the ground truth, which is typically provided as a one-hot encoded representation indicating the presence or absence of the target class for each pixel. Following this initial classification step, the feature maps $\mathbf{X_c}$ undergo subsequent processing through intra-class Transformers and inter-Transformers. Each of these components plays a key role in refining and integrating class-specific information into the feature representations.

4.6.2 Intra-Class Dynamic Transformer

To capture class-specific long-range dependencies, a conventional approach is to apply Transformers that consider global relations for each class. However, traditional Transformers involve a substantial number of keys attending to each query patch, which leads to redundant calculations of correlations with irrelevant keys and high computational demands. Therefore, a dynamic Transformer structure is designed with saliency-aware keys and values focused on the specific class query. This design allows the Transformer to adaptively learn visual patterns for each class, achieving an efficient balance between computational efficiency and effective feature learning.

As depicted in Fig. 4.13, starting with the input feature map $\mathbf{X}_c \in \mathbb{R}^{H\times W\times D}$, the query tokens $\mathbf{q}$ are derived through a linear projection: $\mathbf{q} = \mathbf{X}_c\mathbf{W}_q$. These query tokens are then processed by a Saliency-aware Network θ_{SN} (shown in the top-left corner of Fig. 4.13), which enhances important pixels while suppressing irrelevant ones within the query feature. The input features undergo initial processing via a 3×3

convolutional layer followed by a GELU activation to capture local features. Recognizing that pooling across the channel dimension can effectively emphasize informative regions, max-pooling and average-pooling operations are performed along the channel axis. These pooled results are concatenated and further processed by a 3×3 convolution to yield the saliency-aware map $\mathbf{M} \in \mathbb{R}^{H\times W}$.

To implement Dynamic Key/Value Selection across the entire space Λ, we begin by dividing the saliency-aware map $\mathcal{M}$ into non-overlapping sub-regions using a stride of δ. Each sub-region $\Omega \in \Lambda$ is considered individually. Within each sub-region Ω, we identify the salient position $\hat{p}$ with the highest localization probability:

$$\hat{p} = \arg\max_{p\in\Omega} \mathcal{M}(p) \tag{4.13}$$

where p is the position of each token. In this way, we can obtain a set of salient positions $\phi = \{\hat{p}_i\}_{i=1}^{|\mathcal{N}|}$ for the entire space Λ, where $|\mathcal{N}|$ represents the number of positions and equals to $\lceil H/\delta \rceil \times \lceil W/\delta \rceil$. $\lceil \cdot \rceil$ is the ceiling operation and yields the smallest integer larger than or equal to its input.

Interestingly, we observe that finer sampling with a smaller receptive scale (smaller δ) is effective for capturing details of small objects while maintaining explicit long-range context. Conversely, coarser sampling with a larger receptive field (larger δ) is beneficial for focusing on larger objects with global dependencies. To leverage these insights, a Pyramid Mechanism is employed to adaptively sample salient positions from multiple scales using different δ values. In this mechanism, denoted as $\tilde{\phi}$, salient positions are combined from various scales: $\tilde{\phi} = \{\phi_i\}_{i=1}^{S}$, where S is the total number of scales.

Then, salient tokens $\tilde{\mathbf{X}}_c$ are sampled from the input tokens $\mathbf{X}_c$ based on the salient positions ϕ. The keys $\mathbf{k}$ and values $\mathbf{v}$ are obtained through linear projection matrices:

$$\mathbf{k} = \tilde{\mathbf{X}}_c \mathbf{W}_k, \mathbf{v} = \tilde{\mathbf{X}}_c \mathbf{W}_v \tag{4.14}$$

$$\tilde{\mathbf{X}}_c = \{\mathbf{X}_c^i | i \in \tilde{\phi}\} \tag{4.15}$$

With the linearly projected $\mathbf{q}$, adaptive selected $\mathbf{k}$ and $\mathbf{v}$, the multi-head attention is applied to capture long-range dependency, which is formulated as:

$$A(\mathbf{q}, \mathbf{k}, \mathbf{v}) = \text{softmax}(\mathbf{q}\mathbf{k}^{\mathbf{T}}/\sqrt{d})\mathbf{v} \tag{4.16}$$

Compared to the conventional Transformer with the complexity of $O(DH^2W^2)$, the Dynamic Transformer significantly reduces costs to $O(DHW\sum_{i=0}^{S}|\mathcal{N}_i|)$, where $|\mathcal{N}_i|$ is the number of selected tokens in each scale. The enhanced sequence is then reshaped into 2D lattice and concatenated with the input feature map to produce the final output.

4.6.3 Inter-Class Interactive Transformer

Category-wise dependency is crucial for effective context modeling, involving high-level semantic relationships among local regions. Traditional Transformers focus on pixel-level relations within feature maps, which limits their ability to explicitly model category dependencies among semantic objects. To address this limitation, exploration of object-to-object relations is proposed through interactions among patch tokens with high-level semantics and multiple class-specific tokens. In the following section, we will discuss the semantic attention process in detail. This approach aims to enhance the model's capability to capture and utilize category-specific information effectively.

Given the input $\mathbf{X}_{intra} \in \mathbb{R}^{H\times W\times D}$, at first the max pairing with the stride r is applied, followed by two consecutive 3×3 convolutional layers. This process yields a higher-level representation of size $\frac{H}{r} \times \frac{W}{r} \times D$. Subsequently, this representation is reshaped into a sequence of patch tokens $\mathbf{T} \in \mathbb{R}^{HW/r^2\times D}$ with enriched semantics. To determine the center of each class, the average of all feature pixels belonging to the same class is calculated using $\mathbf{X}_{intra}$. Initially, $\mathbf{X}_{intra}$ is used to predict a coarse segmentation probability map $\mathcal{Y}_c \in \mathbb{R}^{H\times W}$, where each pixel value ranges from 0 to 1, representing the likelihood of belonging to the c-th class. Next, the representations of all pixels are aggregated, weighted by their probabilities of belonging to the c-th object:

$$\mathbf{G}_c = \sum_{i\in I} \mathcal{Y}_c^i \mathbf{X}_{intra}^i \tag{4.17}$$

where $\mathbf{X}_{intra}^i$ is the feature of i-th pixel and $\mathcal{Y}_c^i$ is the normalized probability for i-th pixel belonging to the c-th class. The obtained $\mathbf{G}_c \in \mathbb{R}^D$ is denoted as the c-th class token. As shown in Fig. 4.14, class tokens are concatenated $\mathbf{G} \in \mathbb{R}^{C\times D}$ with patch tokens $\mathbf{T} \in \mathbb{R}^{HW/r^2\times D}$, which then passes through the Transformer to capture the long-range interactions. The linearly projected $\mathbf{Q}$, $\mathbf{K}$ and $\mathbf{V}$ have the same shape of $(C + HW/r^2) \times D$, and attention map $\mathcal{A}$ can be denoted as:

$$\mathcal{A} = \begin{bmatrix} \mathcal{A}^{c\to c} & \mathcal{A}^{c\to p} \\ \mathcal{A}^{p\to c} & \mathcal{A}^{p\to p} \end{bmatrix} \tag{4.18}$$

where $\mathcal{A}^{c\to c}$, $\mathcal{A}^{p\to p}$, $\mathcal{A}^{c\to p}$, $\mathcal{A}^{p\to c}$ are normalized similarity matrices representing class-to-class, patch-to-patch, class-to-patch, and patch-to-class. $\mathcal{A}^{c\to c}$ and $\mathcal{A}^{p\to p}$ perform the self-attentions, which update the class tokens and patch tokens respectively; $\mathcal{A}^{c\to p}$ and $\mathcal{A}^{p\to c}$ learn the complementary information via cross-attention. Therefore, patch tokens enable the exploration of explicit category dependencies through interactions with class tokens, facilitating class-guided feature enhancement. Conversely, class tokens can capture rich semantic information from patch tokens, enhancing the discriminative ability across different categories.

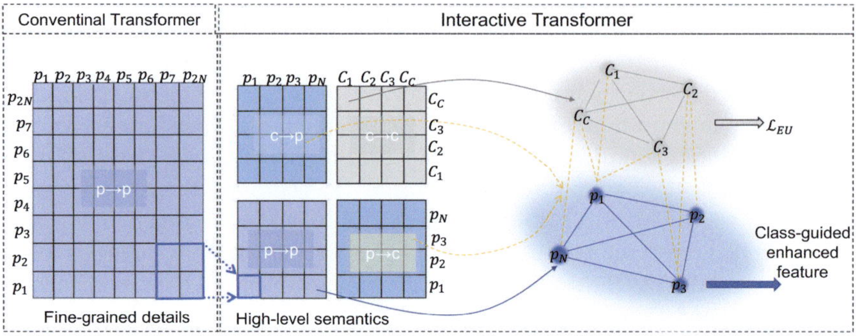

Fig. 4.14 Interactions among patch ($\mathbf{p}_i$) and class tokens ($\mathbf{c}_i$)

After enhancing the sequence with interactions between class tokens and patch tokens, the sequence is reordered back according to the original concatenation order. Additionally, an auxiliary Euclidean distance loss L_{EU} [10] is introduced to increase the distance between the enhanced class tokens, enhancing feature distinctiveness. Next, the enhanced patch tokens are reshaped into 2D features and upsampled to match the original size of $H \times W \times D$. To preserve the original information, the input $\mathbf{X}_{intra}$ is combined using a 1×1 convolutional layer, resulting in the final output $\mathbf{X}_{inter} \in \mathbb{R}^{H \times W \times D}$. This process ensures that the output retains the source information while incorporating enhanced features from both class and patch tokens.

4.6.4 Complexity Analysis

In the first intra-stage, enhanced features initially emphasize the internal structure of objects. These features then provide more condensed features for each object and more representative initial class tokens to the inter-class Transformer. This approach achieves comprehensive exploration of class-specific dependencies. Compared to the conventional Transformer with a complexity of $O(DH^2W^2)$, the computational cost of ClassFormer is reduced. Specifically, the cost can be summarized as:

$$\Omega = \underbrace{O(DHWC\mathcal{N})}_{intra-class} + \underbrace{O(D(C + HW/r^2)^2)}_{inter-class} \tag{4.19}$$

In the intra-class dynamic Transformer, $\mathcal{N} = \sum_{i=1}^{S} N_i$ represents the total number of selected salient keys across S scales. Despite considering multiple classes C and scales S, the computational cost of the intra-class Transformer remains relatively small compared to the conventional approach ($C\mathcal{N} \ll HW$). Furthermore, in the inter-class interactive Transformer, the down-sampling rate r determines the number of high-level patch tokens, significantly reducing the overall computational burden of the Transformer from a global perspective. Both intra-class and inter-class Transformers are designed in a lightweight manner, optimizing efficiency while effectively capturing class-specific and scale-specific dependencies in the data.

In summary, from a class-wise perspective, ClassFormer [5] disentangles the class-agnostic Transformer into intra-class and inter-class components, with the intra-class Transformer focusing on object dependency exploration and the inter-class component capturing object-to-object relations. Being a flexible plug-and-play module, ClassFormer can be seamlessly integrated at multiple scales within existing backbone architectures, offering significant improvements in segmentation performance. This modularity allows it to achieve state-of-the-art results across a wide range of tasks, further enhancing the adaptability and efficiency of medical image segmentation models.

4.7 Performance Validation

4.7.1 Dataset and Implementation Details

We provide an overview of the commonly used experimental datasets used for medical image segmentation tasks, including Automated Cardiac Diagnosis Challenge dataset (ACDC) [11], Multi-organ Synapse dataset (Synapse) [12], and Multi-Organ Nucleus dataset (MoNuseg) [13].

- **ACDC dataset** [11] (4 classes). The Automated Cardiac Diagnosis Challenge includes 200 annotated short-axis cardiac cine-MR images from 100 patients, covering three key organs: the left ventricle (LV), right ventricle (RV), and myocardium (MYO). The dataset is randomly split into 70 cases for training, 10 cases for validation, and 20 cases for testing.
- **Synapse dataset** [12] (9 classes). The Multi-Organ Segmentation Challenge includes 30 abdominal CT scans with 8 organs: the aorta, gallbladder, left kidney, right kidney, liver, pancreas, spleen, and stomach. The dataset is split into 18 cases for training and 12 cases for testing.
- **MoNuSeg dataset** [13] (2 classes). Training data contains 30 images and around 22,000 nuclear boundary annotations. It contains 30 images for training, and 14 images for testing.

To avoid overfitting, two kinds of data augmentations, including random rotation and flipping, are adopted. The image size is all resized into 224×224. Here, we list the batch size (bs), learning rate (lr), maximum training epochs (ep), optimizer (opt) for three datasets as below:

- Synapse: bs=8; lr=3e-3; ep=600; opt=SGD;
- ACDC: bs=8; lr=3e-3; ep=200; opt=SGD;
- MoNuSeg: bs=4; lr=1e-3; ep=200; opt=Adam.

All models are trained with momentum 0.9 and weight decay 1×10^{-4}. For fair comparison, we use the combined cross entropy loss and dice loss as the segmentation losses. Following [5], all 3D volumes are inferenced in a slice-by-slice fashion and the predicted 2D slices are stacked together to reconstruct the 3D prediction for evaluation.

To evaluate the segmentation performance of each method, the Dice Similarity Coefficient (DSC) and the 95th-percentile Hausdorff Distance (HD95) are employed as evaluation metrics for the ACDC and Synapse datasets, while DSC and Intersection over Union (IoU) are used for the MoNuSeg dataset. The Dice score quantifies the overlap between the predicted segmentation and the ground truth, providing a measure of overall accuracy in terms of spatial agreement. HD95, on the other hand, assesses the boundary error by measuring the 95th-percentile distance between the predicted and true boundaries, which is particularly important for evaluating segmentation precision. IoU measures the ratio of the intersection area to the union area between the predicted and ground truth regions, offering another perspective on segmentation accuracy.

4.7.2 Experimental Results and Analysis

For each dataset, following [5], we compare the corresponding methods to provide a thorough evaluation of segmentation performance. Tables 4.1, 4.2, and 4.3 summarize the results of different comparison methods on the Synapse, ACDC, and MoNuSeg datasets, respectively.

Specifically, we evaluate eleven methods on the Synapse dataset, including V-Net [14], U-Net [15], AttUNet [16], TransUNet [2], Swin-UNet [1], UCTransNet [17], MT-UNet [4], AFTerUNet [18], MISSFormer [19], ScaleFormer [3], and Class-Former [5].

For the ACDC dataset, we compare several classical methods based on architectures discussed in this chapter, including TransUNet, Swin-UNet, MT-UNet, UCTransNet, ScaleFormer, and ClassFormer. For the MoNuSeg dataset, following [5], we extend the evaluation by including additional methods such as UNet, AttUNet, MISSFormer, UCTransNet alongside the aforementioned models.

The methods under comparison can similarly be categorized into the types introduced earlier in this chapter:

(1) **CNN-based methods**: UNet, AttUNet and V-Net. U-Net is widely used 2D architectures for medical image segmentation. AttUNet extends U-Net by incorporating attention mechanisms, which enhance feature selection by focusing on the most relevant regions of the feature map. V-Net is a 3D variant of U-Net designed specifically for volumetric data, suitable for handling spatial context in medical images.

Table 4.1 Experiment results on synapse dataset

Method	Params (M)	DSC (%)	HD95 (mm)	Aorta	Gallbladder	Kidney (L)	Kidney (R)	Liver	Pancreas	Spleen	Stomach
V-Net	45.60	68.81	–	75.34	51.87	77.10	80.75	87.84	40.05	80.56	56.98
U-Net	31.04	76.85	39.70	89.07	69.72	77.77	68.60	93.43	53.98	86.67	75.58
AttUNet	34.88	77.77	36.02	89.55	68.88	77.98	71.11	93.57	58.04	87.30	75.75
TransUNet	105.28	77.48	31.69	87.23	63.13	81.87	77.02	94.08	55.86	85.08	75.62
Swin-UNet	27.17	79.12	21.55	85.47	66.53	83.28	79.61	94.29	56.58	90.66	76.60
UCTransNet	66.43	78.23	26.75	88.86	66.97	80.19	73.18	93.17	56.22	87.84	79.43
MT-UNet	79.08	78.59	26.59	87.92	64.99	81.47	77.29	93.06	59.46	87.75	76.81
AFTerUNet	41.50	81.02	–	**90.91**	64.81	87.90	85.30	92.20	63.54	90.99	72.48
MISSFormer	42.46	81.96	18.20	86.99	68.65	85.21	82.00	94.41	65.67	**91.92**	80.81
ScaleFormer	111.58	82.86	16.81	88.73	**74.97**	86.36	83.31	95.12	64.85	89.40	80.14
ClassFormer	35.43	**84.40**	**12.22**	89.04	72.91	**88.95**	**85.68**	**95.55**	**69.25**	90.70	**83.09**

The DSC for each class, the average DSC and HD95, and the number of parameters for each method are reported. The underline indicates the second-highest accuracy

Table 4.2 Experiment results on ACDC dataset

Methods	DSC (%)	RV	Myo	LV
TransUNet	89.63	86.70	86.96	95.33
MT-UNet	89.84	86.06	88.14	95.32
Swin-UNet	90.00	88.55	85.62	95.83
UCTransNet	90.11	88.01	87.29	95.02
ScaleFormer	90.17	87.33	88.16	95.04
ClassFormer	**90.99**	**88.35**	**89.08**	**95.54**

The DSC for each class, the average DSC are reported. The underline indicates the second-highest accuracy

Table 4.3 Comparison with SOTA methods on MoNuSeg dataset

Methods	DSC(%)	IoU (%)
U-Net	73.97	59.42
AttUNet	76.20	62.64
MISSFormer	76.04	61.58
MT-UNet	77.97	64.30
TransUNet	79.20	65.68
Swin-UNet	78.49	64.72
UCTransNet	79.87	66.68
ScaleFormer	80.06	66.87
ClassFormer	**80.75**	**67.88**

The average DSC and IOU are reported. The underline indicates the second-highest accuracy

(2) **Hybrid Transformer methods**: UCTransNet, ScaleFormer. Specially, UCTransNet introduces a channel-wise Transformer module between the U-Net encoder and decoder to better fuse multi-scale features and reduce the semantic gap.
(3) **Pure Transformer methods**: Swin-UNet, MISSFormer. Specially, MISSFormer enhances the traditional Transformer block by designing an advanced mechanism to strengthen long-range dependencies while supplementing local contextual information.
(4) **Advanced Transformer Structures**: MT-UNet, AFTerUNet, ClassFormer. In particular, AFTerUNet introduces an Axial Fusion Transformer, which considers both intra-slice and inter-slice long-range cues to guide segmentation.

Results on Synapse dataset. As illustrated in Table 4.1, compared to traditional CNN methods, TransUNet, Swin-UNet, MT-UNet achieve higher accuracy on multi-organ segmentation by leveraging the Transformer's ability to capture long-range dependencies. ScaleFormer performs better due to its scale-wise structural design, but it also introduces a significant increase in the number of parameters. ClassFormer, as a plug-and-play module, outperforms the existing approaches in both regional measures DSC (84.40%) and boundary-aware measure HD (12.22mm) with fewer model parameters (Params: 35.43M).

Results on ACDC dataset. In Table 4.2, MT-UNet achieves higher average accuracy (DSC: +0.21%) compared to TransUNet by incorporating the improved Transformer module and leveraging inter-sample relationships. While MT-UNet alleviates the reliance on pre-trained model weights, its performance still falls short of Swin-UNet, which demonstrates the effectiveness of the Swin Transformer in feature representation.

UCTransNet further improves the segmentation accuracy by effectively considering multi-scale feature fusion compared with Swin-UNet (DSC:+0.11%). ScaleFormer achieves even better performance by not only addressing feature fusion across scales but also integrating CNN and Transformer modules at each scale, making it a more powerful backbone for segmentation tasks.

ClassFormer further advances the state-of-the-art (DSC: +0.82%) by explicitly modeling inter-class relationships across scales. This approach proves particularly advantageous for multi-class segmentation tasks, where understanding the dependencies between different categories is crucial for achieving higher accuracy.

Results on MoNuSeg dataset. Table 4.3 presents the DSC and IoU results on the MoNuSeg dataset. Traditional method such as U-Net achieves average DSC scores of 73.97%, demonstrating the effectiveness of their encoder-decoder architectures in medical image segmentation. AttUNet further enhances performance by incorporating attention mechanisms, achieving a DSC of 76.20% and an IoU of 62.64%.

Compared to pure CNN-based methods, Transformer-based and hybrid Transformer approaches exhibit superior performance, primarily due to the Transformer's ability to handle variations in segmentation targets more effectively. The strong results achieved by UCTransNet and ScaleFormer highlight the importance of integrating multi-scale information with Transformer or CNN architectures. This combination allows these models to capture fine-grained details and improves segmentation accuracy for challenging tasks like nucleus segmentation, where the targets are both small and numerous.

Finally, ClassFormer demonstrates its exceptional effectiveness by further considering inter-class relationships, which proves particularly advantageous for fine-grained medical image segmentation tasks. By achieving the highest performance among all methods (DSC: 80.75%, IOU:67.88%), ClassFormer underscores its capability in handling complex segmentation scenarios with multiple small targets.

4.8 Summary

In this section, we discuss three types of network architecture designs by integrating Transformer into medical image segmentation models: **Pure Transformer Structure**, **Hybrid Transformer Structure**, and **Advanced Transformer Structures**, evaluating their performance on various benchmarks.

Swin-UNet is a representative pure Transformer-based architecture that utilizes a pre-trained Swin Transformer to construct its Encoder-Decoder framework. Specifically, Patch Merging and Patch Expanding layers are designed for down-sampling and up-sampling, respectively, enabling efficient multi-scale feature representation and reconstruction. Since the model relies entirely on Transformer-based components, it may inevitably overlook some fine-grained local features compared to CNN-based architectures.

TransUNet is the pioneering work in Hybrid Transformer Structure design, while still adhering to the classic Encoder-Decoder framework. In its encoder, a CNN is employed in the shallow layers to extract local features, while a pre-trained Transformer operates in the deeper layers to capture global context and long-range dependencies. The hybrid approach still requires managing the computational complexity of both CNN and Transformer components.

ScaleFormer, another hybrid architecture design, integrates CNN and Transformer within each scale while leveraging the Transformer's ability to capture cross-scale long-range dependencies. However, the full-scale local-global feature extraction in ScaleFormer results in a relatively large number of parameters, leading to increased computational cost and memory usage.

Besides the hybridization of the network architecture, MT-UNet and ClassFormer each make specific improvements to the Transformer structure. MT-UNet improves upon conventional self-attention by introducing a local+global self-attention design, which also considers the dependencies between samples.

ClassFormer separates the class-agnostic Transformer into two distinct components: one that focuses on capturing intra-class dependencies and another that models inter-class relationships between objects. By distinguishing between intra-class and inter-class dependencies, ClassFormer enables more accurate segmentation at the object level. Its modular architecture makes it highly adaptable, facilitating seamless integration with various backbone networks and datasets, and allowing it to be tailored to different segmentation tasks.

References

1. Cao, H., Wang, Y., Chen, J., Jiang, D., Zhang, X., Tian, Q., et al. Swin-unet: Unet-like pure transformer for medical image segmentation. In: *European conference on computer vision* (pp. 205–218). Springer Nature Switzerland, Cham.
2. Chen, J., Lu, Y., Yu, Q., Luo, X., Adeli, E., Wang, Y., et al. (2021). Transunet: Transformers make strong encoders for medical image segmentation. arxiv preprint arxiv:2102.04306.
3. Huang, H., Xie, S., Lin, L., Iwamoto, Y., Han, X., Chen, Y. W., et al. (2022). ScaleFormer: revisiting the transformer-based backbones from a scale-wise perspective for medical image segmentation. In: *Proceedings of the thirty-first international joint conference on artificial intelligence, IJCAI 2022, Vienna, Austria, 23–29 July 2022* (pp. 964–971). ijcai.org, Vienna. https://doi.org/10.24963/IJCAI.2022/135.
4. Wang, H., Xie, S., Lin, L., Iwamoto, Y., Han, X. H., Chen, Y. W., et al. (2022). Mixed transformer u-net for medical image segmentation. In: *ICASSP 2022–2022 IEEE international conference on acoustics, speech and signal processing (ICASSP)* (pp. 2390–2394). IEEE.

5. Huang, H., Xie, S., Lin, L., Tong, R., Chen, Y. W., Wang, H., et al. (2023). ClassFormer: exploring class-aware dependency with transformer for medical image segmentation. *Proceedings of the AAAI Conference on Artificial Intelligence37*(1), 917–925.
6. Liu, Z., Lin, Y., Cao, Y., Hu, H., Wei, Y., Zhang, Z., et al. (2021). Swin transformer: Hierarchical vision transformer using shifted windows. In: *Proceedings of the IEEE/CVF international conference on computer vision* (pp. 10012–10022).
7. He, K., Zhang, X., Ren, S., & Sun, J. (2016). Deep residual learning for image recognition. *Proceedings of the IEEE conference on computer vision and pattern recognition* (pp. 770–778).
8. Dosovitskiy, A., Beyer, L., Kolesnikov, A., Weissenborn, D., Zhai, X., Unterthiner, T., et al. (2020). An Image is worth 16 × 16 words: Transformers for Image Recognition at Scale[C]. *International conference on learning representations* 2021.
9. Ho, J., Kalchbrenner, N., Weissenborn, D., et al. (2019). *Axial attention in multidimensional transformers*. arXiv preprint arXiv:1912.12180.
10. Li, J., Yu, H., Chen, C., Ding, M., et al. (2022). Category guided attention network for brain tumor segmentation in MRI. *Physics in Medicine and Biology, 67*(8), Article 085014.
11. Bernard, O., Lalande, A., Zotti, C., Cervenansky, F., Yang, X., Heng, P. A., et al. (2018). Deep learning techniques for automatic MRI cardiac multi-structures segmentation and diagnosis: is the problem solved? *IEEE Transactions on Medical Imaging, 37*(11), 2514–2525.
12. Landman, B., Xu, Z., Igelsias, J., Styner, M., Langerak, T., et al. (2015). Miccai multi-atlas labeling beyond the cranial vault—workshop and challenge. In: *Proceeding MICCAI multi-atlas labeling beyond cranial vault—Workshop challenge* (vol. 5, p. 12).
13. Kumar, N., Verma, R., Sharma, S., Bhargava, S., Vahadane, A., et al. (2017). A dataset and a technique for generalized nuclear segmentation for computational pathology. *IEEE Transactions on Medical Imaging, 36*(7), 1550–1560.
14. Milletari, F., Navab, N., & Ahmadi, S. A. (2016). V-net: Fully convolutional neural networks for volumetric medical image segmentation. In: *2016 fourth international conference on 3D vision (3DV)* (pp. 565–571). IEEE.
15. Ronneberger, O., Fischer, P., et al. (2015). U-net: Convolutional networks for biomedical image segmentation. In: *Medical image computing and computer-assisted intervention–MICCAI 2015: 18th international conference, Munich, Germany, October 5–9, 2015, proceedings, Part III 18* (pp. 234–241). Springer International Publishing.
16. Schlemper, J., Oktay, O., Schaap, M., Heinrich, M., Kainz, B., Glocker, B., et al. (2019). Attention gated networks: Learning to leverage salient regions in medical images. *Medical Image Analysis, 53*, 197–207.
17. Wang, H., Cao, P., Wang, J., et al. (2022). Uctransnet: Rethinking the skip connections in u-net from a channel-wise perspective with transformer. *Proceedings of the AAAI Conference on Artificial Intelligence, 36*(3), 2441–2449.
18. Yan, X., Tang, H., Sun, S., Ma, H., Kong, D., et al. (2022). After-unet: Axial fusion transformer unet for medical image segmentation. In: *Proceedings of the IEEE/CVF winter conference on applications of computer vision* (pp. 3971–3981).
19. Huang, X., Deng, Z., Li, D., et al. (2021). Missformer: An effective transformer for 2d medical image segmentation [J]. *IEEE Transactions on Medical Imaging, 2022, 42*(5), 1484–1494.

Chapter 5
Deep Learning Networks for 3D Medical Image Analysis

5.1 Introduction

A primary difference between medical images and natural images lies in their physical form and imaging principles. While there are 2D medical images such as X-rays, a significant number of medical images are in 3D format, providing more comprehensive views of patient's organs for clinical evaluation. Deep learning techniques face challenges in analyzing 3D medical images due to their high dimensionality and resolution, leading to significant GPU memory consumption. To reduce computational demands, a common approach is converting 3D images into 2D slices along the z-axis, then applying 2D deep learning models [1]. However, this method fails to preserve inter-slice 3D contextual information, leading to suboptimal results. To overcome this limitation, 2.5D and 2D multi-view methods have been proposed. These approaches enable 2D deep learning models to capture some 3D spatial information by treating adjacent slices as different input channels or by sequentially splitting the 3D image along the x, y, and z-axes. However, they are still limited in fully utilizing the complete 3D spatial information.

In contrast, 3D deep learning models are naturally better suited for processing 3D data. However, to mitigate computational costs, patch-sampling is commonly used in 3D models. This technique involves segmenting the 3D image into smaller subvolumes [2–4], which can result in the loss of global contextual information and reduce inference efficiency. To address these challenges, recent advancements have introduced patch-free 3D methods and lightweight 3D models, enabling efficient 3D analysis with lower computational costs and faster inference.

The following subsections will focus on 3D medical image segmentation and present several contemporary solutions for deep learning in 3D medical image analysis, including 3D CNNs, 3D Transformer-based approaches, and lightweight 3D models.

Y. Chen et al., *Recent Advances in Deep Learning for Medical Image Analysis*, Intelligent Systems Reference Library 278, https://doi.org/10.1007/978-3-031-94791-9_5

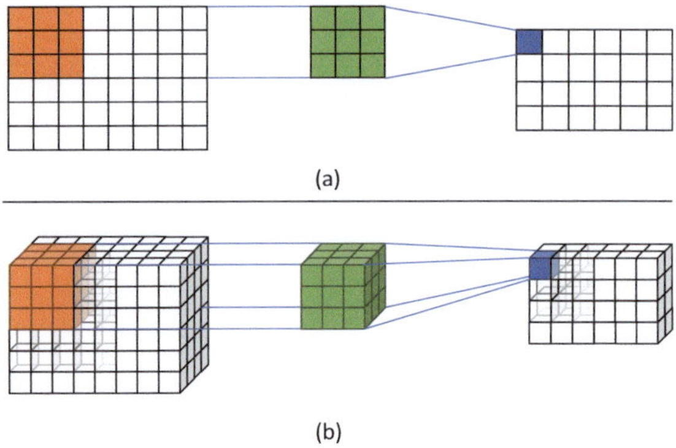

Fig. 5.1 **a** 2D convolution models representations only on a 2D plane, **b** 3D convolution directly performs 3D spatial feature extraction

5.2 3D Convolutional Neural Networks

3D CNNs are natural solutions for processing 3D medical images due to their inherent ability to capture comprehensive 3D spatial information. 3D CNNs utilize 3D convolutional kernels for feature extraction, allowing them to model representations along the x, y, and z axes simultaneously. Different 3D deep learning models can be constructed based on 3D convolutional kernels and their variants, such as 3D U-Net [2], 3D ResUNet [4], and V-Net [3]. An intuitive comparison of 2D and 3D convolution is presented in Fig. 5.1.

However, conventional 3D CNNs face the challenge of high computational cost due to the high dimensionality of the input 3D medical images. As a result, most 3D models are designed with narrower and shallower architectures to reduce GPU memory consumption. Therefore, patch-sampling is widely employed in 3D methods, where the entire 3D medical image is divided into smaller patches for training, further decreasing GPU memory usage. Recently, patch-free CNNs have been developed to improve 3D medical image segmentation. Unlike patch-sampling methods, they retain full semantic information, enabling more efficient training and faster inference.

In this section, we will give brief introductions to classic patch-based and patch-free 3D CNNs. In Sect. 5.2.1, 3D U-Net is discussed first as a basic 3D deep learning model. In Sect. 5.2.2, 3D Residual-U-Net is introduced as an improved version of 3D U-Net. In Sect. 5.2.3, V-Net is presented as a more complex 3D CNN model with learnable pooling layers. In Sect. 5.2.4, Holistic Decomposition-based methods are discussed first as vanilla patch-free methods. In Sect. 5.2.5, PFSeg is presented as a more recent patch-free solution to 3D medical image analysis tasks. In Sect. 5.2.6, performance validation is conducted on these methods to show their pros and cons. Finally, in Sect. 5.2.7, a summary of these 3D CNNs is presented.

5.2.1 3D U-Net

3D U-Net [2] is the 3D version of the 2D deep learning model, U-Net [1]. It uses 3D convolution kernels for feature extraction, which inherently capture the full 3D spatial representations in the images. While maintaining the skip connection design of the original 2D U-Net, 3D U-Net has the following differences:

- 3D U-Net modifies the convolutional layers in the original U-Net into 3D convolutional layers, making it suitable for directly handling 3D medical image inputs such as Computed Tomography (CT) and Magnetic Resonance Imaging (MRI).
- 3D U-Net reduces the width and depth of the network structure to maintain a relatively low GPU memory requirement.

A schematic view of the network structure of 3D U-Net is presented in Fig. 5.2. As shown, 3D U-Net follows an encoder-decoder structure. In the encoder part, each layer comprises a convolutional block for feature extraction, and 3D max pooling is used between layers to reduce the spatial size of the feature map. In the decoder network, 3D transposed convolution is used to enlarge the spatial size of the feature maps for each layer. The skip connection mechanism helps the layers fuse the enlarged feature maps they receive with the low-level features from the corresponding encoder layers. Finally, a point-wise convolution is used to project the output feature maps to the final segmentation mask prediction.

More specifically, for a given 3D input image shaped as $C \times D \times W \times H$, where C, D, W, H are respectively the channel, depth, width and height of the image, the encoder gradually expands the channel dimension through the convolution layers. Each layer's output channel can be configured according to the specific device, yet usually follows a 32-64-128-256 pattern. After each convolutional layer, there is also a normalization layer for stable learning and an activation function layer for adding nonlinearity to the model. Then, a 3D max pooling layer is applied to reduce the spatial size by $2\times$ to further distill the core semantic information for the next convolutional layer. Bottleneck layers are employed throughout the network. The feature size undergoes an $8\times$ reduction in the bottleneck layer, where crucial semantic information is encoded. The decoder stage is subsequently employed to reconstruct the full spatial information. Through the use of transposed convolutional layers, the spatial dimensions of the feature maps are expanded. Following this expansion, the feature maps are concatenated with the low-level features propagated through the skip connection mechanism. Subsequently, an additional convolutional block processes the concatenated feature maps to reduce channel dimensionality and to distinguish the target and background. This process is iterated three times, gradually restoring the feature maps to their original size, finally producing the predicted segmentation mask.

Given that current 3D medical images can often reach dimensions as large as $155 \times 512 \times 512$, even with reduced width and depth, direct processing of these high-resolution (HR) 3D images in one pass remains challenging for 3D U-Net.

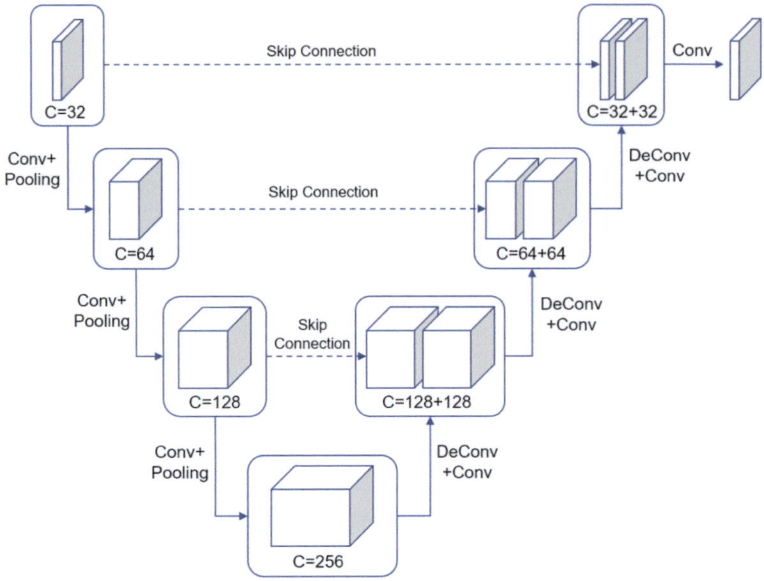

Fig. 5.2 A schematic view of the network structure of 3D U-Net. Max pooling is used between each layer of the encoder to reduce the spatial size of the feature by 2×, and transposed convolution (i.e., deconvolution) is used between each layer of the decoder to generate a 2× larger output. *C* denotes the number of channels in each layer

Consequently, the network must resort to patch-sampling during training. Typically, random cropping is employed during training to extract small patches (e.g., 128 × 128 × 128) from the original 3D image. This approach serves to not only alleviate computational costs during training but also functions as a form of data augmentation, mitigating the challenges associated with small-scale datasets commonly encountered in medical imaging.

During inference, two optional schemes are available. The first one is directly input the entire 3D medical image for prediction. However, this approach is computational-expensive due to the GPU's limited memory capacity, which cannot accommodate the storage of gradients for back-propagation during inference. The second scheme employs a sliding-window strategy to ensure that the inference stage receives the same size of model input as during training. Both solutions have their advantages and disadvantages. The first scheme offers fast inference speed but may suffer from the inconsistency of input sizes between training and testing, which can impact segmentation accuracy. On the other hand, the latter scheme avoids such issue, but it also results in longer processing times since the model needs to be run multiple times to obtain segmentation masks for all patch regions, which are then combined to form the full prediction mask.

In general, 3D U-Net marks a fundamental progress in 3D medical image analysis. It perpetuates the effectiveness of the U-Net structure in medical image analysis field.

However, due to the higher dimensionality of input data and the larger GPU memory consumption during model training, ordinary 3D U-Net is limited to smaller model scales and incomplete perception of the full 3D image, while hinders its performance.

5.2.2 3D Residual-U-Net

Motivated by the residual connection mechanism proposed in ResNet [5], 3D Residual-U-Net (3D ResUNet) incorporates residual connections into the original 3D U-Net architecture to address performance degradation issues [4]. As the residual connection mechanism has been previously introduced in previous chapters, we will only give a brief recap as follows.

In conventional convolutional layers, the model directly learns the mapping function (i.e., with parameters θ) between the input $\mathbf{X}$ and the expected output $\mathbf{Y}$ of the layer, formulated as:

$$\min_{\theta} ||f(\mathbf{X}; \theta) - \mathbf{Y}||^2 \tag{5.1}$$

In contrast, the convolutional layer with residual connection turns to model the residuals between the input $\mathbf{X}$ and the output $\mathbf{Y}$, which is:

$$\min_{\theta} ||f(\mathbf{X}; \theta) - (\mathbf{Y} - \mathbf{X})||^2 \tag{5.2}$$

Residual connections are primarily proposed based on the observation that deeper networks may struggle to surpass simpler ones, even with overfitting addressed. This phenomenon is attributed to the challenge of fitting identity mappings within most deep learning models. Therefore, the explicit introduction of an identity mapping branch within feature extraction layers aids in enhancing the performance of deeper network designs while preserving the overall representation modeling capabilities of the pipeline. 3D ResUNet consistently outperforms the 3D U-Net in various scenarios without significantly increasing computational costs. 3D ResUNet maintains the same encoder-decoder structure, leveraging identical normalization layers, 3D max pooling layers, and skip connection mechanisms. The key distinction lies in the addition of an extra point-wise 3D convolution for residual connections. This facilitates the adjustment of the residual output's shape, ensuring proper addition to the convolution outputs (as illustrated in Fig. 5.3).

The success of residual connections underscores their ability to avoid performance degradation during the representation learning process. However, similar to 3D U-Net, 3D ResUNet relies on patch-sampling for training on most mainstream GPUs. Additionally, its inference speed is marginally slower than that of 3D U-Net due to the inclusion of additional summation operations in the network.

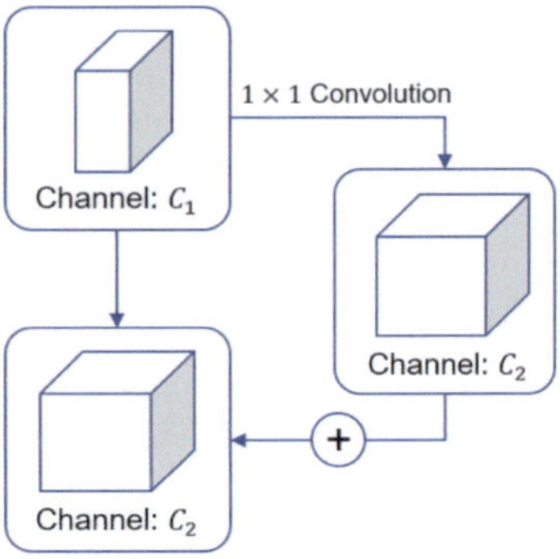

Fig. 5.3 An intuitive illustration of the residual connection mechanism used in 3D ResUNet

5.2.3 V-Net

V-Net [3] is another 3D CNN designed especially for 3D medical image segmentation, which also follows an encoder-decoder structure. Similar to 3D ResUNet, V-Net incorporates skip connections and residual connections into its model architecture. However, notable differences exist between V-Net and 3D ResUNet. In addition to being deeper and wider, V-Net introduces strides convolution to replace traditional pooling layers. Following the normalization and activation layers, V-Net employs another layer of 3D convolution instead of the conventional 3D max pooling to reduce the spatial dimensions of the feature maps. Strided convolution differs from ordinary convolution in that it accepts a larger step size during the sliding window computation, whereas ordinary convolution employs a step size of one by default.

To illustrate, let H represent height, W represent width, D represent depth, C denote the number of channels before convolution, and C' denote the number of channels after convolution. Following ordinary convolution, a $C \times D \times W \times H$ feature map would become $C' \times (D-1) \times (W-1) \times (H-1)$. However, after a s-strided convolutional layer, it becomes $C' \times \frac{D}{s} \times \frac{W}{s} \times \frac{H}{s}$. By employing strided convolutional layers as pooling layers, V-Net renders pooling operations trainable, deepening the model architecture overall. A visual comparison of ordinary convolution and strided convolution is provided in Fig. 5.4.

While V-Net theoretically promises enhanced modeling performance, it may exhibit limited generalization capability on certain datasets. And due to the increased model size, V-Net necessitates patch-sampling before training on mainstream GPUs.

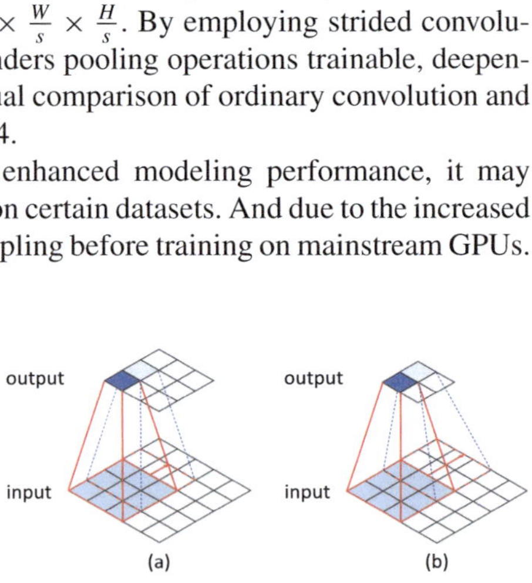

Fig. 5.4 **a** Depicts ordinary convolution with a stride of 1. **b** illustrates strided convolution with a stride of 2. The output size decreases as the sliding window moves more steps at a time in this mode

Furthermore, V-Net demonstrates slower inference speeds compared to 3D ResUNet, due to the introduction of additional convolution operations and a significantly larger parameter count.

5.2.4 *Holistic Decomposition-Based 3D Method*

Although patch-based methods facilitate GPU-memory-efficient training, they often compromise model performance by undermining the global contextual information present in the image. To effectively train 3D models at the entire image level, patch-free methods are increasingly gaining attraction. Among these, G. Zeng, et al. proposed a Holistic decomposition-based method in their work titled "Holistic decomposition convolution for effective semantic segmentation of medical volume images" [6].

Holistic Decomposition is designed as the reverse operation of Pixel-Shuffle. While Pixel-Shuffle generates an output with a larger spatial size by merging several channels of small feature maps, Holistic Decomposition can decompose a large feature map into several channels of small feature maps, as illustrated in Fig. 5.5. Given their simple inverse relationship, Holistic Decomposition and Pixel-Shuffle can be integrated into the front and back of most existing 3D segmentation backbones, transforming them from patch-based to patch-free methods. A typical pipeline is presented in Fig. 5.6.

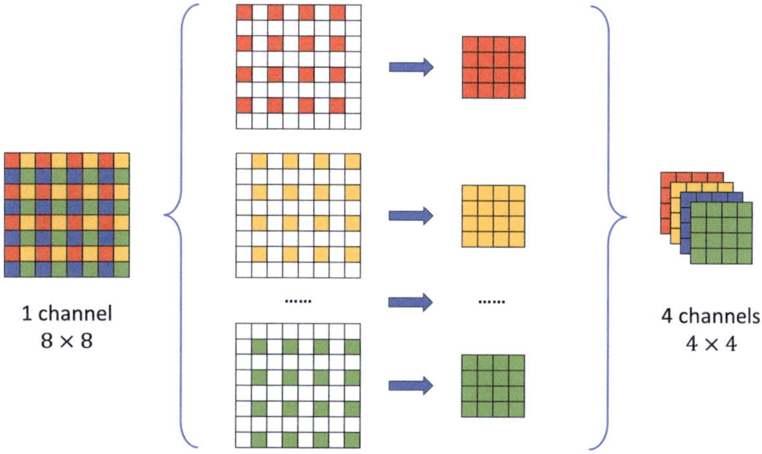

Fig. 5.5 A schematic view of holistic decomposition in 2D, with a down-sampling factor of 2 for each dimension as an example, is illustrated. It shows that each output channel is formed by sampling the pixel at a certain fixed position in those 2 × 2 non-overlapping regions. In the end, the spatial size of the feature is reduced, but all the voxels are still preserved through this method

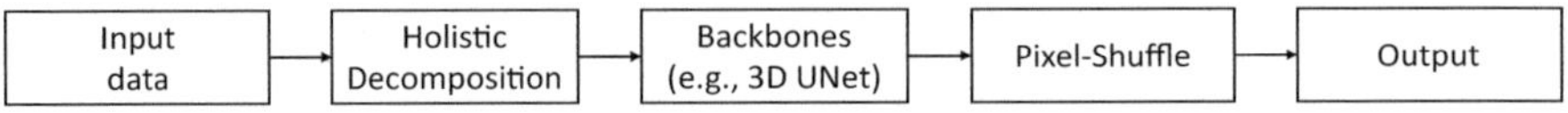

Fig. 5.6 The pipeline of holistic decomposition-based 3D methods. The input data is first decomposed into several small components to reduce the spatial dimensions. Then, ordinary segmentation backbones can be used for segmentation in the lower-dimension space. Finally, pixel-shuffle is used to restore the decomposed components back into the complete output segmentation mask

The reduction in spatial size achieved through Holistic Decomposition significantly alleviates GPU memory requirements during training, facilitating patch-free 3D medical image segmentation. Unlike interpolation algorithms such as bilinear and tricubic [7], Holistic Decomposition realizes downsampling by rearranging voxels for decomposition. Consequently, it retains all voxels post-decomposition, minimizing information loss.

While rearranging voxels from spatial dimensions to channel dimensions reduces the width of the network relatively, it may impact the model's representation learning capability. Additionally, following voxel rearrangement, each channel of the decomposition output contains a blurred 3D image of the original, potentially hindering the representation learning process by losing high-frequency information. However, despite these considerations, Holistic Decomposition remains an effective and user-friendly method for achieving patch-free 3D segmentation based on existing backbones.

5.2.5 Patch-Free 3D Medical Image Segmentation

H. Wang et al. proposed a Patch-Free 3D Medical Image Segmentation (PFSeg) in their work titled "Patch-free 3d medical image segmentation driven by super-resolution technique and self-supervised guidance" [8], which adopts the multi-task learning framework to realize super-resolution-based patch-free segmentation. A schematic view of the PFSeg pipeline is depicted in Fig. 5.7. In PFSeg, the main task is high-resolution (HR) semantic segmentation using low-resolution (LR) input, while the auxiliary task is super-resolution to aid in the restoration of high-frequency information throughout the network. Both tasks share an encoder and are further regulated by the Task-Fusion Module (TFM). During inference, the super-resolution branch can be removed to reduce computational costs, resulting in a highly cost-effective patch-free 3D segmentation solution. PFSeg is compatible with most existing 3D segmentation backbones, offering a patch-free solution for these methods.

PFSeg uses the manually downsampled LR 3D image as patch-free main input. The LR image is fed into the shared encoder to provide global-wise contextual information for the two different tasks. Additionally, a HR patch is also cropped from the original HR image to be the guidance for subsequent high-frequency information restoration process. The extracted features of the two inputs are concatenated to serve as bottleneck features for the two decoders of the tasks. Finally, at the end of the

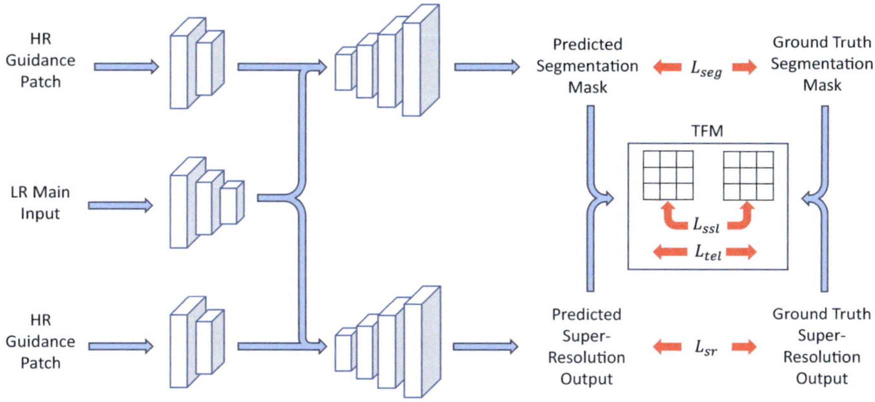

Fig. 5.7 A schematic view of the training pipeline of PFSeg. During inference, only the segmentation branch remains, while other components are removed to reduce GPU memory utilization

pipeline, the segmentation output and the super-resolution output are combined as the input of TFM, where additional similarity evaluations are conducted to ensure a more accurate HR prediction of the target areas in the mask.

The loss functions for PFSeg are carefully designed to ensure good cooperation between different modules in the pipeline. In addition to the conventional Mean Square Error (MSE) L_{sr} for the Super-Resolution branch and Cross-Entropy Loss L_{seg} for the main segmentation branch, specific losses are designed for TFM. TFM evaluates the similarity between the prediction and the ground truth in two streams, aiming at producing a more accurate HR segmentation mask through the LR input. For the first stream, a Target-Enhanced Loss (L_{tel}) is utilized, which calculates the average square Euclidean distance of target-area-only voxels, encouraging more precise segmentation from the segmentation branch; For the second stream, inspired by the Spatial Attention Mechanism in [9], a Spatial Similarity Loss (L_{ssl}) is used to ensure the relative similarity between the prediction and the ground truth. L_{ssl} is computed using a Spatial Similarity Matrix, which primarily describes the pairwise relationships between voxels. Specifically, for an image with dimensions $C \times D \times W \times H$, its Spatial Similarity Matrix $\mathbf{S}$ is computed as follows: first, the image is reshaped into a $C \times V$ matrix, where $V = D \times W \times H$. Subsequently, this matrix is multiplied to its transpose to obtain a $V \times V$ similarity matrix, which is then compared with the ground truth to calculate the loss. Mathematically, the loss function for this module can be defined as follows:

$$L_{tfm} = L_{tel} + L_{ssl} \tag{5.3}$$

$$L_{tel} = \frac{1}{N} \sum_{i=1}^{N} ||p_i \cdot SR(\mathbf{X} \downarrow)_i - y_i \cdot \mathbf{X}_i||^2 \tag{5.4}$$

$$L_{ssl} = \frac{1}{D^2W^2H^2} \sum_{i=1}^{D\cdot W\cdot H} \sum_{j=1}^{D\cdot W\cdot H} ||\mathbf{S}_{ij}^{predict} - \mathbf{S}_{ij}^{gt}||^2 \tag{5.5}$$

$$S_{ij}^{predict} = (p_i \cdot SR(\mathbf{X}\downarrow)_i)(p_j \cdot SR(\mathbf{X}\downarrow)_j)^{\mathrm{T}} \tag{5.6}$$

$$S_{ij}^{gt} = (y_i \cdot \mathbf{X}_i)(y_j \cdot \mathbf{X}_j)^{\mathrm{T}} \tag{5.7}$$

where N is the total number of voxels, $SR(\mathbf{X}\downarrow)$ is the super-resolution restored image based on the downsampled LR main input $\mathbf{X}\downarrow$, p_i denotes the prediction of the i-th voxel after binarization, y_i represents its corresponding ground truth, and $\cdot$ is element-wise multiplication. The overall objective function L of the proposed framework can then be expressed as:

$$L = L_{seg} + \omega_{sr} L_{sr} + \omega_{tfm} L_{tfm} \tag{5.8}$$

where ω_{sr} and ω_{tfm} are hyper-parameters, both set to 0.5 by default.

In summary, PFSeg adopts multi-task learning paradigm to generate HR segmentation tasks from LR inputs, realizing patch-free 3D medical image analysis indirectly. By using the entire 3D image as input, PFSeg conserves the global context for more accurate segmentation, and the introduction of the super-resolution auxiliary task aids in restoring the lost high-frequency information during the input downsampling process, which further enhances the boundaries of the predicted segmentation masks. However, despite its superior performance, PFSeg has complicated training procedures, and may require additional preprocessing steps for the data, which could hinder the clinical application.

5.2.6 *Performance Validation*

Some experimental results comparing the accuracy achieved by the mentioned methods and other existing methods are presented in this subsection.

5.2.6.1 Dataset and Implementation Details

In the experiments, the accuracy and inference speed of different 3D CNN methods are compared. The datasets used in these experiments include BRATS2020 [10–12] and a liver segmentation dataset. BRATS2020 is a brain tumor segmentation dataset containing 369 cases of brain tumors. Each case in BRATS2020 includes four MRI modalities of images (T1, T2, T1ce, and FLAIR) along with three categories of masks (tumor core, enhanced tumor, and whole tumor). All images are sized $155 \times 240 \times$

240 with a spacing of 1 mm × 1 mm × 1 mm. For each image, we removed the edges without brain parts by 24 voxels and resized the remaining part to a resolution of 128 × 192 × 192.

The liver segmentation dataset comprises 347 cases collected from Sir Run Run Shaw Hospital, Zhejiang University School of Medicine. For each case, an MRI image and its corresponding segmentation ground truth mask, labeled by experienced doctors, are provided. These images exhibit varying sizes and spacing settings. As part of the pre-processing pipeline, we initially resample the images, standardizing their spacing to (1.5 mm × 1.5 mm × 1.5 mm. Following resampling, the sizes of the MRI images range from 100 × 230 × 230 to 130 × 250 × 250. Subsequently, we crop the central 128 × 192 × 192 area to serve as the standard model input.

For both datasets, 80% of the data is used for training and 20% of the data is for testing. Patch-based methods utilize a patch size of 64 × 96 × 96, while patch-free methods down-sample the entire image to match this spatial size of 64 × 96 × 96 as their input. During inference, patch-based methods adopt a sliding window strategy with a 50% overlap on all three axes. In the case of the patch-free method PFSeg, 3D ResUNet serves as its backbone. Additionally, a novel patch-free method is included in our experiments, denoted as 3D ResUNet↑ in the experimental results. This method involves training the entire network in the LR space. Specifically, for each original 3D image, both the image and its ground truth are initially down-sampled by a factor of 2 for training. Subsequently, LR segmentation predictions are obtained, and the tricubic interpolation algorithm [7] is employed to upscale the mask to its original size.

All methods are implemented using PyTorch on Ubuntu 16.04.3 LTS, utilizing an Intel Xeon CPU E5-2640 v4, 64GB RAM, and one Nvidia GTX 1080Ti GPU (11GB). Data augmentation techniques such as random cropping (exclusively for patch-based methods), random flipping, random rotation, and random shifting are applied. Optimization is performed using the Adam optimizer [13], with the initial learning rate set to 1e-4. The learning rate is reduced to its 10% if the loss does not decrease for 10 epochs. The training process ends when it reaches $1e^{-7}$. The batch size for all the methods is set to 1. All the methods are trained from scratch using Binary Cross-Entropy (BCE) Loss and Dice Loss.

For both datasets, we employ three quantitative evaluation indices in the experiments, which are Dice Similarity Coefficient (DSC), Jaccard Coefficient (JC) and 95% Hausdorff Distance (HD95). DSC and JC mainly focus on filled segmentation regions whereas HD95 pays more attention to segmentation boundaries. The mathematically illustartion of these metrics are presented as follows.

1. **Dice Similarity Coefficient**. Let $\hat{\mathbf{Y}}$ denotes the predicted output and $\mathbf{Y}$ represents the ground truth, DSC evaluates the similarity between $\hat{\mathbf{Y}}$ and $\mathbf{Y}$ by computing their intersection areas over the average of their sizes. It can be defined as:

$$DSC(\hat{\mathbf{Y}}, \mathbf{Y}) = \frac{2|\hat{\mathbf{Y}} \cap \mathbf{Y}|}{|\hat{\mathbf{Y}}| + |\mathbf{Y}|} \tag{5.9}$$

where $|\cdot|$ calculates the sum of voxels in a region.

2. **Jaccard Coefficient**. JC evaluates the similarity between $\hat{\mathbf{Y}}$ and $\mathbf{Y}$ by computing the size of their intersection areas over the size of their union. It can be defined as:

$$JC(\hat{\mathbf{Y}}, \mathbf{Y}) = \frac{|\hat{\mathbf{Y}} \cap \mathbf{Y}|}{|\hat{\mathbf{Y}}| \cup |\mathbf{Y}|} \tag{5.10}$$

3. **95% Hausdorff Distance**. Before we present the definition of HD95, we first give an introduction to the original Hausdorff Distance (HD).

As a distance-based metric, HD evaluates the similarity between $\hat{\mathbf{Y}}$ and $\mathbf{Y}$ by computing their maximum distance, for all points in $\hat{\mathbf{Y}}$, to the closest point in $\mathbf{Y}$. It can be mathematically defined as:

$$d_H(\hat{\mathbf{Y}}, \mathbf{Y}) = \max_{x \in \hat{\mathbf{Y}}}(\min_{y \in \mathbf{Y}}(d(x, y))) \tag{5.11}$$

$$HD(\hat{\mathbf{Y}}, \mathbf{Y}) = \max(d_H(\hat{\mathbf{Y}}, \mathbf{Y}), d_H(\mathbf{Y}, \hat{\mathbf{Y}})) \tag{5.12}$$

where $d(x, y)$ is the Euclidean distance between voxel x and voxel y. On top of that, HD95 replaces the original maximum distance used for HD with the distance that ranked 95% in the set as its final result, mitigating the negative influence from potential outliers.

5.2.6.2 Results and Comprehensive Analysis

Table 5.1 summarizes the segmentation accuracy of various methods. Among patch-based models, 3D ResUNet achieves the best Dice Similarity Coefficient (DSC): 82.12% on BRATS2020 and 92.36% on the liver segmentation dataset. Despite its complexity, V-Net performs worse, with the lowest DSC and slower inference speed.

For patch-free methods, it is evident that they typically exhibit significantly faster inference speeds compared to patch-based methods, as they require only one run to generate entire segmentation masks. Among these methods, the PFSeg framework stands out for its superior segmentation accuracy. This can be attributed to its retention of global contextual information during training, which enables it to better restore high-frequency details. Through the utilization of the PFSeg framework, 3D ResUNet achieves enhanced performance compared to the patch-based scheme, underscoring the importance of global context in segmentation tasks. However, the failure of 3D ResUNet$\uparrow$ to surpass the performance of the patch-based 3D ResUNet highlights the significance of high-frequency details for accurate segmentation. This further validates the effectiveness of introducing the auxiliary super-resolution task within the PFSeg framework.

Additionally, some visualization results are presented in Fig. 5.8 to provide a more intuitive comparison. As shown, the patch-based results exhibit noticeable

Table 5.1 Experimental results on BRATS2020 and the liver segmentation datasets are presented

Method	Category	BRATS2020			Liver dataset			Inference time (s)
		DSC (%)	JC (%)	HD95 (mm)	DSC (%)	JC (%)	HD95 (mm)	
V-Net [3]	Patch-based	79.91	68.29	13.86	89.49	81.83	14.93	5.26
3D U-Net [2]	Patch-based	81.21	69.89	14.63	90.80	84.11	11.64	4.00
3D ResUNet [4]	Patch-based	82.18	71.19	13.21	92.36	86.46	9.27	4.21
3D ResUNet↑	Patch-free	80.89	70.02	8.56	91.98	85.84	6.40	2.21
3D ResUNet+ Holistic [6]	Patch-free	82.45	72.07	9.21	92.88	87.33	5.58	0.19
PFSeg [8]	Patch-free	83.82	74.01	7.83	93.83	88.70	4.29	0.95

Inference time is evaluated on images of size $128 \times 192 \times 192$, with the patch size being $64 \times 96 \times 96$ for patch-based methods

flaws (highlighted in red): in Case 1, segmentation noise is evident. This issue arises primarily due to the limited context available in patch-based approaches. For instance, when segmenting the upper-right corner patch, the model lacks information about the actual tumor region, increasing the likelihood of misdiagnosing normal areas as lesions.

In Cases 2 and 3, segmentation errors are observed in the corner areas, which are attributed to the use of padding techniques. As noted in [14], padding can introduce artifacts at the edges of feature maps, potentially confusing the network. The issues in Cases 2 and 3 are common when the target region spans multiple patches. Under such circumstances, patch-based models often struggle to accurately predict voxels at the patch edges, leading to inconsistencies during the fusion process.

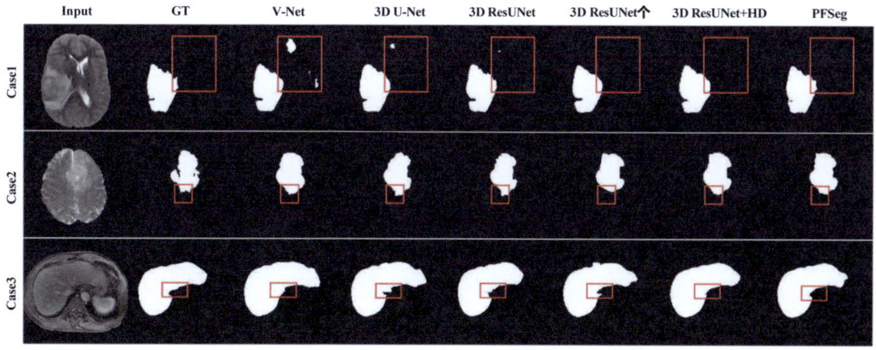

Fig. 5.8 Visualization of the segmentation results. Case1 and Case2 are from BRATS2020, while Case3 is from the liver dataset. The central slice is selected from each case for the visualization

While patch-free segmentation eliminates the aforementioned issues, it may still suffer from performance degradation due to the loss of high-frequency information during down-scaling. However, PFSeg addresses this limitation through the use of an SR auxiliary task and a TFM module, which help to restore HR representations. As a result, PFSeg demonstrates superior performance compared to other patch-free methods in the experiment.

5.2.7 Summary

This section introduces several patch-based and patch-free 3D CNN architectures for medical image segmentation task. The baseline 3D CNN is 3D U-Net, which is essentially the 3D version of the U-Net architecture. Building upon this foundation, 3D ResUNet incorporates residual connections, achieving higher segmentation performance. V-Net, on the other hand, enhances traditional CNNs by replacing pooling layers with strided convolutions, resulting in a deeper network. However, the increased computational cost associated with these 3D CNNs necessitates patch-sampling during training. Also, these approaches have some limitations such as limited global contextual information in the patches, impacting performance.

To address these limitations, patch-free methods have been developed. Holistic Decomposition is one such method, which reduces the spatial size of 3D inputs by increasing their dimensionality through feature decomposition. Another noteworthy patch-free method is PFSeg, which leverages super-resolution techniques to achieve HR segmentation with LR input. In general, patch-free methods offer faster inference speeds as they require only one pass to generate the entire prediction mask. Additionally, patch-free methods also tend to exhibit better performance by retaining the global context of the 3D input in many cases.

5.3 3D Vision Transformer

Although CNNs have demonstrated significant potential in 3D medical image analysis, they are inherently limited in processing long-range dependencies. Convolutional layers operate within a small kernel space, typically 3×3, leading to spatial interactions that are confined to local views of the complete image during training. Consequently, crucial semantic information that may reside within a larger area can be overlooked. To more effectively capture the global-wise information, Transformers [15] have emerged as a prominent solution in the field of medical image analysis. The structure of a typical Transformer block is depicted in Fig. 5.9.

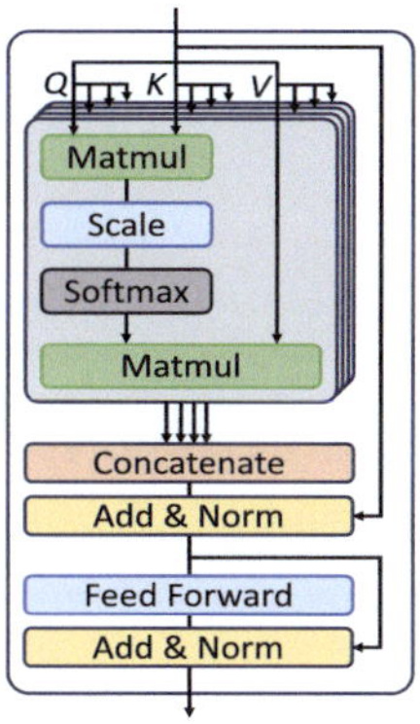

Fig. 5.9 The network structure of a transformer block. For a given sequence input shaped as $L \times C$, transformer will first project it into three different sub-spaces, making it into **K**, **Q**, and **V**. As **K** and $\mathbf{Q}^T$ is multiplied into a self-attention matrix, each token's correlation with all the other tokens is reflected in it. Next, by using matrix multiplication on **V** and the self-attention map, a weighted new sequence is acquired for the subsequent normalization and feed forward layers

Originally developed for Natural Language Processing (NLP), Transformers possess the inherent capability to model global correlations due to the multi-head self-attention (MSA) mechanism. By converting feature maps into sequence data, Transformers can also be applied onto visual tasks to realize global-wise representation modeling. Currently, most existing Transformers are still patch-based due to the high complexity of their structures. However, patch-based methods often compromise the global context in the image, which contradicts the primary goal of leveraging Transformers for medical image analysis. Consequently, there has been a recent surge in the development of carefully designed patch-free Transformers to address this issue.

In this section, we will give brief introductions to classic patch-based and patch-free 3D Vision Transformers. In Sect. 5.3.1, UNETR [16] is discussed as the very first 3D vision Transformer model for medical image analysis. In Sect. 5.3.2, Swin-UNETR [17] is introduced as an improved version of UNETR. In Sect. 5.3.3, VolumeFormer [18] is presented as a more recent patch-free 3D vision Transformer. In Sect. 5.3.4, performance validation is conducted on these methods to demonstrate their strengths and limitations. Finally, in Sect. 5.3.5, these 3D vision Transformers are summarized.

5.3.1 UNETR

UNETR [16] is introduced by A. Hatamizadeh, et al., in their work titled "Unetr: Transformers for 3d medical image segmentation". It is designed specifically for medical image analysis. The architecture of UNETR integrates Transformer blocks with traditional convolutional layers, aiming to utilize the strengths of both approaches. The overall structure is depicted in Fig. 5.10.

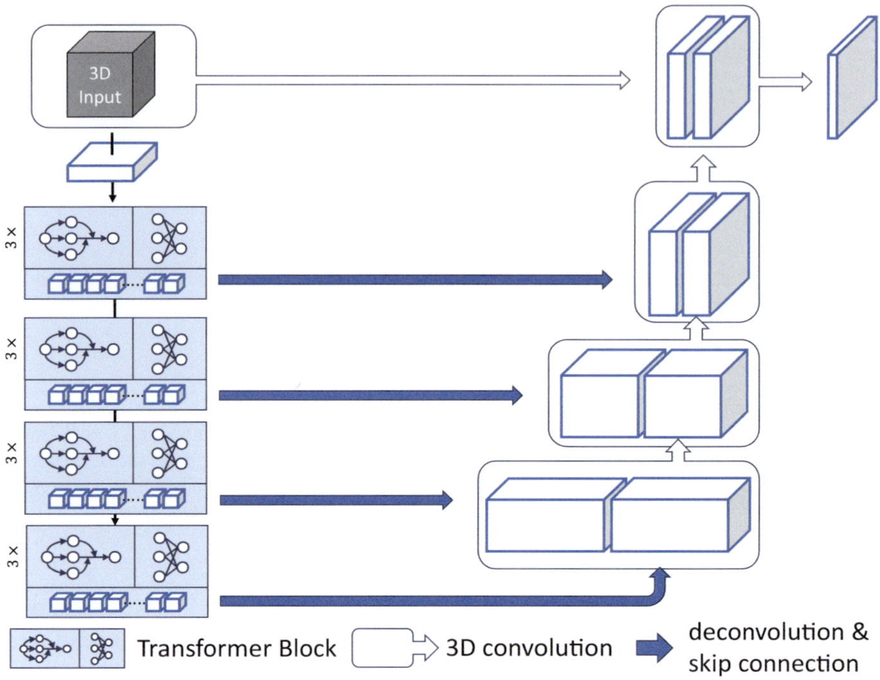

Fig. 5.10 A schematic view of the network structure of UNETR

In general, UNETR employs an asymmetric encoder-decoder architecture, wherein Transformer blocks constitute the encoder and convolutional layers form the decoder. For an input image of dimensions $C \times D \times W \times H$, the network initially transforms it into sequence data of shape $768 \times (\frac{D}{16} \times \frac{W}{16} \times \frac{H}{16})$ using a patch embedding layer. Subsequently, 12 Transformer blocks are utilized for feature extraction from the sequence input, with the outputs of the 3rd, 6th, 9th, and 12th Transformer blocks being upsampled and concatenated with the corresponding decoder feature maps. The decoder consists entirely of several convolutional layers, similar to a 3D U-Net. Due to the discrepancy in shape between the sequence data and the decoder features, skip connections are adapted using transposed convolution layers to upscale the sequence data to match the decoder's feature dimensions. The inclusion of Transformer blocks significantly increases the parameter count of UNETR, rendering it computationally expensive and challenging to deploy on low-end devices.

From another perspective, UNETR can be regarded as a modified 3D U-Net, which replaces the convolutional layers in the encoder with Transformer blocks for feature extraction. The robust representation modeling capability of Transformers enables the extraction of more informative features, enhancing the semantic understanding of the entire model. However, the tokenization process inherent to the Transformer pipeline can result in the loss of 3D spatial position information in the image, and the patch embedding process may omit some crucial high-frequency details from the input. Furthermore, the asymmetric structure can pose challenges for the skip connection mechanism, as low-level features from the encoder differ in form from those in the decoder, leading to potential misalignment. Consequently, UNETR may not consistently achieve optimal performance in 3D medical image segmentation tasks.

5.3.2 *Swin-UNETR*

Inspired by the Swin-Transformer [19], Swin-UNETR [17] was proposed by A. Hatamizadeh, et al., in their work titled "Swin unetr: Swin transformers for semantic segmentation of brain tumors in MRI images". Swin-UNETR retains the asymmetric architecture with a Transformer-based encoder and a convolutional decoder but replaces the original Transformer blocks with 3D Swin Transformer blocks. A schematic view of the 3D Swin Transformer Block is presented in Fig. 5.11. As shown, Transformer Blocks used in Swin-UNETR is composed of two modules: window-MSA and shifted-window MSA. The window-MSA performs multi-head self-attention (MSA) within a small window region to improve the network's locality, while the shifted-window MSA enables information exchange between windows, allowing the Transformer structure to handle long-range dependencies. Although this design enhances the feature extraction capabilities of vision Transformers, its high complexity also necessitates Swin-UNETR being a patch-based method for most GPUs.

Unlike previous Transformer-based methods that adopt model structures designed for NLP tasks, the Swin Transformer enhances the local perception of the model and is specifically designed for computer vision problems. Consequently, Swin-UNETR demonstrates superior visual feature extraction capabilities compared to the previous UNETR. However, Swin-UNETR demands more GPU memory for training and maintains a high parameter count, making it computationally expensive in most cases. Additionally, since Transformer-based methods lack inductive bias

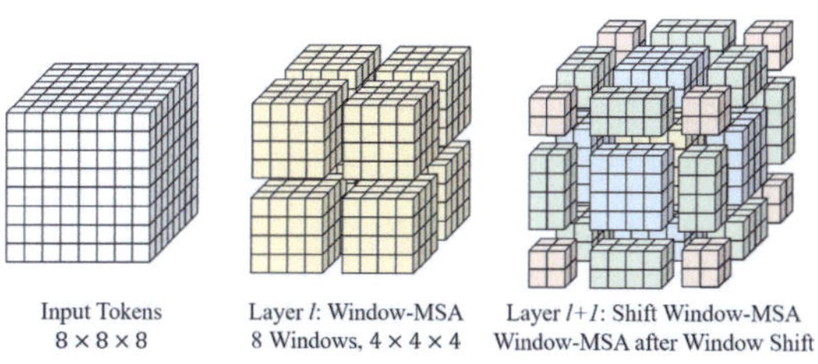

Fig. 5.11 A schematic view of the swin transformer blocks used in Swin-UNETR is depicted. Initially, the transformer block performs Multi-head Self-Attention (MSA) in the local window area to ensure locality for the model. Subsequently, the window is shifted, and MSA is conducted again on these new windows, facilitating information exchange within the entire image

during training, they typically require large-scale pre-training to be fully leveraged. While this requirement is manageable for natural images, which can be captured with various devices, it poses a significant challenge for medical images. These images are costly to obtain and often cannot be publicly shared due to privacy restrictions, complicating the pre-training process.

5.3.3 VolumeFormer

While Transformers have demonstrated effectiveness in 3D medical image segmentation tasks, their performance can be significantly hindered by patch-sampling. The use of patches as input results in loss of long-range dependencies, making global-wise structure of Transformers less meaningful. Consequently, there is a growing desire for patch-free Transformer architectures to address this challenge. H. Wang et al. [18] proposed a VolumeFormer in their work titled "Adaptive decomposition and shared weight volumetric transformer blocks for efficient patch-free 3d medical image segmentation". A schematic representation of VolumeFormer is depicted in Fig. 5.12.

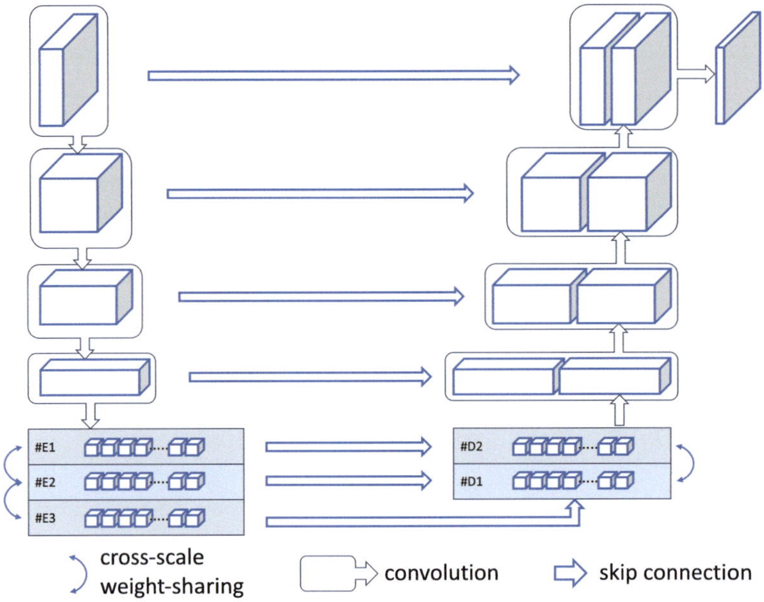

Fig. 5.12 A schematic view of VolumeFormer is presented. The shallow layers of VolumeFormer adopt convolutional layers to efficiently capture local context, while the deeper layers use Shared-Weight Volumetric Transformer Blocks to handle long-range dependencies at a low cost

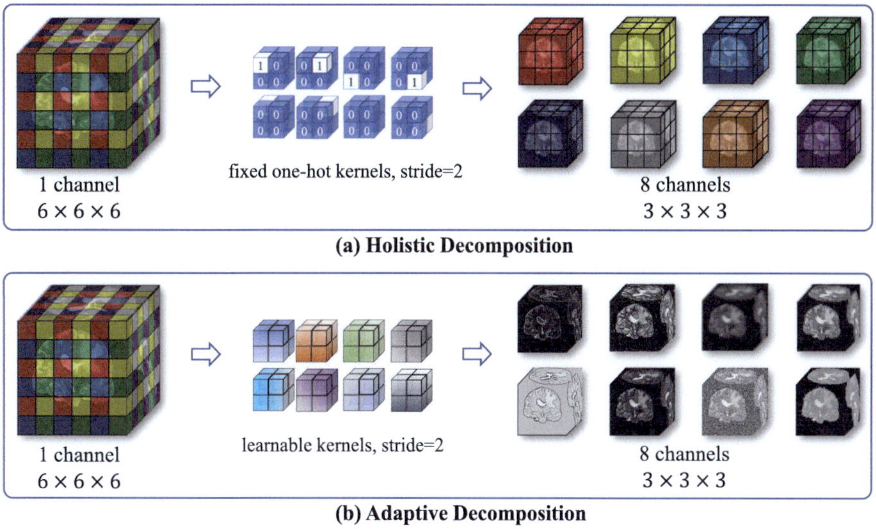

Fig. 5.13 A comparison between **a** Holistic decomposition and **b** Adaptive decomposition. Adaptive decomposition can be viewed as the learnable version of holistic decomposition, and holistic decomposition can be viewed as a specific instance of adaptive decomposition

VolumeFormer makes significant contributions in two key areas. First, it introduces Adaptive Decomposition for feature decomposition, which is a generalized version of Holistic Decomposition. Second, it proposes the Shared-Weight Volumetric Transformer Block, a lightweight Transformer module capable of cross-scale weight-sharing. An intuitive comparison between Holistic Decomposition and Adaptive Decomposition is shown in Fig. 5.13. Employing Holistic Decomposition to downscale the spatial size by a factor of r is equivalent to utilizing r^3 distinct unlearnable one-hot convolution kernels with dimensions $r \times r \times r$ and a stride of r on the original image. However, employing these fixed one-hot kernels restricts the flexibility of the decomposition process. In contrast, Adaptive Decomposition makes these kernels learnable alongside the segmentation backbone, enhancing the flexibility of the decomposition process.

The Shared-Weight Volumetric Transformer Block (SW-VTB) incorporates several design elements to enhance performance and reduce computational cost. These include identity V-mapping, a multi-scale residual branch, and split-head linear layers. Additionally, it is carefully designed to be insensitive to input spatial sizes, enabling cross-scale weight-sharing. The structure of SW-VTB is illustrated in Fig. 5.14.

In this architecture, for a given input sequence, only **Q** and **K** are obtained using linear layers, while **V** is obtained through identity mapping to minimize parameter count. A multi-scale residual branch is then applied to **V**, implemented with two depth-wise convolution layers of different kernel sizes.

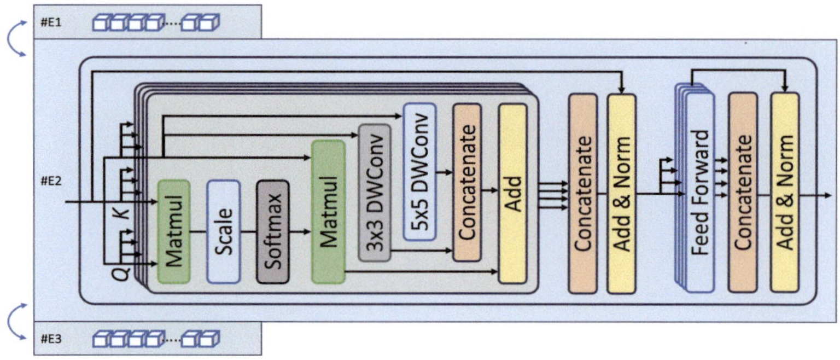

Fig. 5.14 A detailed view of SW-VTBs

The introduction of these design enhancements introduces additional inductive biases within the network, accelerating convergence. Further, the feed-forward layers are modified to adopt a multi-head design, facilitating more efficient modeling. Importantly, the parameter count of the entire SW-VTB remains fixed and invariant to changes in the input sequence length. This characteristic enables cross-scale weight-sharing among SW-VTBs in VolumeFormer. Cross-scale weight-sharing also offers several benefits. It reduces overall computational cost, enables the network to focus on scale-invariant core semantic information, and ultimately leads to more accurate predictions and higher generalization ability for the model.

5.3.4 Performance Validation

5.3.4.1 Dataset and Implementation Details

The dataset used in the experiments are the same of the datasets introduced in the previous part of this chapter, namely BRATS2020 and the liver segmentation dataset. To ensure fair comparison, the same preprocessing procedures as before are followed. For both datasets, 80% of the data is used for training and 20% of the data is for testing. Patch-based methods employ a patch size of $64 \times 96 \times 96$, whereas patch-free methods down-sample the entire image to the same spatial size of $64 \times 96 \times 96$ as their input. During inference, patch-based methods utilize a sliding window strategy with a 50% overlap on all three axes.

In experiments, a more lightweight 3D transformer called VolumeFormer-S has also been introduced, which shares the same structural design as VolumeFormer but with half the width. All methods are implemented using PyTorch on Ubuntu 16.04.3 LTS with an Intel Xeon CPU E5-2640 v4, 64GB RAM, and one Nvidia GTX 1080Ti GPU (11GB). Data augmentation techniques include random cropping (only for patch-based methods), random flipping, random rotation, and random shifting. For optimization, all models are trained using Adam with the initial learning rate set to

1e-4. The learning rate is reduced to 10% if the loss does not decrease for 10 epochs, and training concludes when the loss reaches 1e-7. The batch size for all methods is set to 1, and they are trained with Binary Cross-Entropy (BCE) Loss and Dice Loss.

For both datasets, we employ three quantitative evaluation indices in the experiments, which are the aforementioned Dice Similarity Coefficient (DSC), Jaccard Coefficient (JC) and 95% Hausdorff Distance (HD95). DSC and JC mainly focus on filled segmentation regions whereas HD95 pays more attention to segmentation boundaries. The detailed illustration of these metrics were presented in Sect. 5.2.6.1.

5.3.4.2 Experimental Results and Analysis

Table 5.2 presents the experimental results of the methods discussed. The comparison of these methods reveals several key insights. The patch-based Transformer model, Swin-UNETR, fails to outperform the patch-based 3D ResUNet. This can be attributed to the small patch input size, which compromises the preservation of global-wise information in the original 3D images. As a result, the Transformer's advantage in handling long-range dependencies is diminished. In contrast, the patch-free method VolumeFormer significantly surpasses the 3D ResUNet-based PFSeg. This demonstrates the effectiveness of incorporating a long-range dependency processing unit within the network. Notably, even the lightweight version, VolumeFormer-S, which has half the width of the original VolumeFormer, still outperforms the patch-free 3D CNN by 1.38% in Dice Similarity Coefficient (DSC). This further validates the effectiveness of VolumeFormer's structural design.

Table 5.2 Experimental results on BRATS2020 and the liver segmentation datasets

Method	Category	BRATS2020			Liver dataset			Inference time (s)
		DSC (%)	JC (%)	HD95 (mm)	DSC (%)	JC (%)	HD95 (mm)	
3D ResUNet [4]	Patch-based	82.18	71.19	13.21	92.36	86.46	9.27	4.21
Swin-UNETR [17]	Patch-based	79.69	68.62	16.84	89.74	82.44	16.14	4.82
PFSeg [8]	Patch-free	83.82	74.01	7.83	93.83	88.70	4.29	0.95
UNETR [16]	Patch-free	79.31	67.83	19.00	90.04	82.66	16.77	0.43
VolumeFormer-S [18]	Patch-free	84.20	74.56	7.84	94.87	90.42	3.95	0.66
VolumeFormer [18]	Patch-free	85.58	76.31	7.62	95.11	90.81	3.40	0.87

VolumeFormer-S refers to the simplified version of VolumeFormer with only half the network width

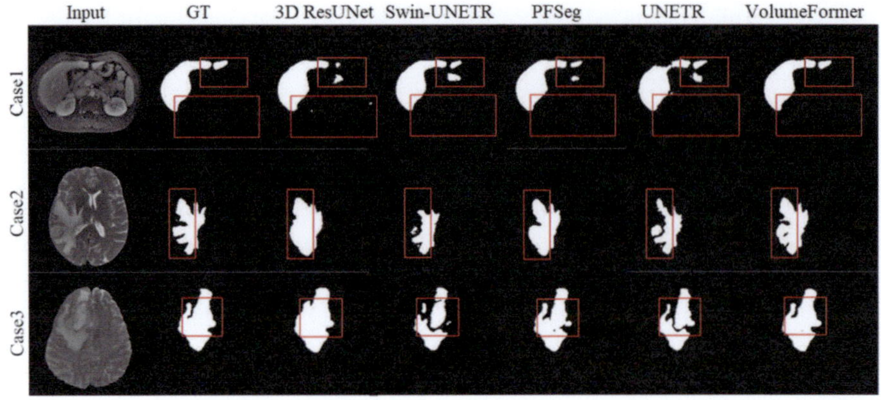

Fig. 5.15 Some segmentation results of our experiment. Case 1 is from the liver segmentation dataset. Case 2 and Case 3 are from BRATS2020. In order to make visualization simple, the central slice is selected from each 3D case for visualization

The performance of UNETR is also noteworthy. Despite being classified as a patch-free method due to its capability to handle full-size 3D inputs within the 11GB GPU memory limit, it does not outperform any other methods, including the patch-based Swin-UNETR. The potential reason for its underperformance could be attributed to the patch embedding layer at the beginning of its pipeline. By embedding a small patch area into a single token, a substantial amount of high-frequency information might be lost, leading to a less accurate final prediction mask. These results highlight the importance of preserving global information in 3D medical image analysis and validate the effectiveness of the proposed methods, particularly VolumeFormer and its lightweight variant.

In addition, Fig. 5.15 presents segmentation results for a more intuitive comparison. In Case 1, patch-based methods exhibit segmentation noise due to their reliance on patches as input, which lack global contextual information. Without a holistic view of the tumor area, these models are prone to misclassifying normal regions as lesions when segmenting patches, such as the bottom-right corner. In contrast, the patch-free method VolumeFormer, leveraging global context, accurately identifies the lesions, as demonstrated in the upper-right corner of Case 1.

In Case 2, CNN-based methods produce over-smoothed segmentation boundaries because they lack mechanisms for global information processing. On the other hand, VolumeFormer effectively handles complex segmentation boundaries, showcasing the benefits of its Adaptive Decomposition and SW-VTB modules in improving performance on challenging samples.

However, in Case 3, some patch-free methods display incorrect segmentation at tumor boundaries. This issue arises because image boundaries can become blurred during embedding decomposition or down-sampling processes in other patch-free methods, leading to reduced segmentation accuracy.

5.3.5 Summary

In this section, various 3D Transformer-based methods are discussed, highlighting their evolution and performance in 3D medical image segmentation. UNETR is the first public study employing 3D Transformers for medical image segmentation and serves as the baseline for subsequent 3D Transformer-based methods. The patch embedding layer, which converts image data into sequence data, has limitations. This conversion can lead to the loss of high-frequency information critical for accurate segmentation, resulting in a less precise segmentation mask. As an improved version of UNETR, Swin-UNETR addresses some issues by incorporating a locality design. This design enhances its capability as a feature extractor for 3D segmentation tasks. Despite these improvements, Swin-UNETR still relies on sequence data input, which retains some limitations of the patch embedding approach. However, its locality-aware design leads to better performance in experimental evaluations compared to UNETR.

VolumeFormer introduces a novel approach by being patch-free. This design preserves the global-wise contextual information from the input data, avoiding the drawbacks of patch embedding. By incorporating Transformer blocks, VolumeFormer efficiently handles global-wise long-range dependencies, crucial for accurate 3D segmentation. VolumeFormer demonstrates superior segmentation accuracy in experiments, showing its effectiveness in maintaining global context and handling complex dependencies in 3D medical images.

References

1. Ronneberger, O., Fischer, P., & Brox, T. (2015). U-Net: Convolutional networks for biomedical image segmentation. In: *International conference on medical image computing and computer-assisted intervention* (pp. 234–241). Springer.
2. Cicek, Ö., Abdulkadir, A., Lienkamp, S. S., Brox, T., & Ronneberger, O. (2016). 3d U-Net: Learning dense volumetric segmentation from sparse annotation. In: *International conference on medical image computing and computer-assisted intervention* (pp. 424–432). Springer.
3. Milletari, F., Navab, N., & Ahmadi, S. A. (2016). V-net: Fully convolutional neural networks for volumetric medical image segmentation. In: *2016 fourth international conference on 3D vision (3DV)* (pp. 565–571). IEEE.
4. Yu, L., Yang, X., Chen, H., Qin, J., & Heng, P. A. (2017). Volumetric convnets with mixed residual connections for automated prostate segmentation from 3d MR images. In: *Proceedings of the AAAI conference on artificial intelligence* (vol. 31).
5. He, K., Zhang, X., Ren, S., & Sun, J. (2016). Deep residual learning for image recognition. In: *Proceedings of the IEEE conference on computer vision and pattern recognition* (pp. 770–778).
6. Zeng, G., & Zheng, G. (2019). Holistic decomposition convolution for effective semantic segmentation of medical volume images. *Medical Image Analysis, 57*, 149–164.
7. Lekien, F., & Marsden, J. (2005). Tricubic interpolation in three dimensions. *International Journal for Numerical Methods in Engineering, 63*(3), 455–471.
8. Wang, H., Lin, L., Hu, H., Chen, Q., Li, Y., Iwamoto, Y., Han, X. H., Chen, Y. W., & Tong, R. (2021). Patch-free 3d medical image segmentation driven by super-resolution technique

and self-supervised guidance. In: *International conference on medical image computing and computer-assisted intervention* (pp. 131–141). Springer.

9. Fu, J., Liu, J., Tian, H., Li, Y., Bao, Y., Fang, Z., & Lu, H. (2019). Dual attention network for scene segmentation. In: *Proceedings of the IEEE/CVF conference on computer vision and pattern recognition* (pp. 3146–3154).
10. Bakas, S., Akbari, H., Sotiras, A., Bilello, M., Rozycki, M., Kirby, J. S., Freymann, J. B., Farahani, K., & Davatzikos, C. (2017). Advancing the cancer genome atlas glioma MRI collections with expert segmentation labels and radiomic features. *Scientific Data, 4*(1), 1–13.
11. Bakas, S., Reyes, M., Jakab, A., Bauer, S., Rempfler, M., Crimi, A., Shinohara, R.T., Berger, C., Ha, S.M., Rozycki, M., et al. (2018). *Identifying the best machine learning algorithms for brain tumor segmentation, progression assessment, and overall survival prediction in the brats challenge*. arXiv preprint arXiv:1811.02629.
12. Menze, B. H., Jakab, A., Bauer, S., Kalpathy-Cramer, J., Farahani, K., Kirby, J., Burren, Y., Porz, N., Slotboom, J., Wiest, R., et al. (2014). The multimodal brain tumor image segmentation benchmark (brats). *IEEE Transactions on Medical Imaging, 34*(10), 1993–2024.
13. Dettmers, T., Lewis, M., Shleifer, S., & Zettlemoyer, L. (2021). *8-bit optimizers via block-wise quantization*.
14. Alsallakh, B., Kokhlikyan, N., Miglani, V., Yuan, J., & Reblitz-Richardson, O. (2021). Mind the pad—CNNs can develop blind spots. *International conference on learning representations*.
15. Vaswani, A., Shazeer, N., Parmar, N., Uszkoreit, J., Jones, L., Gomez, A. N., Kaiser, L., & Polosukhin, I. (2017). Attention is all you need. In: *Advances in neural information processing systems* (pp. 5998–6008).
16. Hatamizadeh, A., Tang, Y., Nath, V., Yang, D., Myronenko, A., Landman, B., Roth, H. R., & Xu, D. (2022). Unetr: Transformers for 3d medical image segmentation. In: *Proceedings of the IEEE/CVF winter conference on applications of computer vision* (pp. 574–584).
17. Hatamizadeh, A., Nath, V., Tang, Y., Yang, D., Roth, H., & Xu, D. (2022). *Swin unetr: Swin transformers for semantic segmentation of brain tumors in MRI images*. arXiv preprint arXiv:2201.01266.
18. Wang, H., Xu, Y., Chen, Q., Tong, R., Chen, Y. W., Hu, H., & Lin, L. (2023). Adaptive decomposition and shared weight volumetric transformer blocks for efficient patch-free 3d medical image segmentation. *IEEE Journal of Biomedical and Health Informatics, 27*(10), 4854–4865.
19. Liu, Z., Lin, Y., Cao, Y., Hu, H., Wei, Y., Zhang, Z., Lin, S., & Guo, B. (2021). *Swin transformer: Hierarchical vision transformer using shifted windows*. arXiv preprint arXiv:2103.14030.

Chapter 6
Multimodal Deep Learning for Medical Image Analysis

6.1 Introduction

Clinical diagnosis based on medical imaging often involves multiple imaging modalities, such as Magnetic Resonance Imaging (MRI), Computed Tomography (CT), Positron Emission Tomography (PET), Ultrasound, and other clinical data such as Electronic Health Records (EHR), laboratory test results, and genetic data. Each modality provides complementary perspectives on a patient's physiological or pathological conditions, and the combination of these diverse data sources holds the potential to offer a more comprehensive understanding of the patient's health status.

Leveraging multimodal information through deep learning models can significantly improve diagnostic accuracy and treatment planning. By integrating data from various modalities, these models can capture complex patterns and interactions that are difficult to detect in a single modality. The key to maximizing the potential of multimodal data lies in effective fusion, which combines the strengths of each modality while addressing their individual limitations. Multimodal fusion allows for a more nuanced and reliable representation of a patient's condition, ultimately improving decision-making in clinical practice.

Multimodal fusion strategies are commonly categorized based on the stage at which information integration occurs in the data processing pipeline. As illustrated in Fig. 6.1, there are three primary types of fusion strategies: (a) early fusion, (b) late fusion, and (c) intermediate fusion. Each strategy possesses distinct characteristics and is suitable for different application scenarios.

Early Fusion: Early fusion integrates raw or low-level features from multiple modalities at the initial stage of the learning pipeline. For example, in medical image and clinical text fusion, it may involve combining image pixel intensities with text embeddings before any further processing [1]. While this strategy offers a unified input space, it often struggles with capturing high-level semantic interactions.

Late Fusion: In late fusion, each modality is processed independently, and their respective predictions or high-level features are combined at a later stage [2, 3]. This

Y. Chen et al., *Recent Advances in Deep Learning for Medical Image Analysis*, Intelligent Systems Reference Library 278, https://doi.org/10.1007/978-3-031-94791-9_6

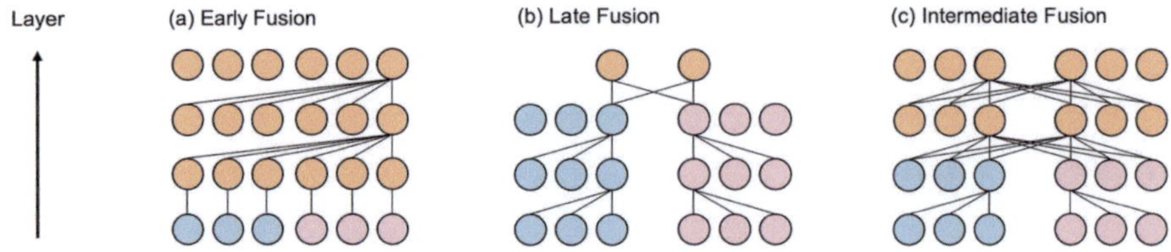

Fig. 6.1 Illustration of multimodal fusion strategies. The circles represent tokens, with the blue ones representing the tokens of modality A, the pink ones representing the tokens of modality B, and the orange ones representing the fused tokens

modular approach allows for flexibility and modality-specific optimization, but may miss subtle cross-modal interactions.

Intermediate Fusion: This strategy performs fusion at an intermediate stage, often within neural network architectures. It aims to balance the benefits of early and late fusion by enabling partial modality-specific processing while still allowing for joint feature learning [4]. Intermediate fusion is especially well-suited to deep learning models and has shown promising results in recent studies due to its ability to capture hierarchical and inter-modality dependencies.

Recent advances have highlighted the growing importance of intermediate fusion methods, particularly in deep neural networks, where flexible fusion mechanisms across multiple layers significantly improve model performance on multimodal tasks.

In the following subsections, we will first introduce various fusion strategies and then delve into specific fusion methods that enable effective integration of multimodal information. In particular, we will focus on vanilla fusion methods, attention mechanisms, and graph-based approaches, which have shown strong potential in recent medical imaging studies.

6.2 Vanilla Fusion Methods

The integration of vectorized features from various sources of information is a fundamental aspect of multimodal fusion. In Fig. 6.2, this integration can be achieved through concatenation that typically involve minimal or no additional parameters. The flexibility of concatenation is derived from the adaptability of deep models during joint training, enabling them to adjust and incorporate the required operations.

6.2.1 *Concatenation*

A commonly employed method for multimodal fusion is concatenation, which involves combining low-level input features or high-level features extracted by pre-

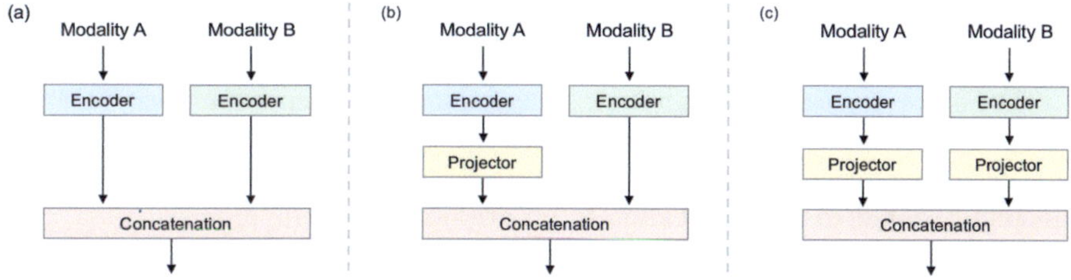

Fig. 6.2 Vanilla fusion architectures: **a** concatenation with no projection, **b** concatenation with projection applied to single modality, and **c** concatenation with projection applied to both modalities

trained models [5, 6]. For example, concatenation can merge low-level features from diverse sources like sensors or modalities, or concatenate high-level features derived from pretrained networks, as shown in Fig. 6.2a. This results in a fused representation that integrates information from multiple sources, facilitating more comprehensive analysis.

6.2.2 *Concatenation with Projection*

Concatenation is a widely used method for fusion in multimodal learning, where features from different sources are combined into a unified representation. However, when features come from diverse sources or modalities, the concatenated features may not always be optimally aligned in a shared semantic space, potentially hindering the fusion process. To address this issue, a common approach is to employ a projector. The projector is designed to map the features from different sources into a shared semantic space, ensuring that the concatenated features are more meaningful and compatible with each other. By aligning features in a common space, the projector helps improve the effectiveness of the concatenation process, allowing for a more accurate and coherent fused representation.

The key advantage of using a projector is that it learns the necessary transformations or mappings for different modalities to align them in a way that enhances their mutual compatibility. This step improves the quality of the concatenated features, making the combined representation more suitable for downstream tasks such as classification, detection, or prediction.

6.2.2.1 Single-Modality Projection

As shown in Fig. 6.2b, single-modality projection refers to mapping features extracted from a single modality into the common semantic space. This approach is often used when features from one modality, such as Magnetic Resonance Imaging (MRI), Computed Tomography (CT), or X-ray, need to be processed and aligned within a unified framework to improve downstream tasks such as segmentation, classification, or disease diagnosis.

The process typically begins by extracting features by the encoder. These features, which capture important patterns and structures, are then projected into a common space using a projector, which is usually a neural network layer or a transformation function designed to align the extracted features into a uniform representation. This projection helps to standardize the feature dimensions, making them more suitable for further processing and fusion with other features or modalities.

This technique simplifies the learning process by focusing on features from one modality at a time and aligning them in a meaningful space. It has been successfully applied in various medical imaging tasks, including the classification of lung diseases from chest X-rays or the detection of brain abnormalities from MRI scans. The effectiveness of single-modality projection lies in its ability to capture and enhance modality-specific features that are critical for accurate diagnosis and prediction. However, its limitation in lacking alignment with other modalities in multimodal learning presents opportunities for further improvement.

6.2.2.2 Dual-Modality Projection

In dual-modality projection, each modality is projected into a shared semantic space using separate projectors designed for each modality, illustrated in Fig. 6.2c. This method is commonly applied when combining features from different data sources, such as combining medical images from different modalities (e.g., MRI and CT scans) or different sensor data. The goal of dual-modality projection is to align the features from both modalities in a shared space, where the features can be effectively fused for further processing. Unlike single-modality projection, dual-modality projection uses two independent projectors. Each projector is responsible for transforming the features of its respective modality into a shared space that preserves modality-specific information but also allows for alignment.

The dual-modality projection method has been widely used in fields like medical image analysis, where different imaging techniques (e.g., MRI and CT) capture complementary information about a subject. For instance, in a typical dual-modality scenario, MRI might capture soft tissue details while CT provides bone structure information. By projecting the MRI and CT features into a shared space, their complementary information can be effectively combined, resulting in a more comprehensive representation of the subject for tasks like segmentation or classification. This approach improves the effectiveness of fusion-based models by ensuring that both modalities are appropriately aligned, making it easier for the model to leverage the unique information from each modality.

6.2.3 *Performance Validation*

Multimodal fusion methods have significant potential in medical image analysis, offering a comprehensive approach to disease diagnosis, treatment planning, and outcome prediction. By integrating different sources of data, such as medical images and clinical information, fusion models can capture complementary features, leading to more accurate and robust predictions. In this section, we introduce concatenation fusion methods and present experimental results to validate their effectiveness in the early recurrence (ER) prediction of hepatocellular carcinoma (HCC) using multi-phase CT images and clinical data. In the case presented here, Modality A refers to medical images (multi-phase CT images), and Modality B refers to clinical data (e.g., patient history, laboratory results, and tumor markers). These two modalities serve as an example of the types of data typically integrated for predictive modeling.

6.2.3.1 Dataset and Implementation Details

Integrating multi-phase CT images with clinical data offers a more holistic view of a patient's condition, enhancing the accuracy of disease diagnosis and treatment planning.

All CT images were obtained with Siemens SOMATON Definition AS scanner (Siemens Healthcare, Forchheim, Germany) and GE LightSpeed VCT scanner (GE Medical systems, Milwaukee, WI). Non-contrast enhanced (NC) phase was routinely performed at first. After injecting contrast agent, arterial (ART) phase, portal venous (PV) phase, and delay (DL) phase was scanned at 30 s, 60 s, 180 s respectively). In this study, NC, ART, and PV phases were selected for the following experiment. The pixels size of these three phases is 512×512, and the thickness of each slice is 5 or 7 mm.

As shown in Table 6.1, the clinical factors collected in our study include sex (male or female), age ($\leq$60 years or $\geq$60 years) tumor size ($\leq$5; $\geq$5 cm), tumor number (single or multiple); portal vein invasion (yes/no), liver cirrhosis (yes/no), HBV infection (yes/no), NLR ($\leq$2.81 or $\geq$2.81), ALT ($\leq$40 or $\geq$40 U/L), AST ($\leq$50 or $\geq$50 U/L), AKP ($\leq$125 or $\geq$125 U/L), GGT ($\leq$45 or $\geq$45 U/L), ALB ($\geq$40 or $\leq$40 U/L), TB ($\leq$20.5 or $\geq$20.5 U/L), AFP ($\leq$9 or $\geq$ 9 μg/L), CP Level (A or B), BCLC stage (0, A, B, C). Clinical data were evaluated through Chi-squared test which is a well-known method used to estimate the dependency between categorical variables. *P* value less than 0.05 was considered as significant differences. Seven clinical factors including age, tumor size, portal vein invasion, N/L ratio, TB, AFP, and BCLC stage were selected and further represented by a binary vector with nine elements of [c1, c2, c3, . . ., c8, c9], where [c1] represents age ([0]: $\leq$60, [1]: $\geq$60); [c2] represents tumor size ([0]: $\leq$5 cm, [1]: $\geq$5 cm); [c3] represents portal vein invasion ([0]: no, [1]: yes); [c4] represents N/L ratio ([0]: $\leq$2.8, [1]: $\geq$2.8); [c5] represents TB ([0]: $\leq$20.5, [1]: $\geq$20.5); [c6] represents AFP ([0]: $\leq$9, [1]: $\geq$9); [c7, c8, c9] represents BCLC stage ([0, 0, 0]: 0, [0, 0, 1]: A, [0, 1, 0]: B, [1, 0, 0]: C).

Table 6.1 Clinical variables of patients with ER and NER

Clinical variables	Total	NER	ER	*P*-value (chi-square test)
	n = 167	n = 102	n = 65	
Sex				0.826
Female	27	17	10	
Male	140	85	55	
Age				0.018
≤ 60	102	55	47	
≥ 60	65	47	18	
Tumor size (mm)				≤ 0.001
≤ 50	102	74	28	
≥ 50	65	28	37	
Tumor number				0.152
Single	155	97	58	
Multiple	12	5	7	
Portal vein invasion				0.001
No	143	95	48	
Yes	24	7	17	
Liver cirrhosis				0.794
No	52	31	21	
Yes	115	71	44	
HBV infection				0.054
No	33	25	8	
Yes	134	77	57	
N/L ratio				0.039
≤ 2.8	106	71	35	
≥ 2.8	61	31	30	
ALT (U/L)				0.417
≤ 40	104	66	38	
≥ 40	63	36	27	
AST (U/L)				0.500
≤ 50	123	77	46	
≥ 50	44	25	19	
AKP (U/L)				0.054
≤ 125	131	85	46	
≥ 125	36	17	19	
GGGT (U/L)				0.065
≤ 45	66	46	20	
≥ 45	101	56	45	

(continued)

Table 6.1 (continued)

Clinical variables	Total	NER	ER	*P*-value (chi-square test)
	n = 167	n = 102	n = 65	
ALB (g/L)				0.256
≥ 40	165	100	65	
≤ 40	2	2	0	
TB (μmol/L)				0.020
≤ 20.5	122	81	41	
≥ 20.5	45	21	24	
AFP (μg/L)				0.003
≤ 9	50	39	11	
≥ 9	117	63	54	
CP level				0.132
A	140	89	51	
B	27	13	14	
BCLC stage				≤ 0.001
0	22	17	5	
A	117	78	39	
B	4	1	3	
C	24	6	18	

Table 6.2 The accuracy and AUC of deep Learning model, clinical model and fusion model A, B, C, D

Model	Average accuracy (%)	Average AUC
Deep learning model	69.52 ± 5.1	0.7233 ± 0.06
Clinical model (RF)	76.03 ± 10.0	0.7532 ± 0.13
Fusion model A	74.80 ± 4.1	0.7880 ± 0.05
Fusion model B	78.05 ± 3.2	0.8045 ± 0.08
Fusion model C	78.66 ± 3.9	0.8248 ± 0.07

The fusion models aim to combine multi-phase CT images with clinical data to improve prediction accuracy. The Visual Branch of the network consists of a convolutional layer, a pooling layer, and 16 residual blocks. The CT images are passed through this pathway to extract high-level features. These features are then flattened and fused with the clinical data to form a mixed-feature vector. The fusion models (A–C) showed in Fig. 6.3 are designed to integrate clinical data and deep image features. Fusion model A combines clinical data directly with image features, while fusion model B reduces the dimension of image features before combining them with clinical data. Fusion models C fuse multimodal features through fully connected layers.

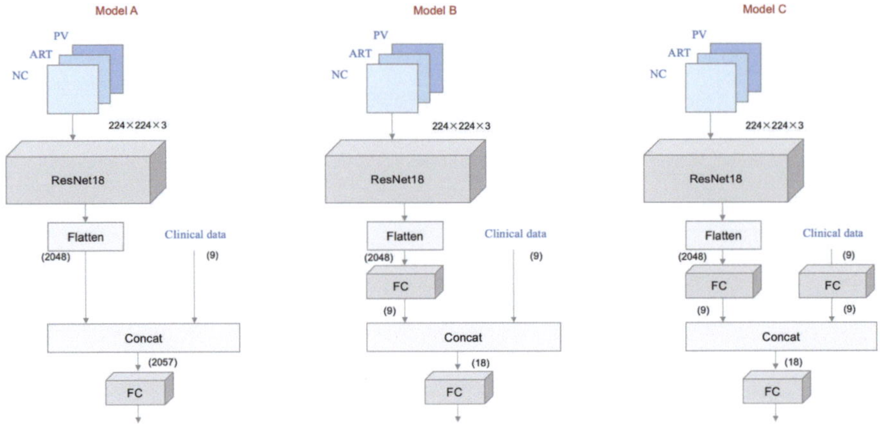

Fig. 6.3 Four deep learning-based fusion models with clinical data and CT image data

6.2.3.2 Results and Comprehensive Analysis

Table 6.2 summarizes the performance of various fusion models. Fusion Model C, which integrates multi-phase CT images and clinical data using fully connected layers, achieved the highest accuracy of 78.66% and an AUC of 0.8248. When a joint loss function was applied to this model, performance further improved to 80.49% accuracy and 0.8331 AUC.

The clinical model based on Random Forest (RF), which uses only clinical data, achieved an accuracy of 76.03% and an AUC of 0.7532. Meanwhile, Fusion Model A, which directly concatenates the clinical data with image features, achieved an accuracy of 74.80% and an AUC of 0.7880. Fusion Model B, which applies single-modality projector for the image features, performed slightly better with an accuracy of 78.05% and an AUC of 0.8045. Fusion Model C, using dual-modality projection for concatenation, showed the best accuracy (78.66%) and an AUC of 0.8248, outperforming the single-modality projection model.

These results clearly demonstrate that incorporating both medical images (Modality A) and clinical data (Modality B) in fusion models leads to better performance than using either modality independently. Fusion Model C particularly stands out, offering the highest prediction accuracy and AUC, showing the value of a multimodal approach in predicting early recurrence of HCC. By combining image and clinical data, the model can better capture the complexity of disease progression, leading to more accurate predictions and improving clinical decision-making.

6.2.4 Summary

A commonly employed method in multimodal analysis is concatenation. This method involves combining low-level or high-level features, which can be extracted from multiple sources, including different sensors or modalities, or derived from pre-trained networks [5–7]. In medical image analysis, for example, concatenation can merge low-level features from clinical data with high-level features derived from images (such as CT and MRI scans). This process results in a fused representation that integrates complementary information from various data sources. By combining the detailed anatomical or functional information from medical images with the contextual information from clinical data, vanilla fusion facilitates more comprehensive analysis, ultimately enhancing prediction accuracy and diagnostic capability.

6.3 Attention-Based Fusion Methods

Vanilla-based fusion methods provide a straightforward and intuitive fusion strategy by directly concatenating multimodal features. However, they lack adaptability to the varying importance of different modalities and fail to capture long-range dependencies and fine-grained associations effectively. This is particularly inadequate for tasks that require precise alignment of multimodal features. In contrast, attention-based methods provide a more flexible and efficient solution. They do not require explicit alignment before fusion. Instead, they use soft selection to dynamically assign learnable weights to features based on their relative importance, selectively highlighting key features and suppressing redundant information. By better capturing nonlinear associations and long-range dependencies across modalities, attention-based methods enable models to deeply exploit useful information within each modality and accurately model complex interactions between modalities.

Among various attention mechanisms, self-attention has been widely adopted to capture feature relationships within individual modalities, focusing primarily on modeling intra-modality dependencies. While these methods help extract intra-modality dependencies and enhance representations from a unified feature space, they do not explicitly model interactions across different modalities. This limitation reduces their ability to exploit cross-modal complementarity and weakens the integration of heterogeneous information sources in multimodal scenarios. To address this challenge, recent research has introduced cross-attention mechanisms designed to facilitate cross-modal interaction and effective exchange of information between modalities. By attending to one modality while conditioning on another, cross-attention promotes fine-grained alignment and the enhancement of complementary features, resulting in more balanced and comprehensive multimodal representations. These mechanisms improve the model's ability to capture inter-modality correlations and are particularly effective in tasks that require precise feature correspondence across modalities.

In this section, we will give brief introductions to classic self-attention-based and cross-attention-based multi-modal deep learning methods. In Sect. 6.3.1 self-attention is discussed focusing on intra-modal feature relationships. Section 6.3.2 extends the discussion to cross-attention, including one-way cross-attention for uni-directional information flow, two-way cross-attention and self-and-mutual attention for bidirectional feature interaction and guidance. In Sect. 6.3.3, performance validation is conducted on these methods to show their applicability in multi-modal biomedical tasks. Finally, in Sect. 6.3.4 Attention-based methods are summarized.

6.3.1 Self-attention

In multimodal medical image analysis, the self-attention mechanism enhances the representation of single-modal features by dynamically assigning weights to the context of the input data, capturing key features within each modality (such as spatial relationships between tissues or abnormal lesions). The core concept is to treat the input data itself as Query (Q), Key (K), and Value (V), enabling each feature vector to aggregate information from all other positions with adaptive weights. This mechanism facilitates flexible global relationship modeling without relying on specific sequence positions or predefined topological structures. By applying self-attention across various modalities, such as images, text, time-series, and genetic data, researchers have achieved more flexible feature representations and richer contextual modeling. The self-attention mechanism has been widely applied in medical image processing and multi-modal fusion tasks, including segmentation, classification, detection, and prediction [8, 9].

Figure 6.4 illustrates a self-attention-based multimodal architecture, which applies attention mechanism within each single modality. Taking modality A and modality B for example, the input feature sequences from modality A are denoted as $X_A = [x_1, x_2, \ldots, x_n] \in \mathbb{R}^{n \times d}$, the input feature sequences from modality B are denoted as $X_B = [x_1, x_2, \ldots, x_n] \in \mathbb{R}^{n \times d}$, where n represents the sequence length and d represents the feature dimension. The self-attention of modality A and modality B first independently map X_A and X_B to their respective Query (Q), Key (K), and Value (V) subspaces through three linear transformations:

$$Q_A = X_A W_{Q_A}, \qquad K_A = X_A W_{K_A}, \qquad V_A = X_A W_{V_A} \tag{6.1}$$

$$Q_B = X_B W_{Q_B}, \qquad K_B = X_B W_{K_B}, \qquad V_B = X_B W_{V_B} \tag{6.2}$$

where $W_{Q_A}, W_{K_A}, W_{V_A} \in \mathbb{R}^{d \times d_k}$ and $W_{Q_B}, W_{K_B}, W_{V_B B} \in \mathbb{R}^{d \times d_k}$ are learnable projection matrices, and d_k represents the feature dimension of the corresponding subspace.

Then, the similarities between Q_A and K_A, Q_B and K_B are separately computed via a dot-product operation, followed by softmax normalization to obtain attention weights. These weights are used to perform weighted summation over the Value

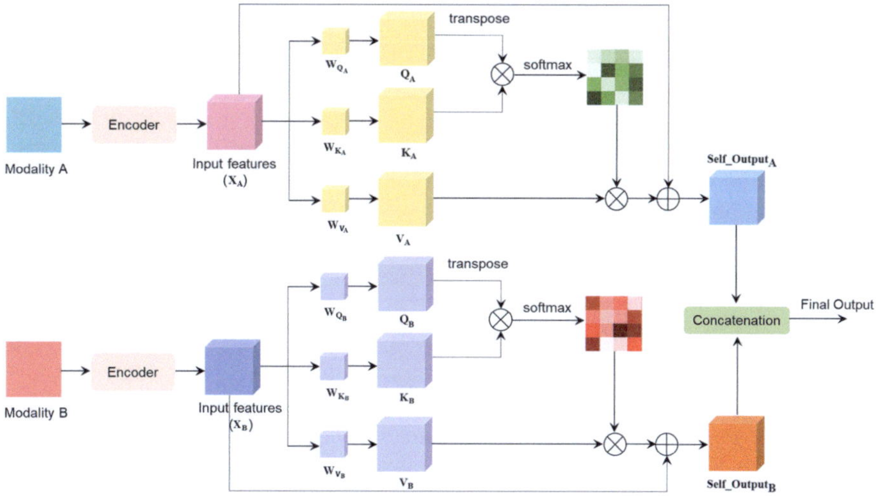

Fig. 6.4 Self-attention-based multimodal architecture. For any modalities A and B, global feature relationships within each modality are modeled by dynamically assigning similarity-based weights through the independent mappings of query (Q), key (K), and value (V). The attention features extracted from each modality are then concatenated to generate the final output

matrix V_A and V_B, producing the enhanced feature representations:

$$Attention\,(Q_A, K_A, V_A) = softmax\left(\frac{Q_A K_A^T}{\sqrt{d_k}}\right) V_A \tag{6.3}$$

$$Attention\,(Q_B, K_B, V_B) = softmax\left(\frac{Q_B K_B^T}{\sqrt{d_k}}\right) V_B \tag{6.4}$$

where $\sqrt{d_k}$ serves as a scaling factor to prevent the dot-product values from becoming excessively large, which could lead to unstable gradients.

Additionally, to preserve the original information of the input features and stabilize gradient propagation, the self-attention module typically incorporates a residual connection. This mechanism adds the input features directly to the enhanced feature representations, producing the outputs:

$$Self\_Output_A = Attention\,(Q_A,\ K_A,\ V_A) + X_A \tag{6.5}$$

$$Self\_Output_B = Attention\,(Q_B,\ K_B,\ V_B) + X_B \tag{6.6}$$

The features $Self\_Output_A$ and $Self\_Output_B$, derived from each modality, are finally concatenated to form the fused feature representation Final Output:

$$Final\_Output = Concat\left(Self\_Output_A, Self\_Output_B\right) \tag{6.7}$$

Although self-attention effectively captures global relationships among multimodal features, it primarily models internal associations within a single set of features, without explicitly distinguishing the directional guidance between modalities. This limitation may hinder its performance in tasks involving significant modality distribution differences or asymmetric inter-modal structures. To address this issue, the next section will explore cross-attention mechanisms that incorporate directional guidance and inter-modal dependencies.

6.3.2 *Cross-attention*

Cross-attention introduces explicit role differentiation in multimodal deep learning. During the computation process, the features of one modality serve as the Query (Q) to actively retrieve relevant features from another modality, which serve as the Key (K) and Value (V). This targeted approach enables more efficient and flexible information extraction and fusion, providing a robust collaborative mechanism for multimodal integration [8, 10, 11].

By assigning different roles to modality A and modality B in attention computation, the model can dynamically select relevant information based on specific task requirements, such as identifying distinctive feature patterns or emphasizing certain key attributes. Cross-attention is generally divided into two types: one-way cross-attention and two-way cross-attention. In practical applications, cross-attention has demonstrated exceptional performance in multimodal tasks, particularly in scenarios requiring modality alignment, feature enhancement, and multi-level information fusion.

6.3.2.1 One-Way Cross-attention

One-way cross-attention is a unidirectional information flow mechanism designed to enable one modality to learn features from another. In one-way cross-attention, one modality is designated as the Q, while the features from the other modality serve as K and V for retrieval and weighting. For instance, the features of modality A can be utilized to retrieve the most relevant features from modality B, and vice versa. The computation process is similar to that of self-attention, except that Q, K, and V are derived from different modalities. As illustrated in Fig. 6.5, taking modality A as the Q_A and modality B as the K_B and V_B, the primary operational steps are as follows:

- Apply a linear projection to the extracted features X_A from modality A to form the Query matrix Q_A.
- Similarly, apply linear projections to the extracted features X_B from modality B to obtain the Key matrix K_B and the Value matrix V_B.

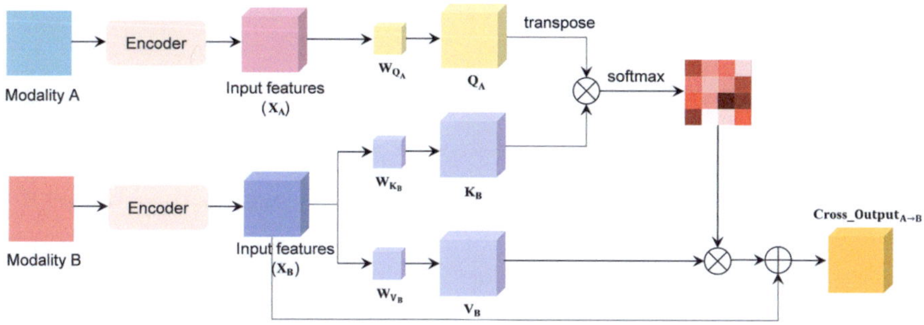

Fig. 6.5 One-way cross-attention-based Architecture. By utilizing features from one modality (Modality A) as the Query (Q_A) and retrieving Keys (K_B) and Values (V_B) from another modality (Modality B), One-Way Cross-attention enables directional information guidance and fusion

- Compute the dot product between Q_A and K_B to obtain the similarity weights, scale the weights, and apply softmax normalization.
- Aggregate V_B using the normalized weights through weighted summation to generate the fused feature representation as follows:

$$Attention_{A\to B}(Q_A, K_B, V_B) = softmax\left(\frac{Q_A K_B^T}{\sqrt{d_k}}\right) V_B \tag{6.8}$$

- Ultimately, input features X_B are directly added to the enhanced feature representation, forming the $Cross\_Output_{A\to \mathrm{B}}$:

$$Cross\_Output_{A\to \mathrm{B}} = Attention_{A\to B}(Q_A, K_B, V_B) + X_B \tag{6.9}$$

One-way cross-attention provides an efficient unidirectional information flow mechanism, enabling the model to retrieve features from one modality based on the specific requirements of another, producing more targeted fused representations. However, its unidirectional nature may result in asymmetric information interaction, potentially overlooking bidirectional dependencies between modalities. This can lead to incomplete information fusion, limiting the model's ability to fully leverage the complementary characteristics of multi-modal data in complex tasks.

6.3.2.2 Two-Way Cross-attention

Two-way cross-attention is an extended interaction-based attention mechanism that differs from the unidirectional information flow of one-way cross-attention. By introducing bidirectional interactions between two modalities, it captures richer feature associations. This mechanism is particularly suitable for tasks requiring symmetrical modeling or deep dependency analysis between two modalities. The core concept of two-way cross-attention lies in computing interaction relationships between two

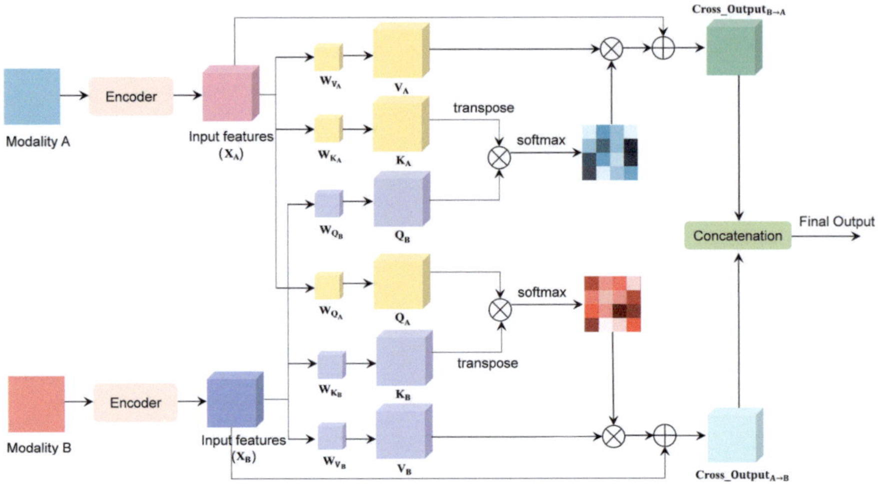

Fig. 6.6 Two-way cross-attention-based architecture. By allowing both modalities (Modality A and Modality B) to serve simultaneously as query and key/value for each other, two-way cross-attention enables bidirectional interaction, guidance, and fusion between multimodal information

modalities, enabling each modality's features to learn useful information from the other. Specifically, by enabling modalities A and B to simultaneously serve as each other's Q, it achieves bidirectional interaction and guided information exchange.

As illustrated in Fig. 6.6, taking modality A and modality B as examples, the features from each modality serve as Q, computing cross-attention matrices with features from the opposite modality. The cross-attention matrix with modality A as the Q_A has already been provided in Eq. (6.8). Similarly, the cross-attention matrix with modality B as the Q_B is computed as follows:

$$Attention_{B\to A}(Q_B, K_A, V_A) = softmax\left(\frac{Q_B K_A^T}{\sqrt{d_k}}\right) V_A \tag{6.10}$$

Finally, the cross-attention results of the two modalities are fused with their respective initial features, resulting in bidirectionally enhanced feature representations that incorporate the most relevant information from both modality A and modality B.

Two-way cross-attention can be regarded as a comprehensive inter-modal collaboration mechanism that captures symmetric relationships and complex dependencies between modalities. It dynamically adjusts the features of each modality to align with the contextual information of the other. This bidirectional interaction mechanism excels in multi-modal fusion, modality alignment, and robustness enhancement. For instance, in multimodal fusion tasks, two-way cross-attention facilitates deep information exchange, improving the quality of fused features. In modality alignment tasks, it effectively captures inter-modal differences and complementarities. Regarding robustness, it strengthens the model's resilience to noise and inconsistent data by simultaneously learning mutual dependencies between modalities. Additionally, the final outputs from both modalities can be concatenated at the feature level to further enrich feature representations. Alternatively, a new cross-attention module can be added on the existing architecture to capture deeper inter-modal relationships, further enhancing overall model performance. The next section will introduce Self-and-Mutual Attention to fully explore and effectively utilize the highly heterogeneous cross-modal information at different levels, modeling the complex relationships between modalities.

6.3.2.3 Self-and-Mutual Attention

Unlike simple fusion and previous attention-based fusion methods, Y. Ai et al. proposed a Self-and-Mutual Attention (SAMA) fusion strategy their work "SAMA: A Self-and-Mutual Attention Network for Accurate Recurrence Prediction of Non-Small Cell Lung Cancer Using Genetic and CT Data" [12]. SAMA effectively models the complex relationships between highly heterogeneous modalities (e.g., CT images and genetic data). As shown in Fig. 6.7, for any two modalities, modality A and modality B, the SAMA module adopts a three-stage information interaction process, including modality self-enhancement, cross-modal guided refinement, and feature aggregation optimized for specificity. This comprehensive approach fully explores specific complementary features and accurately learns complex cross-modal interactions. Drawing on the concept of attention, SAMA combines the advantages of self-attention, one-way cross-attention, and two-way cross-attention to ensure effective interaction at different levels and maximize the extraction of specific complementary information.

Essentially, it first uses two self-attention mechanisms within each model for self-enhancement, improving the quality of single-modal features and ensuring that the internal semantic structures of each modality are fully captured. Furthermore, based on self-attention, a two-way cross-attention is used for each modality, with the enhanced semantic information of modality B being integrated into the enhanced feature representation of modality A, forming a fusion of cross-enhanced modal semantics. Simultaneously, the enhanced semantic information of modality A is integrated into the enhanced feature representation of modality B, promoting cross-modal semantic sharing. One-way cross-attention is used after two-way cross-attention in the optimized feature aggregation to further enhance performance in downstream tasks.

For example, as shown in Fig. 6.8, SAMA is used to fuse and utilize macro CT images and micro genetic data to accurately predict the recurrence of non-small cell lung cancer (NSCLC). The self-enhancement is utilized to enhance modality-specific features of CT images and genes. Gene-guided and CT-guided cross-modality fusion bidirectionally guide the self-enhanced features, aiming to refine each modality with cross-modal complementary information at different levels, enhancing heterogeneity in feature expression for greater discriminative power: for CT images, guided by genetic data to better capture lung structure and tissue details, especially tumor

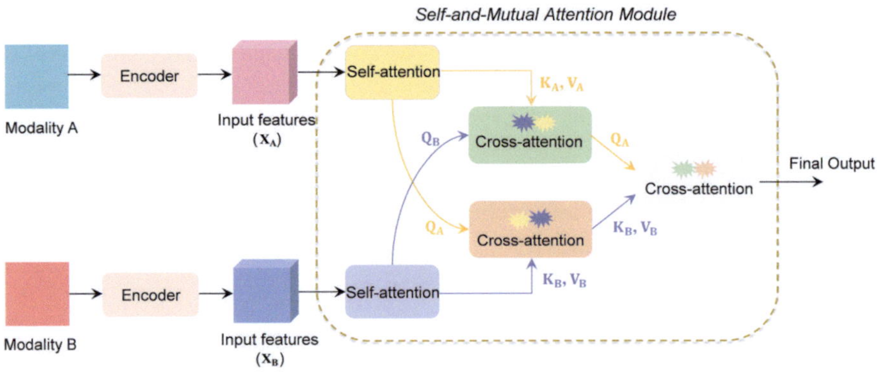

Fig. 6.7 Self-and-mutual attention (SAMA) architecture. SAMA is a three-stage fusion framework for modeling complex macro and micro modality fusion and interaction

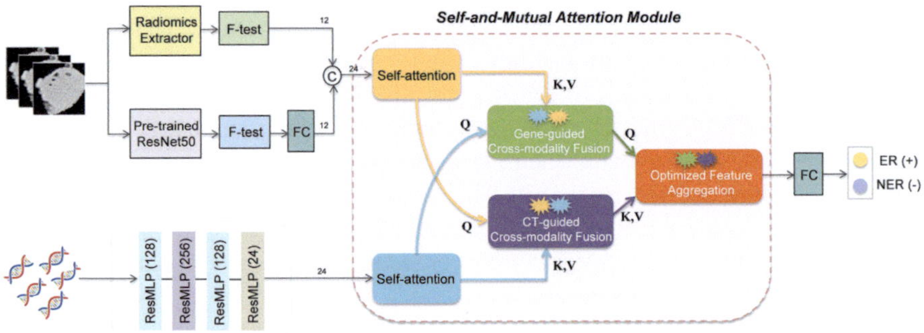

Fig. 6.8 Overview of the self-and-mutual attention (SAMA) network for accurate recurrence prediction of NSCLC. The SAMA network fuses handcrafted and deep features from CT images with gene expression data. The SAMA module employs a three-stage information interaction process involving modality self-enhancement, cross-modality guided refinement, and specificity-optimized feature aggregation to comprehensively mine specific complementary features and learn intricate gene-CT relationships. The numbers displayed on the line represent the counts of extracted or selected features

morphology and locations related to genetic information; for genetic data, guided by CT to better reflect molecular characters tics of tumors and abnormal areas in CT images, such as gene expression patterns. The optimized feature aggregation ensures the refined interactive features are precise in predicting NSCLC recurrence.

6.3.3 Performance Validation

In this section, we introduce self-attention-based and cross-attention-based methods, and present experimental results to validate their effectiveness in predicting preoperative recurrence of NSCLC [12]. In the case presented here, modality A refers to medical imaging data (CT images), while modality B corresponds to gene expression data from the same patient.

6.3.3.1 Dataset and Implementation Details

In our experiments, we compared multiple multimodal fusion strategies. The NSCLC dataset was drawn from the Radiogenomics dataset (R01 cohort) available through The Cancer Imaging Archive (TCIA), funded by the U.S. National Cancer Institute (NCI) to support open science research [13]. This R01 cohort provides data for 162 subjects, each associated with CT images and gene expression data. Following established criteria and the requirements of the recurrence prediction task, subjects lacking recurrence labels, tumor masks, clearly defined gene expression, or those who had unclear causes of death prior to recurrence were excluded. As a result, 88 subjects were selected. To enhance CT data quality, three key preprocessing steps were applied: (1) The intensity values of the CT images were first truncated to the range of − 1000 HU to +400 HU to retain relevant information while removing outliers. (2) The pixel values were then normalized to the range of 0–255 to ensure consistency across imaging devices. (3) For each patient, three slices were selected—one containing the largest tumor cross-section and its two adjacent slices—forming a 2.5D representation that serves as the RGB channels of the input. (4) Using the tumor mask, each image was cropped to the tumor region of interest (ROI) and resized to 224 × 224 pixels for uniformity and compatibility with the model. For the gene expression data, FPKM (Fragments Per Kilobase of transcript per Million mapped reads) values were used for quantification, and a total of 5,587 high-quality genes were selected for analysis [12].

Figure 6.8 illustrates the overall framework for fusing CT images and gene expression data. This network takes handcrafted CT features, deep CT features, and gene features as inputs, and consists of two main components: a SAMA module and a recurrence prediction module. The SAMA module incorporates two self-attention and three cross-attention components to enhance modality-specific features, guide macro- and micro-level refinement, and achieve more targeted feature aggregation.

All experiments were conducted on a workstation equipped with an NVIDIA A100 (40GB) GPU, using Python 3.7 and PyTorch as the software environment. The Adam optimizer was employed with an initial learning rate of 1e-3, trained for 200 epochs, and a weight decay of 1e−4. The batch size was set to 4. Each experiment was repeated 10 times with five-fold stratified cross-validation to obtain fair average performance. In each fold, the training sets had a sample size of 70 or 71, while the validation sets had a sample size of 17 or 18.

Table 6.3 Comparison Results of different fusion models for recurrence prediction of NSCLC on the TCIA dataset

Model	Method	Accuracy (%)	AUC (%)	Sensitivity (%)	Specificity (%)
CT model	–	81.65 ± 7.7	72.08 ± 9.0	70.41 ± 13.5	83.75 ± 9.0
Gene model	–	91.04 ± 5.3	89.09 ± 6.7	87.66 ± 8.5	92.53 ± 5.6
Fusion model 1	Concatenation	92.03 ± 5.6	87.64 ± 6.3	87.76 ± 10.9	93.53 ± 4.2
Fusion model 2	Concatenation with projection	93.14 ± 4.5	90.43 ± 5.8	89.00 ± 7.5	93.87 ± 5.3
Fusion model 3	Self-attention	93.26 ± 4.4	90.15 ± 5.5	89.99 ± 6.3	93.84 ± 4.5
Fusion model 4	CT-guided cross-modality fusion	94.12 ± 3.5	91.30 ± 5.1	89.67 ± 5.5	94.92 ± 4.4
Fusion model 5	Gene-guided cross-modality fusion	93.31 ± 3.8	91.30 ± 5.3	89.33 ± 5.9	96.00 ± 4.0
Fusion model 6	Two-way cross-modality fusion	94.38 ± 3.4	92.39 ± 4.5	91.00 ± 5.3	94.92 ± 4.0
Fusion model 7	SAMA	95.55 ± 2.3	94.17 ± 2.8	91.67 ± 4.9	96.67 ± 3.0

6.3.3.2 Results and Comprehensive Analysis

As shown in Table 6.3, when using only CT images, the model achieved an ACC of 81.65% and an AUC of 72.08%, indicating relatively limited performance. By comparison, the single-modality gene-based model achieved an ACC of 91.04% and an AUC of 89.09%, suggesting that gene expression data offer stronger predictive capabilities for NSCLC recurrence. For the fused models, Model 1 (direct concatenation) showed a significant improvement over single-modality models, achieving an ACC of 92.03% and an AUC of 87.64%. Model 2 introduced a feature alignment mechanism on top of concatenation, further enhancing ACC and AUC to 93.14% and 90.43%, respectively, underscoring the positive impact of feature alignment. Model 3 introduced a self-attention module to capture global feature dependencies within each modality, effectively strengthening feature representations. This approach increased ACC to 93.26%. Models 4 and 5 evaluated CT-guided (CT as Q, genes as K/V) and gene-guided (genes as Q, CT as K/V) cross-modality fusion modules. By introducing cross-modality fusion, single-modality features benefited significantly from guidance and supplementation provided by the other modality. Model 6 integrated both CT-guided and gene-guided cross-modality fusion modules (two-way cross-modality fusion), achieving ACC and AUC scores of 94.38%

Table 6.4 Comparison results of different existing methods for recurrence prediction of NSCLC on the TCIA dataset

Modality	Methods	Accuracy (%)	AUC (%)	Sensitivity (%)	Specificity (%)
CT image	PCA + SVM [15]	67.05 ± 5.8	58.33 ± 7.1	31.03 ± 16.3	85.00 ± 9.8
	Relief-F + SVM [16]	67.56 ± 6.0	61.89 ± 7.5	64.97 ± 13.0	70.80 ± 12.6
	LASSO [17]	68.36 ± 6.0	64.24 ± 7.5	67.00 ± 13.5	73.49 ± 12.8
	Latent + ANN [18]	78.48 ± 7.3	66.38 ± 7.7	55.62 ± 14.7	89.14 ± 8.1
	ResNet50 [19]	79.31 ± 8.1	67.41 ± 7.9	59.00 ± 10.2	89.83 ± 6.4
	DenseNet121 [19]	77.36 ± 8.6	67.88 ± 9.0	38.00 ± 17.0	96.61 ± 3.0
	Radiomics + Deep [20]	82.35 ± 6.7	73.98 ± 8.2	75.63 ± 8.8	88.33 ± 6.0
	GGR [21]	83.28 ± 8.0	76.67 ± 8.0	95.00 ± 5.0	59.00 ± 26.9
	GGR_fusion [22]	84.39 ± 7.6	78.85 ± 7.8	91.21 ± 8.9	65.36 ± 20.7
	ResMLP_GGR [23]	86.38 ± 6.5	82.20 ± 6.8	89.66 ± 8.7	80.75 ± 6.2
	ResMLP_GGR (mixed) [24]	87.78 ± 6.3	83.98 ± 6.6	88.88 ± 8.5	85.07 ± 6.0
Gene data	F-test + ANN [25]	90.28 ± 6.7	86.20 ± 9.4	83.76 ± 14.7	88.64 ± 6.1
	LASSO + F-test + CHI-2 + MLP [26]	90.92 ± 6.7	87.61 ± 7.9	88.00 ± 12.0	91.21 ± 6.3
	KL + DNN [27]	90.70 ± 5.9	84.58 ± 6.8	85.33 ± 12.0	87.83 ± 6.7
CT image and gene data	Cox fusion [28]	91.78 ± 5.8	89.98 ± 6.9	88.66 ± 10.9	91.36 ± 6.4
	DCCAFN [14]	93.06 ± 4.6	91.05 ± 6.2	87.67 ± 12.0	94.35 ± 4.2
	CM-MMF [29]	92.22 ± 5.6	90.09 ± 6.3	88.66 ± 10.9	91.53 ± 5.8
	SAMA [12]	95.55 ± 2.3	94.17 ± 2.8	91.67 ± 4.9	96.67 ± 3.0

and 92.39%, respectively, confirming the value of bidirectional interactive fusion. Building on this foundation, Model 7 incorporated an optimized feature aggregation module, reaching an ACC of 95.55% and an AUC of 94.17%, along with notable improvements in sensitivity and specificity. This results indicate that the optimized feature aggregation module further enhance feature fusion quality while strengthening the model's discriminative capabilities across diverse samples. Additionally, the SAMA method was compared with other widely adopted approaches in the medical domain, including CT-based, gene-based, and CT-gene fusion-based models, for predicting NSCLC recurrence on the TCIA dataset. As shown in Table 6.4, SAMA outperformed existing models, improving ACC, AUC, sensitivity, and specificity by 2.49%, 3%, 4%, and 3%, respectively, and achieved the state-of-the-art [14].

In summary, incorporating self-attention and cross-attention modules significantly elevated feature fusion quality and improved the model's adaptability to complex multimodal tasks. The self-attention module enhances feature diversity and granularity by capturing global dependencies within each modality. The cross-attention module enables dynamic interaction across modalities, allowing them to complement each other and produce more discriminative fused representations. By guiding and aligning features, these modules convert originally difficult-to-integrate modalities into more coherent, correlated representations, effectively handling high-dimensional, heterogeneous data such as medical images and gene expression. Moreover, the cross-attention module emphasizes key features while suppressing irrelevant or redundant information, improving noise robustness. This advantage is particularly evident in multi-modal fusion tasks. The combined use of self-attention and cross-attention in SAMA has achieved the best performance.

6.3.4 Summary

In this section, we introduced attention-based multimodal fusion methods include self-attention and cross-attention. By incorporating self-attention and cross-attention modules, these methods more accurately capture intra-modality feature dependencies and establish dynamic, direction-aware exchanges across modalities, leading to deeper fusion and richer information complementarity. Experiments on the NSCLC dataset—integrating CT imaging and gene data—demonstrate that attention-based methods outperform single-modality models and surpass simpler multimodal fusion techniques. The progressive improvements, from simple concatenation to feature alignment, self-attention, and ultimately bidirectional cross-attention, underscore how these attention mechanisms iteratively refine and strengthen the decision-making process. These findings highlight the potential of attention-based methods in handling complex, heterogeneous clinical data with greater predictive accuracy and robustness, and open opportunities to incorporate even more diverse data sources (e.g., multi-sensor signals, various omics data, and clinical data) within a unified analytical framework.

References

1. Lahat, D., Adali, T., & Jutten, C. (2015). Multimodal data fusion: An overview of methods, challenges, and prospects. *Proceedings of the IEEE, 103*(9), 1449–1477.
2. Chen, T., & Rao, R. R. (1998). Audio-visual integration in multimodal communication. *Proceedings of the IEEE, 86*(5), 837–852.
3. Gadzicki, K., Khamsehashari, R., & Zetzsche, C. (2020, July). Early vs late fusion in multimodal convolutional neural networks. In *2020 IEEE 23rd International Conference on Information Fusion (FUSION)* (pp. 1–6). IEEE.
4. Ahmed, M., Hegazy, M., Klimchik, A. S., & Boby, R. A. (2022). Lidar and camera data fusion in self-driving cars. *Computer Research and Modeling, 14*(6), 1239–1253.
5. Nojavanasghari, B., Gopinath, D., Koushik, J., Baltrušaitis, T., & Morency, L. P. (2016, October). Deep multimodal fusion for persuasiveness prediction. In *Proceedings of the 18th ACM International Conference on Multimodal Interaction* (pp. 284–288).
6. Wang, H., Meghawat, A., Morency, L. P., & Xing, E. P. (2017, July). Select-additive learning: Improving generalization in multimodal sentiment analysis. In *2017 IEEE International Conference on Multimedia and Expo (ICME)* (pp. 949–954). IEEE.
7. Wang, W., Chen, Q., Iwamoto, Y., Aonpong, P., Lin, L., Hu, H., et al. (2020). Deep fusion models of multi-phase CT and selected clinical data for preoperative prediction of early recurrence in hepatocellular carcinoma. *IEEE Access, 8*, 139212–139220.

8. Vaswani, A., Shazeer, N., Parmar, N., Uszkoreit, J., Jones, L., Gomez, A. N., et al. (2017). Attention is all you need, In *Advances in neural information processing systems* (Vol. 30).
9. Ji, Z., Ye, S., & Ma, X. (2024). Sparse Coding Inspired LSTM and Self-Attention Integration for Medical Image Segmentation. *IEEE Transactions on Image Processing, 33*, 6098–6113.
10. Lin, J. W., Lin, J. T., Lu, C., Chen, H., Lin, H., & Zhao, B. C. (2024). CKD-TransBTS: Clinical knowledge-driven hybrid transformer with modality-correlated cross-attention for brain tumor segmentation. *IEEE Transactions on Medical Imaging, 42*(8), 2451–2461.
11. Lin, H., Cheng, X., Wu, X., & Shen, D. (2022, July). CAT: Cross attention in vision transformer. In *2022 IEEE International Conference on Multimedia and Expo (ICME)* (pp. 1–6).
12. Ai, Y., Liu, J., Li, Y., Wang, F., Du, X., Jain, R. K., et al. (2024). SAMA: A self-and-mutual attention network for accurate recurrence prediction of non-small cell lung cancer using genetic and CT data. *IEEE Journal of Biomedical and Health Informatics*. https://doi.org/10.1109/JBHI.2024.3471194
13. Bakr, S., Gevaert, O., Echegaray, S., Ayers, K., Zhou, M., Shafiq, M., et al. (2018). A radiogenomic dataset of non-small cell lung cancer. *Scientific Data, 5*(1), 1–9.
14. Jia, L., Ren, X., Wu, W., Zhao, J., Qiang, Y., & Yang, Q. (2024). DCCAFN: Deep convolution cascade attention fusion network based on imaging genomics for prediction survival analysis of lung cancer. *Complex & Intelligent Systems, 10*(1), 1115–1130.
15. Wang, X., Duan, H. H., & Nie, S. D. (2019, November). Prognostic recurrence analysis method for non-small cell lung cancer based on CT imaging. In *2019 International Conference on Image and Video Processing, and Artificial Intelligence* (Vol. 11321, pp. 411–417).
16. Lee, S., Jung, J., Hong, H., & Kim, B. (2020, June). Radiomic feature-based prediction model of lung cancer recurrence in NSCLC patients. In *International Workshop on Advanced Imaging Technology (IWAIT) 2020* (Vol. 11515, pp. 106–110).
17. Christie, J. R., Abdelrazek, M., Lang, P., & Mattonen, S. A. (2021). February). A multi-modality radiomics-based model for predicting recurrence in non-small cell lung cancer. In *Medical imaging 2021: Biomedical applications in molecular. Structural, and functional imaging* (Vol. 11600, pp. 143–149).
18. Song, F., Tian, J., Zhang, P., Ma, C., Sun, Y., Feng, Y., et al. (2023). A novel feature engineering method based on latent representation learning for radiomics: Application in NSCLC subtype classification. *IEEE Journal of Biomedical and Health Informatics, 28*(1), 31–41.
19. Aonpong, P., Iwamoto, Y., Wang, W., Lin, L., & Chen, Y. W. (2020). Hand-crafted and deep learning-based radiomics models for recurrence prediction of non-small cells lung cancers. In *Innovation in Medicine and Healthcare: Proceedings of 8th KES-InMed 2020* (pp. 135–144).
20. Ai, Y., Li, Y., Jain, R. K., & Chen, Y. W. (2023, October). A self-attention based fusion model of radiomics and deep features for early recurrence prediction in NSCLC. In *2023 IEEE 12th Global Conference on Consumer Electronics (GCCE)* (pp. 833–837).
21. Aonpong, P., Iwamoto, Y., Han, X. H., Lin, L., Chen, Y. W., et al. (2021). Genotype-guided radiomics signatures for recurrence prediction of non-small cell lung cancer. *IEEE Access, 9*, 90244–90254.
22. Aonpong, P., Iwamoto, Y., Han, X. H., Lin, L., Chen, Y. W., et al. (2021, November). Improved genotype-guided deep radiomics signatures for recurrence prediction of non-small cell lung cancer. In *2021 43rd Annual International Conference of the IEEE Engineering in Medicine & Biology Society (EMBC)* (pp. 3561–3564).
23. Ai, Y., Aonpong, P., Wang, W., Li, Y., Iwamoto, Y., et al. (2022, July). Residual multilayer perceptrons for genotype-guided recurrence prediction of non-small cell lung cancer. In *2022 44th Annual International Conference of the IEEE Engineering in Medicine & Biology Society (EMBC)* (pp. 447–450).
24. Ai, Y., Li, Y., Chen, Y. W., Aonpong, P., & Han, X. (2023). ResMLP_GGR: Residual multilayer perceptrons-based genotype-guided recurrence prediction of non-small cell lung cancer. *Journal of Image and Graphics, 11*(2).
25. Aonpong, P., Iwamoto, Y., Wang, W., Lin, L., & Chen, Y. W. (2021). Genomics-based models for recurrence prediction of non-small cells lung cancers. In *Innovation in Medicine and Healthcare: Proceedings of 9th KES-InMed 2021* (pp. 41–49).

26. Bhattacharjee, S., Saha, B., & Saha, S. (2023, January). Prediction of recurrence in non small cell lung cancer patients with gene expression data using machine learning techniques. In *2023 International Conference on Computer, Electrical & Communication Engineering (ICCECE)* (pp. 1–8).
27. Liu, S., & Yao, W. (2022). Prediction of lung cancer using gene expression and deep learning with KL divergence gene selection. *BMC Bioinformatics*, *23*(1), 175.
28. Subramanian, V., Do, M. N., & Syeda-Mahmood, T. (2020, April). Multimodal fusion of imaging and genomics for lung cancer recurrence prediction. In *2020 IEEE 17th International Symposium on Biomedical Imaging (ISBI)* (pp. 804–808).
29. Deng, R., Shaikh, N., Shannon, G., & Nie, Y. (2024, April). Cross-modality attention-based multimodal fusion for non-small cell lung cancer (NSCLC) patient survival prediction. In *Medical Imaging 2024: Digital and Computational Pathology* (Vol. 12933, pp. 46–50).

Chapter 7
Semi-Supervised Learning for Medical Image Analysis

7.1 Introduction

In the realm of natural image analysis, the availability of large-scale, high-quality annotated datasets has significantly propelled the advancement of deep learning models. However, different from the nature image, acquisition of high-quality annotated medical images is a more challenging and resource-intensive task, primarily due to the necessity of expert knowledge for accurate labeling. Consequently, the scarcity of annotated data in medical imaging poses a substantial barrier to the development and deployment of robust deep learning models.

To alleviate the burden of annotation in medical imaging, researchers have explored various strategies that leverage partially annotated data. These strategies can be broadly categorized into two main scenarios:

- **Sample-based Partial Annotations**: it involves annotating only a portion of the tissues within each sample. It involves learning from the annotated regions within each sample and propagating these annotations to adjacent, similar regions.
- **Dataset-based Partial Annotations**: it entails providing complete annotations for a subset of dataset, which allows for the extraction of rich information about tissues or organs. Through the comprehensive annotation of a complete sample, one can learn detailed characteristics such as the shape, texture, and density of organs or lesions.

Given the prevalence of unannotated data in medical imaging, a critical question arises: How can we effectively utilize these unannotated data to enhance model performance? Semi-Supervised Learning (SSL) methods have been proposed to address this challenge. SSL techniques aim to leverage both labeled and unlabeled data to improve learning efficiency and model accuracy. The primary SSL approaches can be classified into the following categories:

- **Co-Training based Semi-Supervised Learning**: This method involves training multiple models simultaneously, where each model is trained on different views or

Y. Chen et al., *Recent Advances in Deep Learning for Medical Image Analysis*, Intelligent Systems Reference Library 278, https://doi.org/10.1007/978-3-031-94791-9_7

subsets of the data. The models iteratively share their predictions on the unlabeled data, which are then used as pseudo-labels to further refine the training process.

- **GAN-based Semi-Supervised Learning**: Generative Adversarial Networks (GANs) are employed in this approach to generate synthetic data that resemble the real data distribution. The discriminator in the GAN framework is trained to distinguish between real and synthetic data. This dual objective helps in improving the model's generalization capability.
- **Consistency based Semi-Supervised Learning**: This category focuses on enforcing consistency in the model's predictions under different perturbations or augmentations of the input data. The key techniques within this category include: (**1**) *Consistency based Semi-Supervised Learning with Model Ensemble*: Multiple models are trained, and their predictions are aggregated to enforce consistency. This ensemble approach helps in reducing the variance and improving the robustness of the predictions. (**2**) *Mean-Teacher based Semi-Supervised Learning*: This technique involves maintaining two models: a student model and a teacher model. The teacher model's weights are updated as an exponential moving average of the student model's weights. The student model is trained to match the teacher model's predictions, promoting consistency and stability in learning.
- **Pseudo-Labeling Techniques**: These methods generate pseudo-labels for the unlabeled data based on the model's predictions. The pseudo-labeled data are then used to augment the training set. The key techniques within this category include: (**1**) *Pseudo-Labeling Baseline Techniques*: Simple approaches where the model's predictions with high confidence are directly used as pseudo-labels. (**2**) *Pseudo-Labeling Techniques with High Confidence*: More sophisticated methods that incorporate confidence thresholds or uncertainty estimation to select the most reliable pseudo-labels, thereby improving the quality of the augmented training data.

The following subsections will focus on medical image segmentation and provide a detailed description of each SSL method, offering a comprehensive overview of their mechanisms, advantages, and applications in medical imaging.

7.2 Co-Training Based Semi-Supervised Learning

Co-Training algorithm [1] is a renowned semi-supervised learning method, which assumes that each data x in the dataset $\mathcal{D}$ has two distinct views, represented as $x = (v_1, v_2)$, with each view v_i being adequate for developing an effective model. For instance, the views may originate from different data sources or utilize distinct representations. Let $\mathcal{X}$ denotes the distribution from which $\mathcal{D}$ is sampled. The Co-Training framework posits that the models f_1 and f_2, trained on views v_1 and v_2 respectively, produce consistent predictions on $\mathcal{X}$, meaning that:

$$f(x) = f_1(v_1) = f_2(v_2), \forall x \in (v_1, v_2) \sim \mathcal{X} \tag{7.1}$$

According to this assumption, Co-Training develops a dual-view self-training framework, which trains a distinct classifier for each view using the labeled set $\mathcal{S}$ in the first step, and then incorporates the predictions made by both classifiers on the unlabeled set $\mathcal{U}$ into $\mathcal{S}$ to further enhance the training process.

Considering the outstanding performance of deep neural networks in various vision tasks, Qiao et al. [2] further extends the Co-Training framework to a deep Co-Training method for Semi-Supervised Image Recognition, termed as DCT. To prevent the neural networks from collapsing into one another, DCT imposes a view difference constraint on the traditional Co-Training framework by training each network to be resilient to the adversarial examples $\mathcal{X}'(\mathcal{X}' \cap \mathcal{X} = \emptyset)$ generated by the other. This can be formulated as follows:

$$\exists \mathcal{X}' : f_1(v_1) \neq f_2(v_2), \forall x \in (v_1, v_2) \sim \mathcal{X}' \quad (7.2)$$

In this way, the key contribution of DCT lies in its differentiable modeling, which incorporates both the Co-Training assumption and the view difference constraint. Based on the above assumptions, we will provide a detailed introduction to the DCT method in the following.

Supervised Loss. In the semi-supervised learning scheme, the dataset $\mathcal{D}$ can be divided into a small annotated subset $\mathcal{S}$ and a large unannotated subset $\mathcal{U}$. Let $v_1(x)$ and $v_2(x)$ represent the two views of the data x, where these views are convolutional representations of x prior to the final fully-connected layer $f_i(\cdot)$. The author utilizes the standard cross-entropy loss on the supervised subset $\mathcal{S}$, which can be formulated as:

$$L_{sup}(x, y) = H\left(y, f_1\left(v_1(x)\right)\right) + H\left(y, f_2\left(v_2(x)\right)\right) \quad (7.3)$$

where y is the corresponding ground truth of x, and $H(p, q)$ denotes the cross entropy between the distribution of p and q.

Co-Training Loss. Additionally, DCT aims for the networks $p_1(x) = f_1(v_1(x))$ and $p_2(x) = f_2(v_2(x))$ to produce similar predictions on unannotated subset $\mathcal{U}$. To quantify this similarity, the author employs the Jensen-Shannon divergence between $p_1(x)$ and $p_2(x)$, defined as follows:

$$L_{cot}(x) = H(\frac{1}{2}\left(p_1(x) + p_2(x)\right)) - \frac{1}{2}\left(H\left(p_1(x)\right) + H\left(p_2(x)\right)\right) \quad (7.4)$$

where $x \in \mathcal{U}$ and $H(p)$ represents the entropy of p. Training neural networks under the Co-Training assumption aims to minimize the expected loss $L_{cot}]$ on the unannotated subset $\mathcal{U}$. For the annotated set $\mathcal{S}$, minimizing the supervised loss L_{sup} inherently promotes similar predictions on $\mathcal{S}$ due to the presence of labels. Consequently, it is unnecessary to minimize L_{cot} on $\mathcal{S}$, and thus DCT focuses solely on its implementation for $\mathcal{U}$.

Difference Constraint Loss. A crucial requirement for the success of Co-Training is that the two views must be distinct and offer complementary information regarding each data point x. However, minimizing Eqs. (7.3) and (7.4) merely leads the neural networks to produce identical predictions on the whole dataset $\mathcal{D}$. Thus, it is essential to promote differences and complementarity between the networks. To accomplish this, we generate an additional set of images $\mathcal{D}'$ ($\mathcal{D} \cap \mathcal{D}' = \emptyset$) such that $p_1(x) \neq p_2(x), \forall x \in \mathcal{D}'$, which will be created using adversarial example.

DCT explores a straightforward generative method $g(x)$ that utilizes data x from $\mathcal{D}$ to construct $\mathcal{D}'$, specifically $\mathcal{D}' = \{g(x)|x \in \mathcal{D}\}$. Co-Training operates under the assumption that $p_1(x) = p_2(x)$ for all $x \in \mathcal{D}$, while we desire $p_1(g(x)) \neq p_2(g(x))$. Given that $p_1(g(x)) = p_1(x)$, it follows that $p_2(g(x)) \neq p_2(x)$. This indicates that $g(x)$ serves as an adversarial example for p_2, successfully deceiving network p_2 while leaving network p_1 unaffected. To prevent the deep networks from converging to similar outputs, we suggest training network p_1 (or p_2) to withstand the adversarial examples $g_2(x)$ from p_2 (or g1(x) from p_1) by minimizing the cross-entropy loss between $p_2(x)$ and $p_1(g_2(x))$ (or between $p_1(x)$ and $p_2(g_1(x))$), formulated as:

$$L_{dif}(x) = H(p_1(x), p_2(g_1(x))) + H(p_2(x), p_1(g_2(x))) \tag{7.5}$$

Objective function. In summary, the total loss function of DCT can be formulated as:

$$L = \mathbb{E}_{(x,y)\in\mathcal{S}} L_{sup}(x, y) + \lambda_{cot}\mathbb{E}_{x\in\mathcal{U}} L_{cot}(x) + \lambda_{dif}\mathbb{E}_{x\in\mathcal{D}} L_{dif}(x) \tag{7.6}$$

where λ_{cot} and λ_{dif} are two hyperparameters to regulate the balance between different losses.

7.3 GAN-Based Semi-Supervised Learning

In recent years, Generative Adversarial Networks (GANs) [3] have shown remarkable success across various applications. These include image generation [4], image completion [5], super-resolution [6], object detection [7], domain adaptation [8], and semantic segmentation [9]. Impressively, GAN-based methods have opened up numerous possibilities for unsupervised and semi-supervised learning, showcasing their impressive capabilities. In this section, we focus on discussing the application and research of the GAN-based methods in semi-supervised segmentation [10, 11].

Figure 7.1 illustrates the overview of GAN-based semi-supervised learning [11], which is composed of two main networks, *i.e.*, the segmentation network (SN) and the evaluation network (EN). The former one is designed for yielding final segmentation prediction. That is, given an input image with the size of $H \times W \times 3$, the segmentation network generates the pixel-wise probability maps with the size of $H \times W \times C$, where C is the number of semantic categories. The latter one, *i.e.*, evaluation network, processes the segmentation probability maps along with the corresponding

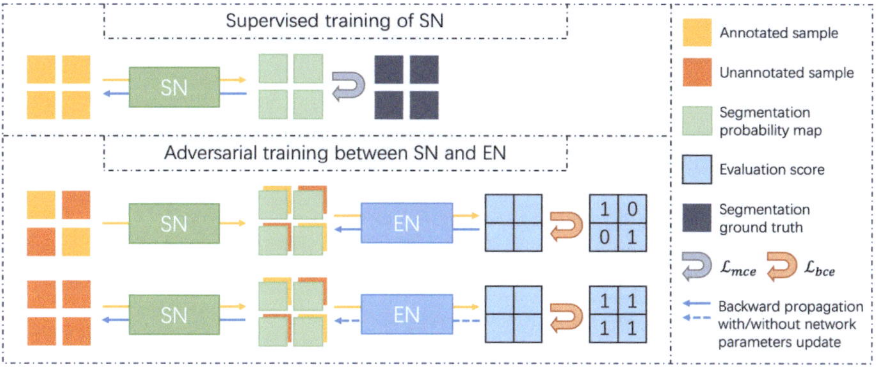

Fig. 7.1 An overview of GAN-based semi-supervised learning

input image I to assign a score that reflects the quality of the segmentation, where a score of 1 indicates good quality and a score of 0 indicates poor quality.

Specifically, EN is incentivized to assign high scores (1) to the segmentation of annotated images and low scores (0) to that of unannotated images, during the training process. Meanwhile, SN is trained on annotated images and also facilitates to generate segmentation results for unannotated images that could receive high scores from EN. The following section provides details about our adversarial training model.

Given M annotated training images X_m with corresponding pixel-level ground truth Y_m, and N unannotated images U_n, the total loss function can be defined as follows:

$$\begin{aligned}\ell(\theta_S, \theta_E) &= \sum_{m=1}^{M}(\ell_{mce}(S(X_m), Y_m)) \\ &- \lambda[\sum_{m=1}^{M} \ell_{bce}(E((X_m), X_m), 1) + \sum_{n=1}^{N} \ell_{bce}(E((U_n), U_n), 0)]\end{aligned} \tag{7.7}$$

where θ_S and θ_E represent the parameters of SN and EN, respectively. The term ℓ_{mce} denotes the multi-class cross-entropy loss, while ℓ_{bce} refers to the binary-class cross-entropy loss. The first component of the loss function is designed for the supervised training of SN using annotated images, whereas the second component contributes to the adversarial training aspect. During the model training, part of the loss is minimized concerning the parameters θ_S of SN, while the loss is maximized concerning the parameters θ_E of EN.

Specifically, the goal of training EN is to optimize its parameters θ_E by minimizing Eq. (7.8):

$$\lambda[\sum_{m=1}^{M} \ell_{bce}(E((X_m), X_m), 1) + \sum_{n=1}^{N} \ell_{bce}(E((U_n), U_n), 0)] \tag{7.8}$$

Correspondingly, the goal of training SN is to minimize the following Eq. with respect to the parameters θ_S of SN:

$$\sum_{m=1}^{M}(\ell_{mce}(S(X_m), Y_m)) - \lambda \sum_{n=1}^{N} \ell_{bce}(E((U_n), U_n), 0) \tag{7.9}$$

when updating SN, the authors substitute the term $-\lambda\sum_{n=1}^{N}\ell_{bce}(E((U_n), U_n), 0)$ with $\lambda\sum_{n=1}^{N}\ell_{bce}(E((U_n), U_n), 1)$. A standard stochastic gradient descent method is utilized to optimize this loss function.

7.4 Consistency-Based Semi-Supervised Learning

Consistency-based methods have gained significant popularity in semi-supervised segmentation tasks, as they enhance the robustness of learned models against various perturbations. These methods can be broadly categorized into two main types: (**1**) The first type maintains input invariance while altering the model itself—this includes approaches such as distinct decoders (i.e., Sect. 7.4.1.1), distinct initialized networks (i.e., Sect. 7.4.1.2), and distinct structured networks (i.e., Sect. 7.4.1.3)—ensuring consistency in the outputs. (**2**) The second type enforce an invariance of the model's predictions over small perturbations imposed to the inputs, exemplified by the classic Mean Teacher model [12, 13], which we will discuss in detail in the next section (i.e., Sect. 7.4.2).

7.4.1 Consistency-Based Semi-Supervised Learning with Model Ensemble

7.4.1.1 Consistency Over Distinct Decoders

As some works [12, 13] have proven that the efficacy of consistency training is profoundly influenced by the characteristics of the data distribution, particularly the cluster assumption, which posits that classes should be delineated by regions of low density. In the context of semantic segmentation, Ouali et al. [14] reveals that low-density regions separating classes are not evident within the input data itself, but rather manifest in the outputs of the encoder. Building on this insight, Ouali et al. [14] design a novel approach, termed as CCT, aiming to enforce consistency across various perturbations applied to the encoder's outputs.

Specifically, CCT considers a shared encoder and a main decoder that are trained using the labeled samples. To fully capitalize on the potential of unlabeled samples, CCT then considers multiple auxiliary decoders that analyze perturbed variations of the output generated by the shared encoder. The consistency is imposed between the main decoder's predictions and that of the auxiliary decoders (refer to Fig. 7.2). In this manner, the representation produced by the shared encoder is augmented through the incorporation of supplementary training signals derived from the unlabeled data. The auxiliary decoders introduced possess a minimal number of parameters in comparison to the encoder.

Problem Definition. Let $\mathcal{D}_l = \{(\mathrm{x}_1^l, y_1), \ldots, (\mathrm{x}_n^l, y_n)\}$ denotes the set of n labeled examples, while $\mathcal{D}_u = \{\mathrm{x}_1^u, \ldots, \mathrm{x}_m^u\}$ signifies the collection of m unlabeled examples. Here, x_i^u represents the i-th unlabeled input image, and x_i^l denotes the i-th labeled input image, both possessing spatial dimensions of $H \times W$. The corresponding pixel-level label for each labeled image is denoted as $y_i \in \mathbb{R}^{H \times W \times C}$, where C indicates the number of classes.

The goal of CCT is to leverage the significantly larger pool of unlabeled examples (where $n \ll m$) to facilitate the segmentation network f, for performing effectively on the test dataset drawn from the same distribution as the training dataset. CCT comprises a shared encoder h and a primary decoder g, together forming the segmentation network $f = g \circ h$. Additionally, CCT introduces a set of K auxiliary decoders g_a^k, where $k \in [1, K]$. While the segmentation network f is trained on the labeled dataset $\mathcal{D}_l$ using a conventional supervised approach, the auxiliary networks $g_a^k \circ h$ are trained on the unlabeled dataset $\mathcal{D}_u$ by enforcing consistency in predictions between the main decoder and the auxiliary decoders. Each auxiliary decoder pro-

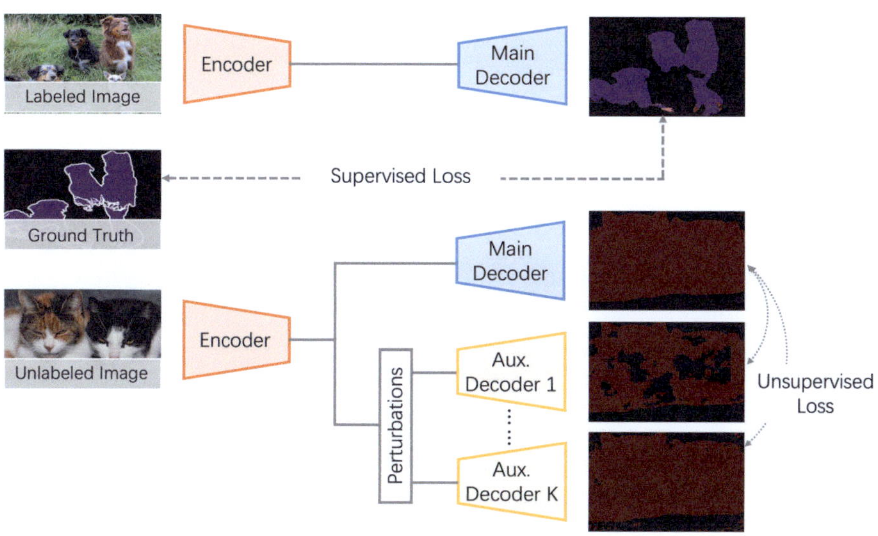

Fig. 7.2 An overview of CCT

cesses a perturbed version of the encoder's output, while the main encoder receives the unaltered intermediate representation. This methodology enhances the representation learning capabilities of the encoder h through the utilization of unlabeled examples, thereby improving the overall performance of the segmentation network f.

Supervised Loss. For the labeled data x_i^l with corresponding pixel-level ground truth y_i, a Cross-Entropy (CE) loss L_s is employed, which is instrumental in quantifying the discrepancy between the predicted segmentation outputs $f(\mathrm{x}_i^l)$ generated by the network and the actual ground truth labels. The process can be formulated as:

$$L_s = \frac{1}{|\mathcal{D}_l|} \sum_{\mathrm{x}_i^l, y_i \in \mathcal{D}_l} \mathrm{H}(y_i, f(\mathrm{x}_i^l)) \tag{7.10}$$

where $\mathrm{H}(\cdot)$ is represented as CE.

Unsupervised Loss. For the unlabeled example x_u^i, CCT derives an intermediate representation of the input through the shared encoder, denoted as $\mathrm{z}_i = h(\mathrm{x}_i^u)^*$. Then, CCT introduces R stochastic perturbation functions, labeled as p_r for $r \in [1, R]$, with each perturbation function potentially linked to multiple auxiliary decoders. By employing a variety of perturbation configurations, K perturbed versions of the intermediate representation $\widetilde{\mathrm{z}}_i^k$ are generated, where the k-th perturbed version is specifically allocated to the k-th auxiliary decoder. To maintain consistency, CCT incorporates the perturbation function as an essential element of the auxiliary decoder, with expressing g_a^k as $g_a^k \circ p_r$. Thus, the training objective is to minimize the unsupervised loss L_u, aiming to measure the divergence between the outputs of the main decoder and those of the auxiliary decoders, which can be formulated as:

$$L_u = \frac{1}{|\mathcal{D}_u|} \frac{1}{K} \sum_{x_i^u \in \mathcal{D}_u} \sum_{k=1}^{K} \mathrm{d}(g(z_i), g_a^k(z_i)) \tag{7.11}$$

where $\mathrm{d}(\cdot)$ denotes the distance metric that quantifies the divergence between two output probability distributions, specifically the results obtained from applying a softmax function across the channel dimension. In this study, CCT has opted to utilize Mean Squared Error (MSE) as their chosen distance measure.

Combined Loss. The overall loss L for CCT in the semi-supervised setting can be calculated as follows:

$$L = L_s + \omega_u L_u \tag{7.12}$$

where ω_u serves as a weighting function for the unsupervised loss.

7.4.1.2 Consistency Over Distinct Initialized Networks

In contrast to the CCT method, which relies on multiple, structurally distinct decoders, Chen et al. propose a consistency-based regularization technique called cross pseudo supervision (CPS). Rather than architectural variation, CPS employs two segmentation networks with identical architectures but different random initializations. Both labeled and unlabeled images are input to these parallel networks. For labeled data, each network is trained individually using the corresponding ground-truth segmentation masks. For each image, both networks generate predictions referred to as pseudo segmentation maps; these pseudo labels are then leveraged to supervise the counterpart network, creating a cross-supervision dynamic.

This cross pseudo supervision strategy offers two main advantages. Firstly, akin to other consistency regularization frameworks, CPS encourages agreement between network predictions for the same image, pushing decision boundaries into low-density regions and enhancing generalization. Secondly, as optimization progresses, the pseudo segmentation outputs become increasingly reliable, often surpassing the accuracy achievable through standard supervised learning limited to ground-truth data. Consequently, the pseudo labels effectively supplement the training set, allowing the model to learn from a broader data distribution and achieve superior segmentation performance.

Problem Definition. Consider two image datasets: a labeled set D_l containing N annotated samples, and an unlabeled set D_u with M samples lacking annotations. The goal of semi-supervised semantic segmentation is to develop a segmentation model that effectively leverages both the annotated and unannotated data to improve performance.

To address this challenge, CPS method employs two segmentation networks operating in parallel:

$$\mathrm{P}_1 = f(\mathrm{X}; \theta_1) \tag{7.13}$$

$$\mathrm{P}_2 = f(\mathrm{X}; \theta_2) \tag{7.14}$$

Both segmentation networks share an identical architecture, but their parameters, denoted as θ_1 and θ_2, are initialized independently. Each network receives the same augmented input X, and generates a segmentation confidence map—P_1 for the first network and P_2 for the second—obtained via softmax normalization of the output logits. The overall framework of the proposed approach is depicted as follows:

$$\begin{aligned} \mathrm{X} \rightarrow \mathrm{X} \rightarrow\ & f(\theta_1) \rightarrow \mathrm{P}_1 \rightarrow \mathrm{Y}_1 \\ \searrow\ & f(\theta_2) \rightarrow \mathrm{P}_2 \rightarrow \mathrm{Y}_2 \end{aligned} \tag{7.15}$$

In this formulation, Y_1 and Y_2 represent the predicted one-hot label maps, serving as pseudo segmentation maps. For each spatial location i, the label vector y_{1i} (or

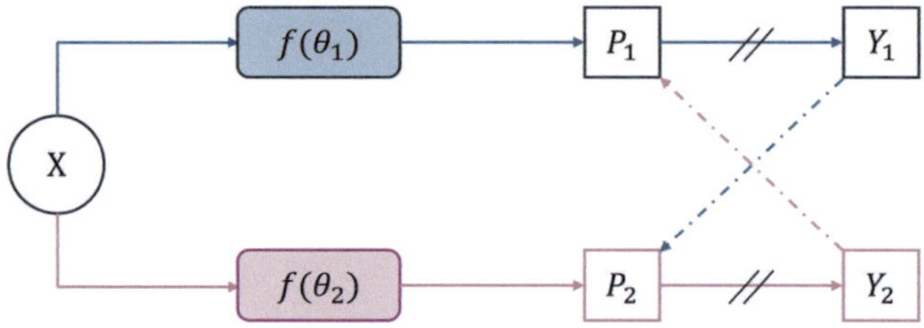

Fig. 7.3 An overview of CPS

y_{2i}) is constructed as a one-hot encoding derived from the corresponding confidence vector p_{1i} (or p_{2i}). Fig. 7.3 provides a comprehensive illustration of the method; for clarity, the supervision via loss functions is omitted from the Eq. (7.15).

Supervised Loss. To ensure effective learning from labeled data, the supervised loss $\mathcal{L}_s$ is defined via a standard pixel-wise cross-entropy function, applied independently to predictions from both segmentation networks. The loss is calculated over the entire set of labeled images as follows:

$$\mathcal{L}_s = \frac{1}{|\mathcal{D}_l|} \sum_{X \in \mathcal{D}_l} \frac{1}{W \times H} \sum_{i=0}^{W \times H} (l_{ce}(p_{1i}, y_{1i}^*) + l_{ce}(p_{2i}, y_{2i}^*)) \tag{7.16}$$

Here, l_{ce} denotes the cross-entropy loss, while p_{1i} and p_{2i} are the predictions from network 1 and network 2 for the i-th pixel, matched against their respective ground-truth annotations y_{1i}^* and y_{2i}^*. W and H correspond to the width and height of the input image.

Unsupervised Loss. To harness unlabeled data, the cross pseudo supervision mechanism introduces a bidirectional consistency loss between networks. For each unlabeled image, one network produces a one-hot pseudo-label map, which is then used to supervise the pixel-wise output of the other network, and vice versa. The unsupervised CPS loss is therefore given by:

$$\mathcal{L}_{cps}^u = \frac{1}{|\mathcal{D}_u|} \sum_{X \in \mathcal{D}_u} \frac{1}{W \times H} \sum_{i=0}^{W \times H} (l_{ce}(p_{1i}, y_{2i}) + l_{ce}(p_{2i}, y_{1i})) \tag{7.17}$$

This approach is analogously extended to the labeled set, generating a cross pseudo supervision loss $\mathcal{L}_{cps}^l$ based on labeled data. The overall CPS loss aggregates both components:

$$\mathcal{L}_{cps} = \mathcal{L}_{cps}^l + \mathcal{L}_{cps}^u \tag{7.18}$$

Combined Loss. The final optimization objective incorporates both the supervised and cross pseudo supervision losses. This unified objective is formulated as:

$$\mathcal{L} = \mathcal{L}_s + \lambda \mathcal{L}_{cps} \tag{7.19}$$

where λ serves as a balancing parameter to modulate the influence of the cross pseudo supervision term during training.

7.4.1.3 Consistency Over Distinct Structured Networks

Deep learning approaches utilizing Convolutional Neural Networks (CNNs) and Transformers have demonstrated promising results in fully supervised medical image segmentation. However, achieving strong performance with limited training annotations remains a significant challenge. As illustrated in Fig. 7.4, the CTCT [15] framework offers a straightforward yet effective solution for semi-supervised medical image segmentation by leveraging cross-teaching between CNNs and Transformers. CNNs and Transformers represent distinct paradigms for visual recognition—CNNs depend on local convolutional operations, while Transformers utilize long-range self-attention mechanisms. Consequently, their predictions inherently differ in their output characteristics.

For an input image x_i, the proposed framework produces two predictions: $p_i^c = f_\phi^c(x_i)$ and $p_i^t = f_\phi^t(x_i)$, where p_i^c and p_i^t represent the prediction of a CNN $f_\phi^c(\dot{)}$ and a Transformer $f_\phi^t(\dot{)}$ respectively. Based on the predictions, the pseudo labels for the cross teaching strategy are generated by this way:

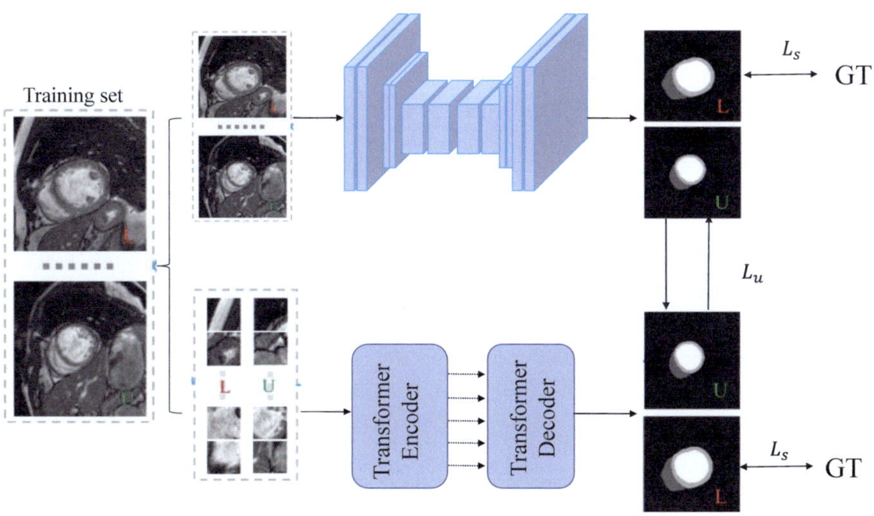

Fig. 7.4 An overview of CTCT

$$\begin{aligned} pl_i^c &= argmax(p_i^t) = argmax(f_\phi^t(x_i)) \\ pl_i^t &= argmax(p_i^c) = argmax(f_\phi^c(x_i)) \end{aligned} \tag{7.20}$$

where pl_i^c and pl_i^t are generated pseudo labels for the CNN and the Transformer respectively. It's worthy to point that pl_i^c and pl_i^t are pseudo segmentation results, and there is no gradient back-propagation between p_i^c and pl_i^c, and between p_i^t and pl_i^t in each mini-batch. Then, the cross teaching loss for the unlabeled data is defined as

$$L_{ctl} = \underbrace{L_{dice}(p_i^c, pl_i^c)}_{supervision\ for\ CNNS} + \underbrace{L_{dice}(p_i^t, pl_i^t)}_{supervision\ for\ trans} \tag{7.21}$$

The overall training objective function is a joint loss with two parts, a supervised loss on the labeled data and an unsupervised loss for the unlabeled data.

$$\begin{aligned} L_{sup} &= L_{ce}(p_i, y_i) + L_{dice}(p_i, y_i) \\ L_{total} &= L_{sup} + \lambda L_{ctl} \end{aligned} \tag{7.22}$$

where L_{ce}, L_{ce} are the cross-entropy loss and dice loss, respectively. p_i and y_i represent the prediction and label of image x_i. λ is a weight factor, which is defined by a time-dependent Gaussian warming up function commonly: $\lambda(t) = 0.1 \cdot e^{(-5(1-\frac{t_i}{t_{total}})^2)}$, where t_i denotes the current training iteration and t_{total} is the total iteration number.

7.4.2 Mean-Teacher Based Semi-Supervised Learning

7.4.2.1 Mean-Teacher Based Baseline Method

Mean-Teacher based methods [12] has demonstrated its effectiveness and shown promising results in various vision tasks, such as segmentation, classification, and object detection. By leveraging both labeled and unlabeled data, the algorithm improves segmentation accuracy and enhances the generalization capabilities of deep learning models. This is achieved by enforcing consistency between predictions generated on different variations of the same input image. By training both the Teacher and Student models, the algorithm effectively utilizes information from unlabeled data, improving the model's ability to generalize to unobserved samples.

Specifically, as depicted in the Fig. 7.5, the input image is imposed with two different noises η and η' for the computation of the student and the teacher models, respectively. For the labeled data, the output of the student model is compared with the ground truth label using the classification loss, and with the teacher prediction by applying the consistency loss. After updating the weights of the student model using gradient descent, an exponential moving average scheme is utilized to update the weights of the teacher model. For the unlabeled data, the optimization during

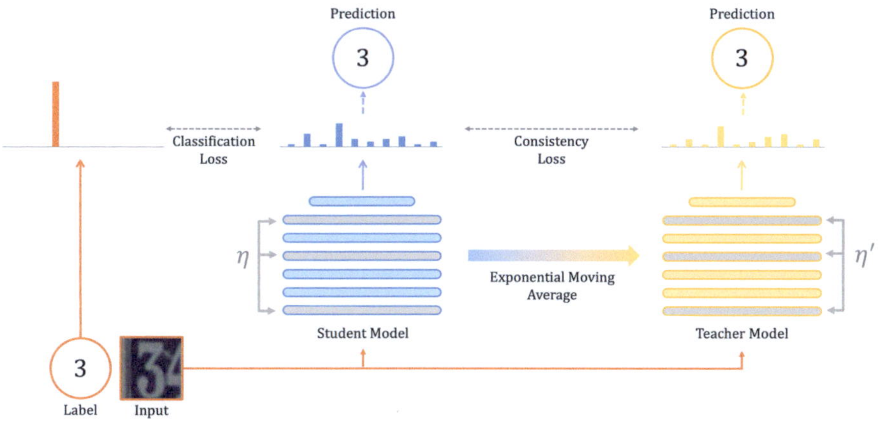

Fig. 7.5 Architecture of mean-teacher semi-supervised classification scheme

the training phase is similar, except that no supervised classification loss is applied. This section discusses the application and research of the Mean-Teacher methods in semi-supervised medical image segmentation.

7.4.2.2 Mean-Teacher Based Method with Uncertainty

In this section, we focus on semi-supervised methods based on Mean-Teacher (MT) learning for application in the context of 3D medical image segmentation. The most well-known work in this area is Uncertainty-Aware (UA) MT [13], an novel semi-supervised learning framework based on the MT scheme for left atrium segmentation. Following the same principles as the MT scheme [12], UA-MT comprises a teacher model and a student model. The teacher model guides the student model by minimizing two types of losses. As depicted in Fig. 7.6, one loss is the segmentation loss L_s on the labeled data, and the other is the consistency loss L_c on all input data, which aims to ensure consistency between the predictions from the student model and the targets provided by the teacher model under different perturbations. Considering that the predicted target from the teacher model may be unreliable and noisy, the authors introduce the uncertainty scheme to enable the student model to gradually derive knowledge from the meaningful and reliable regions by capturing the uncertainty-aware signals from the teacher model. The following section will elaborate on training details of UA-MT.

Overall Loss Function: In the semi-supervised setting, there are generally two kinds of data: the labeled set $\mathcal{D}_L = \{(x_i, y_i)\}_{i=1}^{N}$ and the unlabeled set $\mathcal{D}_U = \{(x_i)\}_{i=N+1}^{N+M}$, where the input volume x_i and the ground truth $y_i \in \{0, 1\}$ has the same size of $H \times W \times D$. The overall loss function is designed by simultaneously minimizing a segmentation loss L_s and a consistency loss L_c, which can be formulated as:

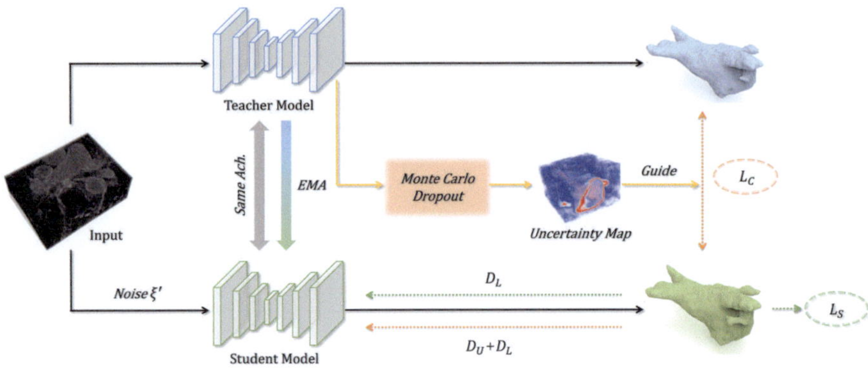

Fig. 7.6 The pipeline of uncertainty-aware scheme for semi-supervised medical image segmentation

$$\min_{\theta} \sum_{i=1}^{N} L_s(f(x_i;\theta), y_i) + \lambda \sum_{i=1}^{N+M} L_c(f(x_i;\theta', \xi'), f(x_i;\theta, \xi)) \tag{7.23}$$

where $f(\cdot)$ represents the segmentation network, (θ, ξ) and (θ', ξ') denote the model parameters and different perturbation operations (such as imposing noise to input and applying dropout to network) of the student and teacher models, respectively. λ is a coefficient that balances the supervised L_s (i.e., cross-entropy loss) and unsupervised loss L_c. To prevent significant loss of previously learned information, we apply an exponential moving average (EMA) to consolidate insights from various stages of the training process. This can be formulated as:

$$\theta'_t = \alpha\theta'_{t-1} + (1-\alpha)\theta_t \tag{7.24}$$

where α is the decay parameter for the exponential moving average (EMA), which determines the rate of updating.

Uncertainty-Aware Consistency Loss: Motivated by Bayesian networks, the authors apply Monte Carlo Dropout [6] to estimate uncertainty. Through T stochastic forward passes on the teacher model, employing random dropout and Gaussian noise for each input, a collection of probability vectors $\{p_t\}_{t=1}^{T}$ is generated, and the entropy can be computed as follows:

$$\mu_c = \frac{1}{T}\sum_t p_t^c \quad \text{and} \quad u = -\sum_c \mu_c \log \mu_c \tag{7.25}$$

where p_t^c is the probability of the c-th class at the t-th time prediction. Benefiting from the estimated uncertainty U, the reliable region with relatively lower uncertainty can be selected, ensuring reliable and meaningful guidance. Subsequently, the uncertainty-aware consistency loss L_c is designed as the voxel-level mean squared error (MSE) loss, which maintains coherence between the teacher and student models in the most certain regions:

$$L_c(f', f) = \frac{\sum_v I(u_v < H) \left\| f'_v - f_v \right\|^2}{\sum_v I(u_v < H)} \tag{7.26}$$

where $I(\cdot)$ denotes the indicator function. f'_v and f_v are the predictions of teacher and student model at the v-th voxel, respectively. u_v represents the corresponding uncertainty and H is a threshold to control the selection of the most certain regions.

7.4.2.3 Mean-Teacher Based Method with Joint CNN-Transformer Network

Semi-supervised learning offers an efficient approach to leverage large-scale data without annotations and reduce reliance on labeled data. The Mean-Teacher (MT) scheme, employing a teacher-student model architecture, has proven effective in semi-supervised medical image segmentation. Here, the student network learns from the teacher by minimizing pixel-wise consistency loss. However, existing MT-based SSL methods still face two primary concerns: (1) Limited learning ability of the student network: Current approaches often overlook the integration of local features and global cues extraction, which are crucial for robust representation learning of diverse objects. (2) Limited knowledge-transferring ability of the teacher network: Teachers typically rely solely on pixel-level consistency regularization, which may lead to insufficient and unstable guidance for the student.

Addressing these concerns is crucial for advancing semi-supervised medical image segmentation methods, ensuring more comprehensive and stable learning from both local and global contexts. To overcome these challenges, MedFCT [16] introduces two innovative approaches: (1) Dual Student Architecture: MedFCT employs a dual student architecture featuring parallel CNN and Transformer branches. This design enables simultaneous extraction of local and global features. The interaction between CNN and Transformer is enhanced using a frequency domain cross-fusion (FDCF) module. This module facilitates exploration of the complementary aspects of both paradigms, leveraging their respective strengths effectively. (2) Comprehensive Multi-level Consistency Regularization: MedFCT introduces a multi-level consistency regularization approach that considers pixel-wise, feature-wise, and class-wise information. This regularization scheme aims to enhance the guidance and knowledge transfer from the teacher network to the student network. By incorporating these multiple levels of consistency, MedFCT improves the stability and effectiveness of semi-supervised learning in medical image segmentation.

Frequency Domain Cross-Fusion Module: As shown in Fig. 7.7, given two feature maps from CNN branch ($f_{cnn} \in \mathbb{R}^{D \times H \times W}$) and Transformer branch ($f_{trans} \in \mathbb{R}^{D \times H \times W}$) in spatial domain, the first step is to convert them into frequency domain. It can be defined as:

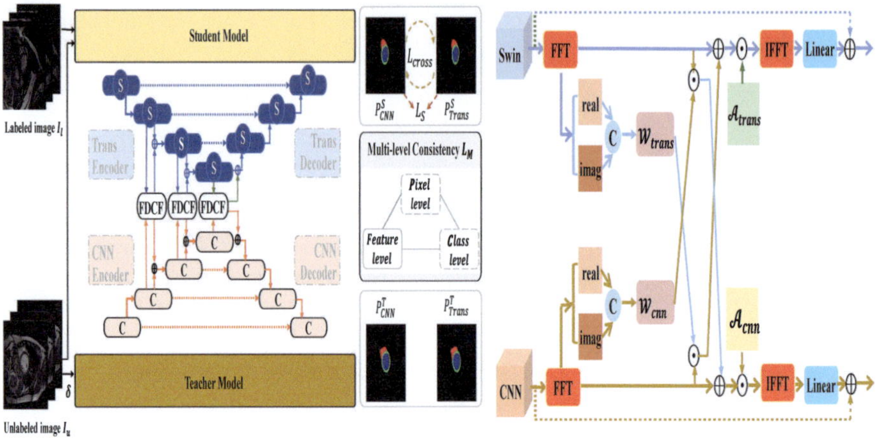

Fig. 7.7 An overview of MedFCT. **a** The overall architecture of MedFCT. **b** The design of frequency domain cross-fusion module

$$\mathcal{F}(u, v) = \sum_{x=0}^{H-1} \sum_{y=0}^{W-1} f(x, y) e^{-i2\pi(\frac{ux}{H} + \frac{vy}{W})} \tag{7.27}$$

where $f(x, y)$ is the pixel value at position (x, y) of an image in spatial domain, while (u, v) is the coordinate of a spectrum in frequency domain. $\mathcal{F}(u, v)$ is the complex frequency value, which depends on all the image pixels in spatial domain. To efficiently compute 2D DFT, the author utilizes the a faster version, 2D fast Fourier transform (**FFT**), to obtain the converted $\mathcal{F}_{cnn}$ and $\mathcal{F}_{trans}$, which can be defined as:

$$\mathcal{F}_{trans} = FFT(f_{trans}); \; \mathcal{F}_{cnn} = FFT(f_{cnn}) \tag{7.28}$$

Benefitting from the conjugate symmetric property of the Fast Fourier Transform (FFT), only half of the spatial dimensions are retained, reducing the resolution of $\mathcal{F}_{cnn}$ (and $\mathcal{F}_{trans}$) to $H \times \lceil W/2 \rceil \times D$. Taking $\mathcal{F}_{cnn}$ as an example, the author concatenates its **real** and **imaginary** parts along the channel dimension to form $\mathcal{F}'_{cnn} = \left[\mathcal{F}^{real}_{cnn}, \mathcal{F}^{imag}_{cnn}\right]$. Next, $\mathcal{F}'_{cnn}$ is passed through a learnable weight module $\mathcal{W}$ to generate the global weight, formulated as:

$$\mathcal{W}_{cnn} = \sigma(Conv_{1\times1}(RELU(BN(Conv_{3\times3}(\mathcal{F}'_{cnn}))))) \tag{7.29}$$

Finally a sigmoid activation σ is applied to generate the global weight $\mathcal{W}_{cnn}$. Similarly, the global weight $\mathcal{W}_{trans}$ can be obtained in the same way.

To further learn the complementary information from two branches, the learned global weight $\mathcal{W}_{trans}$ or $\mathcal{W}_{cnn}$ is leveraged to multiply with the converted spectrum

$\mathcal{F}_{cnn}$ / $\mathcal{F}_{trans}$. The updated spectrum is further added with corresponding original one to using a residual path:

$$\begin{aligned} \hat{\mathcal{F}}_{cnn} &= (\mathcal{W}_{cnn} \odot \mathcal{F}_{trans} + \mathcal{F}_{cnn}) \\ \hat{\mathcal{F}}_{trans} &= (\mathcal{W}_{trans} \odot \mathcal{F}_{cnn} + \mathcal{F}_{trans}) \end{aligned} \tag{7.30}$$

This process can be interpreted as follows: one branch initially generates a spatial mask using its own features. This mask is then multiplied with the features from another branch, facilitating the extraction of complementary frequency information. To further mitigate redundant and noisy frequency components, a learnable parameter matrix $\mathcal{A} \in \mathbb{R}^{1\times H \times \lceil \frac{W}{2} \rceil}$ is employed to modulate the enhanced spectrum of each branch. This modulation can be defined as:

$$\tilde{\mathcal{F}}_{cnn} = \hat{\mathcal{F}}_{cnn} \odot \mathcal{A}_{cnn}; \ \tilde{\mathcal{F}}_{trans} = \hat{\mathcal{F}}_{trans} \odot \mathcal{A}_{trans} \tag{7.31}$$

The final step of our Frequency-Domain Class Feature (FDCF) approach is to convert the enhanced spectrum back to the spatial domain. This reverse operation of Eq. (7.27), 2D inverse DFT can be formulated as:

$$f(x, y) = \frac{1}{H \cdot W} \sum_{u=0}^{H-1} \sum_{v=0}^{W-1} \mathcal{F}(u, v) e^{i2\pi(\frac{ux}{H} + \frac{vy}{W})} \tag{7.32}$$

Accordingly, the 2D inverse FFT (**IFFT**) is utilized for simpler computation. After IFFT, the converted feature in the spatial domain is further processed through a linear layer ($Linear$) for channel mixing. Then, it is added to the original input features f_{cnn} and f_{trans} to obtain the enhanced features in the spatial domain $\tilde{f}_{cnn}$ and $\tilde{f}_{trans}$. This can be formulated as:

$$\begin{aligned} \tilde{f}_{cnn} &= Linear(IFFT(\tilde{\mathcal{F}}_{cnn})) + f_{cnn} \\ \tilde{f}_{trans} &= Linear(IFFT(\tilde{\mathcal{F}}_{trans})) + f_{trans} \end{aligned} \tag{7.33}$$

Since the fast version of Fourier transform is utilized for concrete implementation, the proposed FDCF has the complexity of $O(NlogN)$, where $N = H \times W$. Moreover, FDCF possesses a global receptive field capable of capturing both long-term and short-term interactions. This makes it more effective compared to self-attention mechanisms designed specifically for modeling long-distance dependencies, which typically have a complexity of $O(N^2)$. Additionally, FDCF offers flexibility because the key FFT (Fast Fourier Transform) and IFFT (Inverse Fast Fourier Transform) operations do not require learnable parameters. As a result, FDCF can process images of arbitrary sizes efficiently.

Multi-level Consistency Regularization: In addition to the typical pixel-level constraints, ClassFormer introduces consistency constraints at both the feature-level and class-level. As shown in Fig. 7.8, intermediate representations from the teacher net-

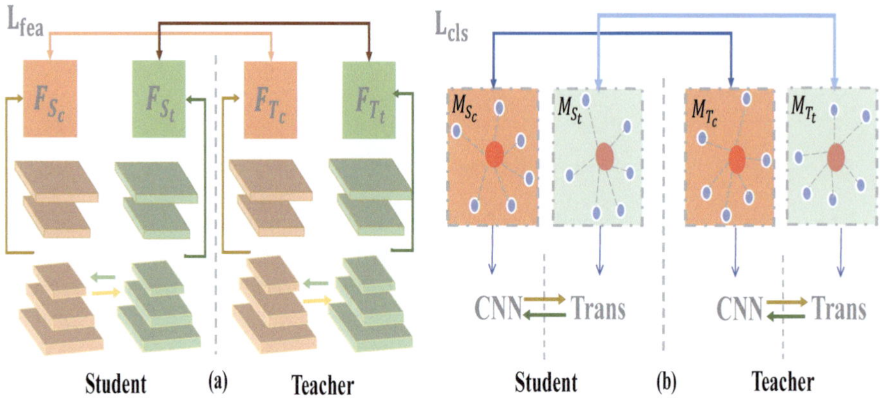

Fig. 7.8 Illustration of **a** Feature-level consistency **b** Class-level consistency. In **b** pink circle denotes class center G_c of c-th class, while blue circles represent features in feature map F

work can enhance the training process and improve the final performance of the student network. Particularly in deeper networks, relying only on pixel-level prediction consistency at the final stage of the teacher network may not sufficiently guide the earlier layers of the student network. To assist the student in learning a better initial position in the feature space, feature-level consistency is proposed using mean square error loss (L_{mse}):

$$L_f = \frac{1}{2} L_{mse}(F_{S_{cnn}}, F_{T_{cnn}}) + \frac{1}{2} L_{mse}(F_{S_{trans}}, F_{T_{trans}}) \tag{7.34}$$

where $F_{S_{cnn}}$ ($F_{T_{cnn}}$) and $F_{S_{trans}}$ ($F_{T_{trans}}$) denotes the feature from deepest encoder layer of CNN and Transformer branch from student (teacher).

As shown in Fig. 7.8, to transfer the important class distribution from the teacher to the student, the author first extracts the output of the deepest encoder layer from the parallel branches ($F_{S_{cnn}}$ and $F_{S_{trans}}$ from the student, $F_{T_{cnn}}$ and $F_{T_{trans}}$ from the teacher). The class distribution $\mathcal{M}$ is defined as the cosine similarity between each feature $F(i) \in \mathbb{R}^{1\times D}$ in the feature map F and the class centers $G \in \mathbb{R}^{C\times D}$, where C is the number of classes. The class center for each class can be formulated as:

$$G_c = \sum_{i \in F} P_i^c F(i) \tag{7.35}$$

where P is the prediction map. Then, $\mathcal{M}$ can be represented as:

$$\mathcal{M}(i) = \cos(F(i), G) \in \mathbb{R}^{1\times C} \tag{7.36}$$

The goal is to make the student's class distribution mimic the teacher's class distribution. This class-wise consistency can be formulated as:

$$L_{cls} = L_{mse}(\mathcal{M}_{T_{trans}}, \mathcal{M}_{S_{trans}}) + L_{mse}(\mathcal{M}_{T_{cnn}}, \mathcal{M}_{S_{cnn}}) \tag{7.37}$$

7.5 Pseudo-Labeling Techniques

7.5.1 Pseudo-Labeling Baseline Techniques

In the most direct form of pseudo-labeling, the model is trained on a small set of labeled data. Once the initial model is trained, it is used to predict labels for the unlabeled data. These predicted labels, known as pseudo-labels, are then treated as true labels, and the model is retrained using both the original labeled data and the newly pseudo-labeled data. This process can be iterated multiple times to progressively improve the model's performance.

For example, Anthimopoulos et al. introduce a deep purely convolutional network specifically designed to tackle the challenge of lung tissue semantic segmentation [17]. By carefully designing the network architecture, they aim to enhance the quality of pseudo-labels, thereby improving the overall performance and reliability of the segmentation process. This network is constructed entirely from convolutional layers that employ dilated kernels to expand the receptive field, thereby avoiding the downsampling of feature maps. Such an architecture has proven effective for similar dense prediction tasks that demand high-resolution precision.

Network structure. The proposed network, as depicted in Fig. 7.9, consists of 13 convolutional layers and achieves a total receptive field of 287 × 287. Specifically, each of the first ten layers contains 32 kernels of size 3 × 3 with dilation rates of 1, 1, 2, 3, 5, 8, 13, 21, 34, and 55, respectively. The author deliberately refrains from exponentially increasing the dilation rates, a common practice, to avoid the extreme gridding issues that have been previously documented. Instead, the first terms of the Fibonacci sequence are utilized as dilation rates, which alleviates the gridding problem by providing a more gradual increase in dilation rates, thus enabling denser sampling.

The output from the first ten layers, is concatenated with the network's input, resulting in a total of 1 + 10 × 32 = 321 feature maps. These feature maps are sub-

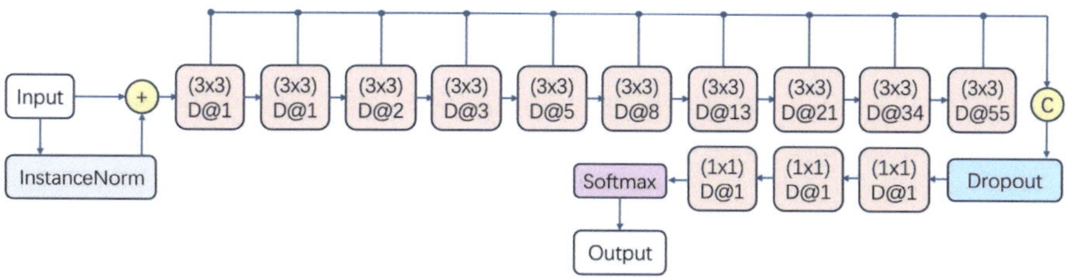

Fig. 7.9 An overview of Pseuso-labeling baseline techniques

sequently processed through a dropout layer with a rate of 0.5 before being directed to the remainder of the network. This concatenation is facilitated by the absence of pooling layers and the appropriate zero padding applied to each convolution, yielding several advantages. It allows for the aggregation of features across various scales and levels of abstraction while enhancing the gradient flow throughout the network, thereby promoting faster training.

The final three layers utilize 1×1 kernels and function as locally dense layers, reducing the feature dimensionality for each pixel from 321 to 128, then to 32, and ultimately to 6, which corresponds to the number of classes considered. The output is transformed into a probability distribution using the softmax function.

A batch normalization (BN) layer follows each convolution, relying on batch statistics during both training and testing phases. This approach is feasible since the batch size is set to one, ensuring a complete batch is available during inference. This methodology has been previously recognized under the term instance normalization, which has demonstrated effective performance in various applications. Instance normalization provides invariance to intensity and contrast variations, making the features adaptable for each slice and potentially mitigating issues arising from different CT scanners.

Additionally, the author observes that incorporating the normalized activations with the non-normalized ones prior to passing them through the ReLU function, as illustrated in Fig. 7.4, significantly enhances the results. This instance normalization skip connection effectively cancels the mean normalization of activations when the trainable parameters remain untrained, while simultaneously performing a form of feature contrast enhancement that reduces the impact of variance shifts without achieving complete invariance to them.

Semi-supervised Strategy. The dense nature of the classification problem under consideration, coupled with the limited availability of annotations, has led to two significant challenges. Firstly, substantial portions of the dataset remain unannotated, rendering them unusable for both supervised training and testing. Secondly, the distribution of the classes within the dataset is highly imbalanced, which poses difficulties for any classification method.

To address these issues, the author implemented a strategy that involves scaling the loss and accuracy metrics using appropriate weighting schemes calculated for each dataset subset. Each pixel corresponding to an annotated area was assigned a weight that is inversely proportional to the number of samples belonging to its respective class within that subset. This approach ensures that all classes contribute equally to the evaluation metrics.

Additionally, a semi-supervised learning technique was employed to further leverage the unannotated areas of the data. An extra term was incorporated into the supervised loss function, representing the entropy of the network's output in regions not involved in supervised learning. This entropy minimization technique has been shown to yield significant performance improvements in various applications.

Furthermore, the concept of pseudo-labeling was introduced, wherein the network classifies non-annotated regions and subsequently uses these classifications as ground truth for fine-tuning. Semi-supervised learning techniques of this nature are grounded in the cluster assumption, which posits that samples from the same class tend to form compact clusters. By minimizing the entropy of the network's output, the decision boundaries are pushed away from areas densely populated by labeled samples. If the cluster assumption holds true and there is minimal overlap between classes, this method can enhance the network's generalization capabilities.

This approach functions similarly to manifold learning and encompasses self-learning as a specific case, as it increases the classifier's confidence. The influence of the semi-supervised term is regulated by an appropriate weight, which is scaled relative to the proportion of unlabeled regions compared to annotated ones. Consequently, the loss for a pixel x with output $\hat{y}$ can be expressed as follows:

$$L(x, \hat{y}) = \begin{cases} -\sum_{i=1}^{C} w_i^s y_i \log(\hat{y}_i), & \text{when } y \text{ is } given \\ -\alpha w_u \sum_{i=1}^{C} \hat{y}_i \log(\hat{y}_i), & \text{otherwise} \end{cases} \tag{7.38}$$

In this context, let y represent the true label in one-hot encoding, C denote the number of classes, w_i^s signify the supervised weight for class i (which is inversely proportional to the number of samples in that class), α be a scaling factor, and w_u indicate the unsupervised weight.

The training process is designed to terminate when the network's performance on the validation set does not show significant improvement over a span of 50 epochs. Performance evaluation is conducted using weighted (balanced) accuracy, with a significant improvement defined as a relative increase of at least 0.5%.

To artificially augment the training dataset and mitigate the risk of overfitting, the author employed transformations on the images, including flips and rotations, which are deemed label-preserving within this domain. This augmentation process is executed online; for each training image in every epoch, one operation from a set of eight possible combinations of flips and rotations is randomly selected and applied.

7.5.2 *Pseudo-Labeling Techniques with High Confidence*

This approach investigates how to generate and select high-quality pseudo-labels for incorporation into the training process [18]. Figure 7.10 provides a comprehensive overview of the proposed methodology. The core concept involves simultaneously training an end-to-end segmentation network with both labeled and unlabeled lung regions, where the labels for the unannotated regions are derived from the annotated ones. In essence, this approach is compatible with nearly all existing end-to-end segmentation networks. The label propagation process comprises three key steps: calculating the center, generating a similarity map, and assigning pseudo-labels. The training procedure is detailed in the following sections.

Center Calculation. Given an image, $f(x, y)$ is denoted as the feature at the spatial coordinate (x, y) within the feature map. In this study, the final feature maps is lever-

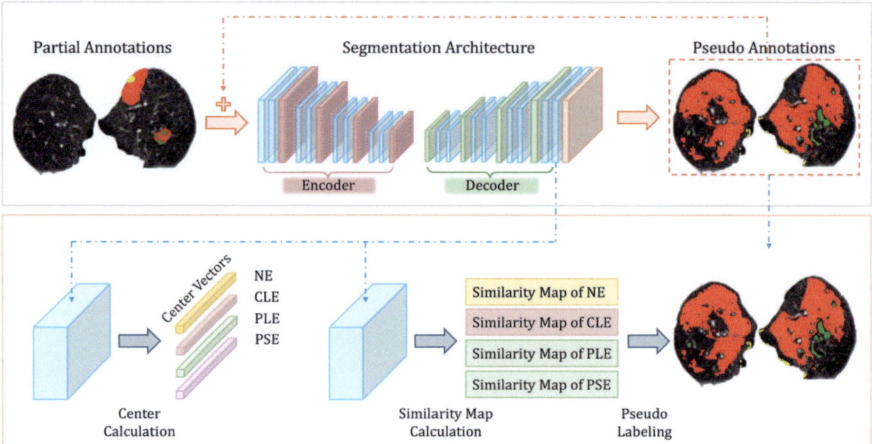

Fig. 7.10 An overview of Pseudo-labeling techniques with high confidence

aged to precede the logits. Typically, these feature maps retain the same dimensions as the input image, meaning that each pixel corresponds to a feature vector $f \in \mathbb{R}^d$, where d represents the number of channels. The term $C_k \in \mathbb{R}^d$ signifies the center of the k-th class in the learned feature space. For each image, the class centers are computed by averaging the features associated with their respective classes, which can be expressed mathematically as:

$$C_k = \frac{1}{N_k} \sum_{f \in D_k \cap L} f \tag{7.39}$$

where N_k denotes pixel number of class k, D_k indicates feature space of class k, and L is labeled regions.

Similarity Map Calculation. Let a_j^k denote the similarity between the j-th pixel and the k-th class center. Hence, a_j^k can be represented as:

$$a_j^k = \frac{f_j^T C_k}{\left\| f_j \right\|_2 \|C_k\|_2} \tag{7.40}$$

where f_j is the feature vector of j-th pixel. The similarity map for the k-th class is subsequently generated by computing a_j^k for every pixel in the input image. Figure 7.11 illustrates several visual representations of these similarity maps.

Pseudo Labeling. Pseudo labels act as target classes for unlabeled data, effectively treating them as if they were authentic labels. The technique of pseudo labeling is founded on the cluster assumption, which suggests that samples from the same category typically form compact clusters. the center of each cluster can be estimated via using the labeled regions. Intuitively, each pixel is associated with the cluster that

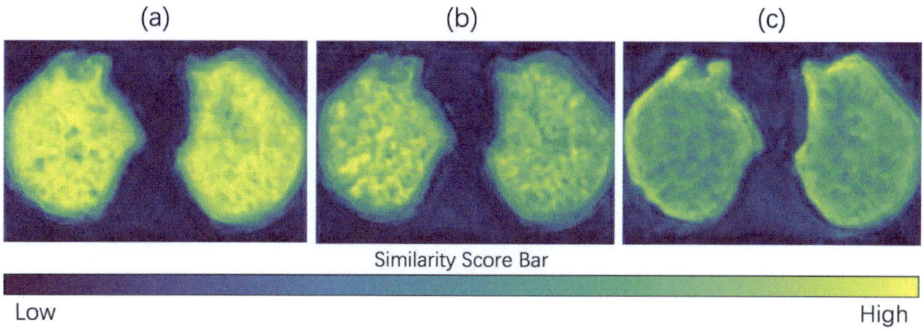

Fig. 7.11 Examples of similarity maps of NE **a**, CLE **b**, and PSE **c**

exhibits the highest similarity. Therefore, the author assigns $y_j{}'$ (the pseudo label) to each unlabeled pixel, which can be expressed mathematically as:

$$y_j{}' = argmax_k(a_j^k) \tag{7.41}$$

In this manner, every pixel in the unannotated areas will be assigned a pseudo label. However, certain unlabeled regions may not be suitable for model training due to their low-confidence (low-similarity) pseudo labels. To address this issue, the author focuses exclusively on confident pseudo-labeled regions (high-similarity areas) for training purposes. After computing the similarity map for each class, the author can readily identify the confident regions by applying a threshold (denoted as γ) to the similarity map. These confident regions are then utilized as masks to isolate the pseudo-labeled areas that exhibit high confidence.

Semi-Supervised Learning. Consider an input image x. Let y represent its partially annotated label map, y' denote the pseudo label map, and $\hat{y}$ signify the outputs generated by the segmentation model. It is important to note that both y and y' are expressed in one-hot encoding format. The semi-supervised loss proposed in this context can be formulated as follows:

$$L_{semi}(x, \hat{y}) = -\sum_i \sum_k \sum_{j \in L} y_{i,k,j} \log \hat{y}_{i,k,j} - \alpha \sum_i \sum_k \sum_{j \in U} y'_{i,k,j} \log \hat{y}_{i,k,j} \tag{7.42}$$

where i represents the index of the image, j denotes the index of the pixel, and k indicates the class index. Here, L refers to the labeled regions, while U signifies the confident pseudo-labeled regions that have been selected from the unlabeled areas. The term $\alpha(t)$ serves as a time-dependent scaling factor designed to balance the loss. This scaling factor can be mathematically expressed as:

$$\alpha(t) = \begin{cases} 0, & t < T_1 \\ \frac{t - T_1}{T_2 - T_1}, & T_1 \le t < T_2 \\ \lambda, & t \ge T_2 \end{cases} \tag{7.43}$$

In this context, t denotes the iteration index, T_1 marks the time point at which the unsupervised term is incorporated into the loss function, and T_2 indicates the time point when $\alpha(t)$ stabilizes. The parameter λ represents the weight assigned to the unsupervised loss. The function $\alpha(t)$ gradually increases from zero to λ. This implies that prior to T_1, the loss and the associated learning gradients are predominantly influenced by the first term, which relies solely on the labeled data. Following T_1, the unsupervised loss is integrated into the training process.

To facilitate effective optimization and prevent the model from becoming trapped in suboptimal local minima, it is essential that the ramp-up of the second term in the loss function occurs at a controlled pace. Specifically, $\alpha(t)$ is designed to increase gradually. Moreover, selecting an appropriate value for $\alpha(t)$ is critical for achieving optimal model performance. If $\alpha(t)$ is set too low, the model may fail to leverage the unlabeled data effectively; conversely, if it is set too high, the training process could be adversely affected. Based on empirical trials, T_1 and T_2 have been heuristically determined to be 5000 and 10,000 iterations, respectively.

Class imbalance is a prevalent challenge in the realm of medical image segmentation. In this study, weight balancing is utilized, a straightforward and widely adopted technique, to address the problem of class imbalance. Weight balancing adjusts the contribution of each training sample to the overall loss calculation by modifying their respective weights. By incorporating these additional weights, the proposed semi-supervised loss can be defined as follows:

$$L_{semi}(x, \hat{y}) = -\sum_{i}\sum_{k}\sum_{j \in L} \omega_k y_{i,k,j} \log \hat{y}_{i,k,j} - \alpha \sum_{i}\sum_{k}\sum_{j \in U} \omega'_k y'_{i,k,j} \log \hat{y}_{i,k,j} \tag{7.44}$$

where ω_k denotes the supervised balancing weight of class k and ω'_k represents the unsupervised balancing weight of class k. ω_k can be formulated as

$$\omega_k = \frac{1}{n_k} \tag{7.45}$$

where n_k indicates pixel number of class k in labeled areas. ω'_k can be formulated as

$$\omega'_k = \frac{1}{n'_k} \tag{7.46}$$

where n'_k represents the number of pixels belonging to class k within the selected unlabeled regions. Prior to initiating the semi-supervised training of our network, the author conducted multiple training sessions in a fully supervised manner to accurately estimate n'_k.

Fisher Loss. As previously highlighted, pseudo labeling serves as a pivotal component of our proposed semi-supervised architecture. In this phase, the author leverages the similarities in deeply learned features between labeled and unlabeled regions to facilitate label propagation, making the discriminative strength of these features essential for ensuring accurate label assignment. In this section, the author first introduces an effective loss function known as Fisher loss, which is designed to enhance the discriminative capability of the learned features. Subsequently, we will detail the integration of Fisher loss into our semi-supervised framework.

The fundamental principle behind our approach is to maximize the "inter-class distinction" while simultaneously bolstering "intra-class consistency". This dual focus is crucial for crafting a loss function that effectively enhances the discriminative power of the learned features. Inspired by this concept, the author proposes the Fisher loss function, which is formally defined in Eq. (7.47).

$$L_{Fisher} = \frac{\sum_k S(C_k, C)}{\sum_k \sum_f \in D_k} S(f, D_k) \tag{7.47}$$

where S denotes the cosine similarity (rather than distance) between two feature vectors, k represents the category index, and C signifies the centroid of all class centers, which can be expressed mathematically as follows:

$$C = \frac{1}{K} \sum_{K=1}^{K} C_K \tag{7.48}$$

where K denotes the total number of classes. The Fisher loss function compels the feature centroids of distinct categories to remain separated while simultaneously attracting the features of the same category towards their respective centroids.

Combination of Fisher Loss and the Proposed Semi-Supervised Architecture. By combining the Fisher loss with the semi-supervised loss, the overall loss function for training the segmentation network can be articulated as shown in Eq. (7.49).

$$\begin{aligned} L_{total}(x, \widehat{y}) = &- \sum_i \sum_k \sum_{j \in L} \omega_k y_{i,k,j} \log \hat{y}_{i,k,j} \\ &+ \beta \frac{\sum_i \sum_k S(C_{i,k}, C_i)}{\sum_i \sum_k \sum_{f \in D_k \cap L} S(f, C_{i,k})} \\ &- \alpha \sum_i \sum_k \sum_{j \in U} \omega'_k y'_{i,k,j} \log \hat{y}_{i,k,j} \end{aligned} \tag{7.49}$$

where β serves as the coefficient to balance the losses. The training methodology is detailed in Algorithm 1, with the only modification being the substitution of Eq. (7.44) with Eq. (7.49).

Figure 7.12 illustrates examples of similarity maps produced by different models: one trained without the Fisher loss (i.e., $\beta = 0$) and another trained with the Fisher

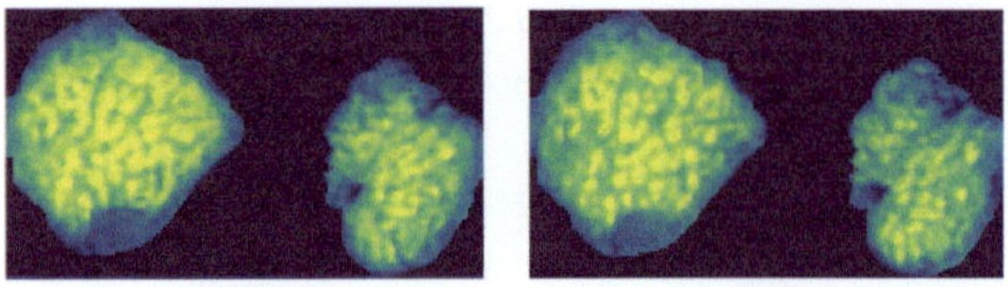

Fig. 7.12 The similarity maps generated from different models. **a** $\beta = 0$. **b** $\beta \neq 0$

loss (i.e., $\beta \neq 0$). It is evident that the model incorporating the Fisher loss yields clearer delineations between non-emphysema and emphysema tissues. This indicates an enhancement in the discriminative capability of the deep features.

7.6 Performance Validation

7.6.1 Dataset and Implementation Details

To validate the effectiveness of the existing semi-supervised segmentation methods, this subsection presents several experiments on two public datasets, including ACDC dataset for automated cardiac diagnosis [19], and ISIC dataset for skin lesion segmentation [20].

- **ACDC Dataset**: This dataset comprises 200 annotated short-axis cardiac cine-MR images from 100 patients, covering three organs: the left ventricle (LV), right ventricle (RV), and myocardium (MYO). We divided the dataset into 140 images from 70 patients for training and 60 images from 30 patients for validation. For unsupervised training, we evaluated our method using 68 images from 3 labeled patients and 136 images from 7 labeled patients out of the 70 training cases, with the remaining data being unlabeled.
- **ISIC Dataset**: This skin lesion segmentation dataset includes 2594 dermoscopy images categorized into 2 classes. We allocated 1838 images for training and the remaining 756 images for validation. For unsupervised settings, 3% (91 images) and 10% (181 images) of the training data were labeled.

In line with the methodology described in [15], we employed standard augmentation techniques to prevent overfitting, such as random cropping to a 224×224 patch, random rotation, and random flipping. To ensure a fair comparison, we maintained the same experimental settings and data splits as [15]. Specifically, all experiments were conducted with a batch size of 16, with half of the batch consisting of labeled data. We optimized our network using the SGD optimizer, training for a total of 30,000 iterations. The learning rate was initially set to 0.01.

7.6.2 *Experimental Results and Analysis*

We evaluate current state-of-art semi-supervised methods includes MT [12], UA-MT [13], EM [21], DCT [2], CCT [14], CPS [22], ICT [23], DAN [11], URPC [24], CTCT [15], SSNet [25], ICT-Med [26] and MedFCT [16], on 3% labeled and 10% labeled of the two datasets separately, and the rest of training samples are unlabeled data.

7.6.2.1 Quantitative Comparison

As demonstrated in Table 7.1, we performed an extensive comparison of existing semi-supervised learning (SSL) techniques using the ACDC dataset under two distinct settings. MedFCT consistently outperforms existing methods, achieving superior results in both area error metrics, with Dice Similarity Coefficient (DSC) scores of 72.6% and 88.5%, and boundary error metrics, with HD95 scores of 10.5 mm and 4.1 mm, for 3 and 7 labeled data points respectively.

Table 7.2 showcases the performance of MedFCT model on the ISIC dataset, where it continues to excel, surpassing 12 other state-of-the-art methods by a significant margin, with DSC improvements of + 1% and + 2%. These quantitative results across two datasets, evaluated under different partition protocols, underscore the robust performance and effectiveness of MedFCT approach.

7.6.2.2 Qualitative Comparison

Figure 7.13 provides a visual comparison of the qualitative results obtained by several state-of-the-art methods on the ACDC dataset, specifically under the condition of using 7 labeled volumes with annotations. The methods compared include SSNet [25], CTCT [15], ICT-Med [26] and MedFCT [16]. The figure clearly demonstrates the superior performance of MedFCT in accurately and comprehensively segmenting organs, regardless of their diverse sizes, shapes, and anatomical locations.

7.7 Application to Emphysema Quantification

This subsection presents an application of Pseudo-Labeling based SSL techniques (Sec.7.5) to emphysema quantification. Emphysema is a type of Chronic Obstructive Pulmonary Disease (COPD), a common chronic respiratory condition characterized by progressive airflow limitation. With its high prevalence and mortality rates, COPD affectes approximately 65 million people worldwide [27]. Emphysema, a major subtype of COPD, can be further classified into three subtypes: Centrilobular Emphysema (CLE), Paraseptal Emphysema (PSE), and Panlobular Emphysema (PLE) [28]. Centrilobular Emphysema is characterized by abnormal enlargement of the central

Table 7.1 Comparison of SSL methods on 3% labeled cases and 10% labeled cases of ACDC dataset

Method	3% labeled		10% labeled	
	DSC	95HD	DSC	95HD
MT	0.566	34.5	0.810	14.4
UA-MT	0.610	25.8	0.815	14.4
EM	0.602	24.1	0.791	14.5
DCT	0.582	26.4	0.804	13.8
CCT	0.586	27.9	0.816	13.1
CPS	0.603	25.5	0.833	11.0
ICT	0.581	22.8	0.811	11.4
DAN	0.528	32.6	0.795	14.6
URPC	0.567	31.4	0.829	10.6
CTCT	0.704	12.4	0.864	8.6
SSNet	0.705	17.4	0.853	10.6
ICT-Med	0.563	22.6	0.837	13.1
MedFCT	**0.726**	**10.5**	**0.886**	**4.3**

Table 7.2 Comparison of SSL methods on 3% and 10% labeled data of ISIC dataset respectively

Method	3% labeled		10% labeled	
	DSC	95HD	DSC	95HD
MT	0.728	37.4	0.734	34.0
UA-MT	0.730	38.6	0.734	33.2
EM	0.723	**36.3**	0.727	39.3
DCT	0.729	40.6	0.760	35.7
CCT	0.677	42.2	0.723	31.7
CPS	0.686	44.4	0.743	35.7
ICT	0.732	37.2	0.753	34.6
DAN	0.695	39.5	0.724	30.4
URPC	0.703	39.3	0.758	32.8
CTCT	0.713	43.2	0.760	37.3
SSNet	0.728	40.8	0.758	32.8
ICT-Med	0.714	39.2	0.749	33.1
MedFCT	**0.742**	39.0	**0.780**	**29.6**

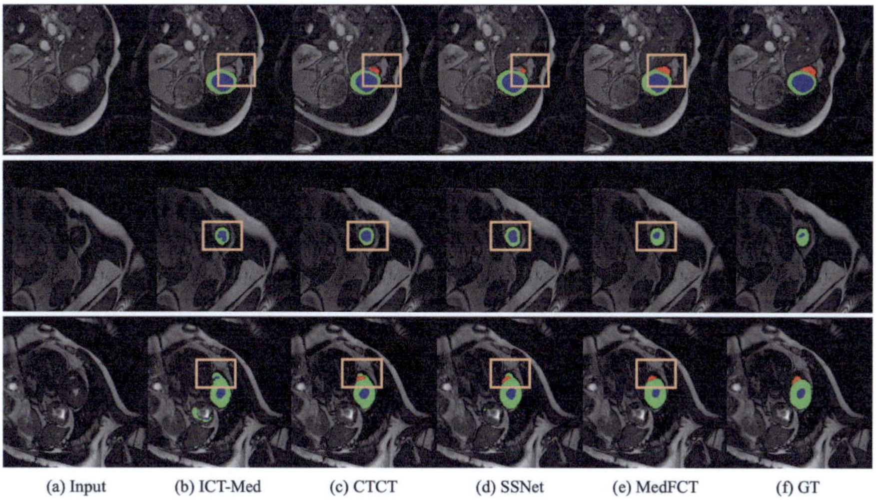

Fig. 7.13 The visualization of various segmentation methods is presented, where red indicates the left ventricle (LV), blue represents the right ventricle (RV), and green denotes the myocardium (MYO)

airspaces of the respiratory bronchioles; Paraseptal Emphysema involves emphysematous changes adjacent to the pleura; and Panlobular Emphysema is defined by diffuse abnormal enlargement of the entire secondary pulmonary lobule [29].

According to relevant literature [30], different subtypes of emphysema exhibit distinct pathogenesis. Centrilobular Emphysema is closely associated with smoking; Panlobular Emphysema shows no significant correlation with smoking in the general population but is linked to congenital alpha-1 antitrypsin deficiency; the pathogenesis of Paraseptal Emphysema remains unclear. Furthermore, studies [31] have shown that different subtypes of emphysema have varying clinical significance: Centrilobular and Panlobular Emphysema are correlated with symptom exacerbation and reduced exercise tolerance, whereas Paraseptal Emphysema generally does not lead to respiratory obstruction. Therefore, qualitative and quantitative analysis of emphysema subtypes is crucial for personalized diagnosis, treatment, and prognosis of emphysema. The core technology to achieve these goals is the classification and semantic segmentation of emphysema subtypes based on CT imaging.

Figure 7.14 illustrates the typical manifestations of different emphysema subtypes on HRCT images. From left to right, the red dashed boxes in the figure show lesions of Centrilobular Emphysema, Paraseptal Emphysema, and Panlobular Emphysema, respectively. Centrilobular Emphysema appears as scattered, small-scale low-density areas; Paraseptal Emphysema presents as single or clustered cystic low-density areas adjacent to the pleura; Panlobular Emphysema is characterized by larger-scale low-density regions. These three subtypes of emphysema may coexist in a single case. In addition to the varying scales of lesions, emphysema also exhibits a diffuse nature, as indicated by the blue dashed lines in the figure, showing that emphysematous lesions

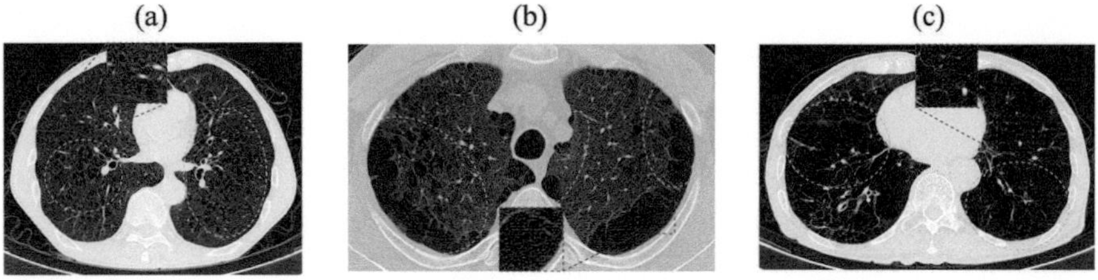

Fig. 7.14 Typical manifestations of emphysema subtypes on HRCT images

typically spread across both lungs. Consequently, it is impractical to annotate all emphysematous lesions in CT images in clinical practice. Given the manageable workload, radiologists can only annotate a portion of the lung tissue in CT images.

Emphysema is a typical diffuse lung disease, and compared to other diseases (such as tumors), annotating its images is more challenging. According to our statistics, a radiologist requires 36 working hours to fully annotate a single emphysema case and an additional 3 to 5 h for verification. Therefore, it is impractical to fully annotate the entire dataset in clinical practice. Moreover, for diffuse diseases, numerous similar lesions repeatedly appear in the CT images of the same patient. If complete annotation is required, radiologists would need to repeatedly annotate a large number of similar lesions, which is a tedious and monotonous task, severely wasting expert resources. To alleviate the annotation burden on radiologists, partial annotation can be considered a more efficient approach. Compared to full annotation (annotating every pixel in the image), partial annotation involves annotating only a portion of the tissue, significantly reducing the workload of annotators. However, deep learning methods generally require a large amount of annotated data to drive them, and relying solely on partially annotated data for training may lead to insufficient model generalization, thereby limiting model performance. In partially annotated medical images, a large number of unannotated regions still contain rich information (such as lesion shape, texture, and density), incorporating them into training can help increase the diversity of training samples, thereby improving model generalization to some extent and helping the model learn more robust features. Additionally, existing semantic segmentation work on emphysema generally adopts patch-based segmentation methods, which have low testing efficiency, necessitating the development of a more efficient emphysema segmentation method. This chapter focuses on training the network using partially annotated regions to achieve accurate and comprehensive segmentation of emphysema.

7.7.1 Dataset and Implementation Details

The dataset employed in this study comprises 91 HRCT volumes from patients who underwent pulmonary function tests (PFTs) confirming emphysema. These scans

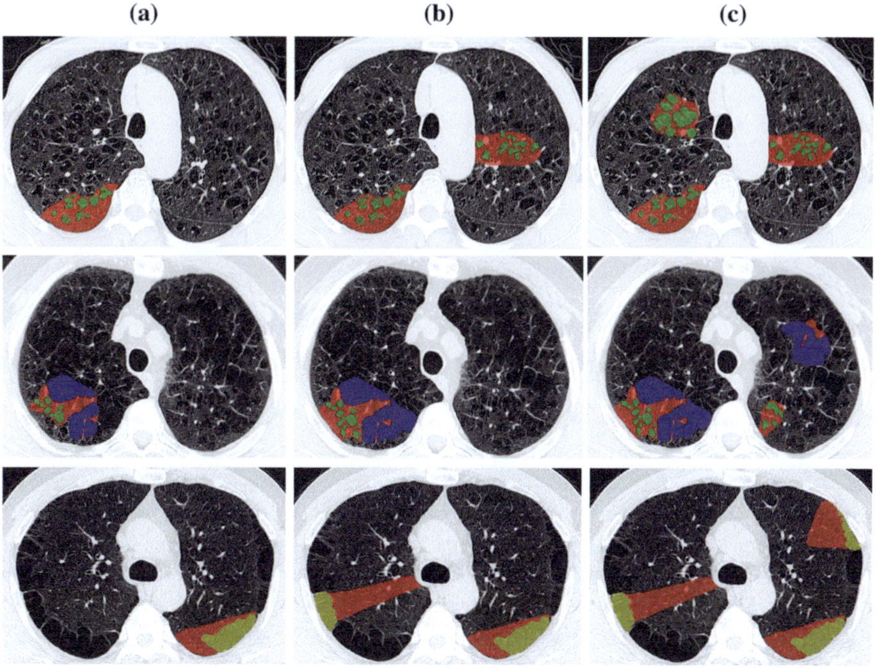

Fig. 7.15 Illustrations of annotated data at varying annotation levels are provided. Each row of images represents a single subject. The color-coded masks indicate different tissue types: green for CLE lesions, yellow for PSE lesions, blue for PLE lesions, and red for non-emphysema tissue. The annotation levels are as follows: **a** Level one (5–10%), **b** Level two (10–15%), and **c** Level three (15–20%)

were collected from two hospitals using three different CT machines, featuring an in-plane resolution of 0.62 mm to 0.69 mm, slice collimation of 1 mm to 2 mm, and a matrix size of 512 × 512 pixels. Among the 91 subjects, 63 are male and 28 are female. Two experienced radiologists annotated the data, with a third expert chest radiologist verifying the annotations. The dataset is partially annotated; radiologists selected specific lung regions and manually delineated masks. Four tissue types were identified: PLE, PSE, CLE, and non-emphysema (NE), the latter indicating tissue without emphysema but potentially with other pulmonary conditions.

To evaluate the performance across datasets with varying annotation levels, which require different manual efforts, radiologists annotated CT images at three levels: level one (5–10% of lung regions labeled), level two (10–15%), and level three (15–20%). These levels were determined through discussions among radiologists, lung specialists, and computer scientists, balancing training needs and radiologists' workload. Figure 7.15 illustrates examples of CT lung slices with different annotation levels.

For preprocessing, all CT slices were cropped and resized to 256 × 256 pixels using bilinear interpolation. The training set images and corresponding label masks

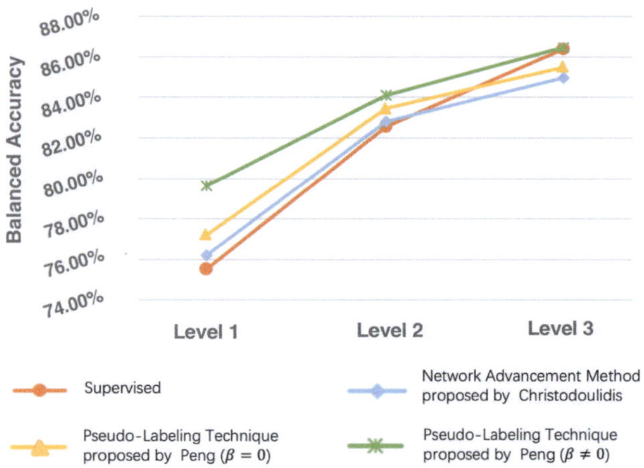

Fig. 7.16 Comparison on network advancement method [17] and Pseudo-labeling technique [18]

were further augmented through rotation and mirror-image reflection. Given that our inputs are 2D images, augmentation was applied to individual slices. Specifically, each training slice was randomly rotated, reflected, or both. This process increased the number of slices to 6,000 in the training set, though the total number of volumes remained unchanged. The dataset was divided into three subsets: training (59 subjects), validation (14 subjects), and test (18 subjects), ensuring balanced category representation. Augmentation was performed post-splitting.

7.7.2 Experimental Results and Analysis

In the conducted experiments, datasets with varying levels of annotation were utilized. As depicted in Figure 7.16, when datasets with fewer annotated samples were used, all semi-supervised methods demonstrated superior performance compared to the purely supervised approach. Among these, the method based on pseudo-labeling techniques with high confidence [18], which incorporates Fisher loss, achieved the most favorable results. Additionally, this method outperformed the approach based on network advancement [17].

The method in [17] exhibited relatively lower performance, which may be attributed to its inclusion of pseudo-labels with low confidence, increasing the likelihood of errors. The authors of [17] did not selectively filter the pseudo-labeled data, potentially incorporating unlabeled data with incorrect labels into the training set. This oversight could have adversely affected the overall performance of their method.

7.8 Summary

In this chapter, we focused on semi-supervised learning for the medical image analysis, whose primary goal was to alleviate the annotation burden on radiologists while achieving accurate and comprehensive segmentation results. Firstly, from the perspective of data annotation, we identified two current annotation scenarios designed to reduce the workload of medical professionals: (a) Sample-based partial annotations by labeling only a portion of tissues within each sample. (b) Dataset-based partial annotations with providing complete labels for a subset of dataset. Secondly, we explored various semi-supervised learning methods, categorizing them into the following types: Co-Training based Semi-Supervised Learning, GAN-based Semi-Supervised Learning, Consistency based Semi-Supervised Learning, and Pseudo-Labeling Techniques. Through the experimental evaluations, we compared the performance of these different semi-supervised methods. The results demonstrated the effectiveness of each approach in leveraging partially annotated data to enhance the segmentation accuracy of emphysema, thereby providing valuable insights into the practical application of semi-supervised learning in medical imaging.

References

1. Blum, A., & Mitchell, T. (1998). Combining labeled and unlabeled data with co-training. In: *Proceedings of the eleventh annual conference on computational learning theory* (pp. 92–100).
2. Qiao, S., Shen, W., Zhang, Z., Wang, B., et al. (2018). Deep co-training for semi-supervised image recognition. In: *Proceedings of the European conference on computer vision* (eccv) (pp. 135–152).
3. Goodfellow, I., Pouget-Abadie, J., Mirza, M., Xu, B., Warde-Farley, D., Ozair, S., et al. (2014). Generative adversarial nets. In: *Advances in neural information processing systems*.
4. Radford, A., Metz, L., & Chintala, S. (2015). Unsupervised representation learning with deep convolutional generative adversarial networks. arXiv preprint arXiv:1511.06434.
5. Li, Y., Liu, S., Yang, J., & Yang, M. H. (2017). Generative face completion. In: *Proceedings of the IEEE conference on computer vision and pattern recognition* (pp. 3911–3919).
6. Ledig, C., Theis, L., Huszár, F., Caballero, J., Cunningham, A., Acosta, A., et al. (2017). Photo-realistic single image super-resolution using a generative adversarial network. In: *Proceedings of the IEEE conference on computer vision and pattern recognition* (pp. 4681–4690).
7. Wang, X., Shrivastava, A., & Gupta, A. (2017). A-fast-rcnn: Hard positive generation via adversary for object detection. In: *Proceedings of the IEEE conference on computer vision and pattern recognition* (pp. 2606–2615).
8. Hoffman, J., Wang, D., Yu, F., & Darrell, T. (2016). Fcns in the wild: Pixel-level adversarial and constraint-based adaptation. arXiv preprint arXiv:1612.02649.
9. Luc, P., Couprie, C., Chintala, S., & Verbeek, J. (2016). Semantic segmentation using adversarial networks. arXiv preprint arXiv:1611.08408.
10. Huang, H., Zheng, H., Lin, L., Cai, M., Hu, H., Zhang, Q., et al. (2021). Medical image segmentation with deep atlas prior. In: *IEEE transactions on medical imaging* (pp. 3519–3530).
11. Zhang, Y., Yang, L., Chen, J., Fredericksen, M., Hughes, D. P., et al. (2017). Deep adversarial networks for biomedical image segmentation utilizing unannotated images. In: Medical image computing and computer assisted intervention: MICCAI 2017: 20th international conference, Quebec City, QC, Canada, September 11–13, 2017, Proceedings, Part III 20 (pp. 408–416). Springer International Publishing.

12. Tarvainen, A., & Valpola, H. (2017). Mean teachers are better role models: Weight-averaged consistency targets improve semi-supervised deep learning results. In: *Advances in neural information processing systems*.
13. Yu, L., Wang, S., Li, X., Fu, C. W., & Heng, P. A. (2019). Uncertainty-aware self-ensembling model for semi-supervised 3D left atrium segmentation. In: *Medical image computing and computer assisted intervention* (pp. 605–613).
14. Ouali, Y., Hudelot, C., & Tami, M. (2020). Semi-supervised semantic segmentation with cross-consistency training. In: *Proceedings of the IEEE/CVF conference on computer vision and pattern recognition* (pp. 12674–12684).
15. Luo, X., Hu, M., Song, T., Wang, G., et al. (2022). Semi-supervised medical image segmentation via cross teaching between CNN and transformer. In: *International conference on medical imaging with deep learning* (pp. 820–833). PMLR.
16. Xie, S., Huang, H., Niu, Z., Lin, L., et al. (2023). MedFCT: A frequency domain joint CNN-transformer network for semi-supervised medical image segmentation. In: *2023 IEEE international conference on multimedia and expo (ICME)* (pp. 1913–1918). IEEE.
17. Anthimopoulos, M., Christodoulidis, S., Ebner, L., Geiser, T., Christe, A., & Mougiakakou, S. (2018). Semantic segmentation of pathological lung tissue with dilated fully convolutional networks. *IEEE Journal of Biomedical and Health Informatics, 23*(2), 714–722.
18. Peng, L., Lin, L., Hu, H., Zhang, Y., Li, H., Iwamoto, Y., & Chen, Y. W. (2019). Semi-supervised learning for semantic segmentation of emphysema with partial annotations. *IEEE Journal of Biomedical and Health Informatics,24*(8), 2327–2336.
19. Bernard, O., Lalande, A., Zotti, C., Cervenansky, F., Yang, X., Heng, P. A., et al. (2018). Deep learning techniques for automatic MRI cardiac multi-structures segmentation and diagnosis: is the problem solved? *IEEE Transactions on Medical Imaging, 37*(11), 2514–2525.
20. Codella, N., Rotemberg, V., Tschandl, P., Celebi, M. E., Dusza, S., Gutman, D., et al. (2019). Skin lesion analysis toward melanoma detection 2018: A challenge hosted by the international skin imaging collaboration (ISIC). arXiv preprint arXiv:1902.03368.
21. Vu, T. H., Jain, H., Bucher, M., Cord, M., & Pérez, P. (2019). Advent: Adversarial entropy minimization for domain adaptation in semantic segmentation. In: *Proceedings of the IEEE/CVF conference on computer vision and pattern recognition* (pp. 2517–2526).
22. Chen, X., Yuan, Y., Zeng, G., et al. (2021). Semi-supervised semantic segmentation with cross pseudo supervision. In: *Proceedings of the IEEE/CVF conference on computer vision and pattern recognition* (pp. 2613–2622).
23. Verma, V., Kawaguchi, K., Lamb, A., Kannala, J., Solin, A., Bengio, Y., et al. (2022). Interpolation consistency training for semi-supervised learning. *Neural Networks, 145*, 90–106.
24. Luo, X., Liao, W., Chen, J., Song, T., Chen, Y., Zhang, S., et al. (2021). Efficient semi-supervised gross target volume of nasopharyngeal carcinoma segmentation via uncertainty rectified pyramid consistency. In: Medical image computing and computer assisted intervention-MICCAI 2021: 24th international conference, Strasbourg, France, September 27-October 1, 2021, Proceedings, Part II 24 (pp. 318–329). Springer International Publishing.
25. Wu, Y., Wu, Z., Wu, Q., Ge, Z., et al. (2022). Exploring smoothness and class-separation for semi-supervised medical image segmentation. In: *International conference on medical image computing and computer-assisted intervention* (pp. 34–43). Springer Nature Switzerland, Cham
26. Basak, H., Bhattacharya, R., Hussain, R., et al. (2022). An embarrassingly simple consistency regularization method for semi-supervised medical image segmentation. arxiv preprint arxiv:2202.00677.
27. World Health Organization. (2010). *World health statistics 2010*. World Health Organization.
28. Takahashi, M., Fukuoka, J., Nitta, N., Takazakura, R., Nagatani, Y., Murakami, Y., & Murata, K. (2008). Imaging of pulmonary emphysema: A pictorial review. *International Journal of Chronic Obstructive Pulmonary Disease,3*(2), 193–204.
29. Lynch, D. A., Austin, J. H., Hogg, J. C., Grenier, P. A., Kauczor, H. U., Bankier, A. A., & Crapo, J. D. (2015). CT-definable subtypes of chronic obstructive pulmonary disease: A statement of the Fleischner society. *Radiology,277*(1), 192–205.

30. Anderson, A. E., & Foraker, A. G. (1973). Centrilobular emphysema and panlobular emphysema: two different diseases. *Thorax, 28*(5), 547–550.
31. Smith, B. M., Austin, J. H., Newell, J. D., Jr., D'Souza, B. M., Rozenshtein, A., Hoffman, E. A., & Barr, R. G. (2014). Pulmonary emphysema subtypes on computed tomography: The MESA COPD study. *The American Journal of Medicine,127*(1), 94–e7.

Chapter 8
Domain Adaptation and Generalization for Medical Image Analysis

8.1 Introduction

Deep learning-based methods have achieved state-of-the-art performance in various medical image analysis tasks. Most of these models are trained under the assumption that the training data (source domain) and testing data (target domain) are independent and identically distributed (i.i.d.). In practice, variations in image acquisition, such as differences in medical centers, equipment manufacturers, scanning protocols, and imaging modalities, render this assumption unrealistic, often resulting in significant performance degradation. This challenge is commonly referred to as domain shift or domain gap. An example of this is illustrated in Fig. 8.1. The figure displays two contrast-enhanced CT images from different phases: (a) the PV phase (source domain), (b) the ART phase (target domain) and (c) their distributions based on reduced two-dimensional features. A notable difference can be observed between the two distributions. This domain shift issue will significantly compromise the model's performance.

To address domain shift, transfer learning is commonly employed as an effective solution. Transfer learning leverages knowledge from a source domain to improve model performance in a target domain, even when the two domains differ in data distribution. Within transfer learning, common strategies for addressing domain shift include fine-tuning (FT), domain adaptation (DA), and domain generalization (DG). Fine-tuning (FT) involves leveraging knowledge from a pre-trained model and adapting it to a new, typically related task through additional training on the corresponding dataset. This approach leverages the extensive knowledge embedded in the model from its initial training, allowing it to perform effectively on the target domain with less data and reduced training time. However, the limitations of fine-tuning are obvious: It requires access to labeled datasets for the target domain, which limits the model's scalability and deployment. Additionally, it is often sensitive to domain discrepancies and may struggle to bridge significant gaps when the source and target domains differ substantially. Furthermore, users need to prepare labeled target

Y. Chen et al., *Recent Advances in Deep Learning for Medical Image Analysis*, Intelligent Systems Reference Library 278, https://doi.org/10.1007/978-3-031-94791-9_8

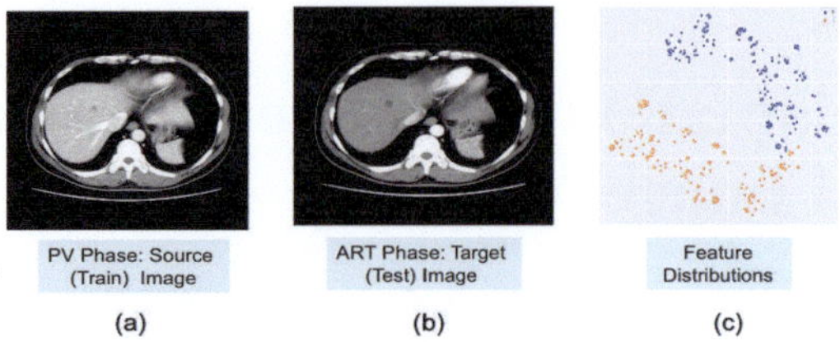

Fig. 8.1 Comparison of PV and ART phase distributions. **a** Contrast-enhanced CT images from the PV phase (source domain) and **b** ART phase (target domain). **c** Two-dimensional feature distributions (orange: source domain; blue: target domain) highlighting the notable differences between the two phases

domain data, which can be a labor-intensive task. To overcome these limitations, many efforts have been made such as Domain Adaptation (DA). DA aims to reduce the distributional differences between the source and target domains by aligning their feature spaces, allowing the model to perform effectively on the target data without extensive retraining. Techniques like adversarial training and feature alignment improve model robustness across diverse domains, making DA particularly valuable when domain shift is substantial.

However, DA assumes that target data are accessible for model adaptation, which is unknown before model deployment, requiring the applicability of DA in scenarios where reliable performance on unseen domains is required. To address these constraints, domain generalization (DG) has emerged as an alternative. Unlike DA, DG trains the generalized model only on source domain(s), enabling it to perform well in any unseen target domains without any prior adaptation. By encouraging the model to learn general representations or domain-invariant features, DG enhances robustness and is well-suited for medical applications where the target environment is unpredictable or inaccessible before deployment.

The following subsections will focuses on the application of various DA and DG approaches in medical image analysis. By addressing the challenges posed by domain shift, DA and DG techniques enhance model robustness and adaptability, making them valuable methods for accurate medical image interpretation across diverse clinical settings.

8.2 Domain Adaptation

Domain Adaptation leverages unlabeled target-domain data to adapt the model, reducing the need for extensive labeling in new domains. To address the domain shift problem and prevent performance degradation, domain adaptation techniques have recently been applied in medical imaging [1–10]. These methods aim to bridge the gap between differing data distributions in style, characteristics, and feature representations. By leveraging deep learning frameworks to minimize discrepancies between various data distribution, domain adaptation offers a promising solution. Specifically, these techniques align the distributions and synchronize the feature spaces of the source (training) and target (test) datasets. This alignment is achieved by generating extracted image feature maps within a common shared space or by learning target-like features, ensuring that the extracted characteristics are robust and invariant to variations across features. In the following subsection, we discuss different learning schemes employed in domain adaptation.

8.2.1 Source and Target Domain Descriptions

Consider the source domain (training), which consists of N^s images with corresponding ground truths (labels), denoted as $\{(\mathbf{x}_i^s, y_i^s)\}_{i=1}^{N^s}$, where $\mathbf{x}_i^s \in \mathcal{X}_S$ and $y_i^s \in \mathcal{Y}_S$. Here, $\mathcal{X}_S$ represents the set of input images, and $\mathcal{Y}_S$ represents the set of corresponding labels. Specifically, $\mathbf{x}_i^s \in \mathbb{R}^{H \times W \times 3}$ is an input image, and y_i^s denotes its corresponding label. For detection tasks, $y_i^s = (B_i^s, C_i^s)$, where B_i^s represents the bounding box coordinates and C_i^s represents the associated category labels. For segmentation tasks, $y_i^s \in \mathbb{R}^{H \times W \times C}$, where C is the number of classes at each pixel. The target domain (test) consists of N^t images without annotations, denoted as $\{\mathbf{x}_i^t\}_{i=1}^{N^t}$, where $\mathbf{x}_i^t \in \mathcal{X}_T$. Here, $\mathcal{X}_T$ represents the set of target domain images, and $\mathbf{x}_i^t$ represents an input target image.

8.2.2 Domain Adaptation Using Adversarial Learning

Adversarial learning-based methods, known as Domain-Adversarial Neural Networks (DANN) [11, 12], utilize adversarial training techniques to align the feature distributions of the source and target domains. Typically, during training, an additional domain discriminator is integrated into the deep learning network, and both are trained in a competitive manner. The core idea is to train the model so that the domain discriminator network struggles to distinguish between features from the source and target domains.

8.2.2.1 Adversarial Loss-Based Domain Adaptive Training Process

The domain discriminator, usually implemented as a convolutional neural network, is specifically designed to identify whether features originate from the source or target domain. Meanwhile, the task-specific deep learning network consists of two main components: a feature extractor and a task-specific head. The task-specific head could be a FC layer or a MLP for classification tasks, a decoder for segmentation

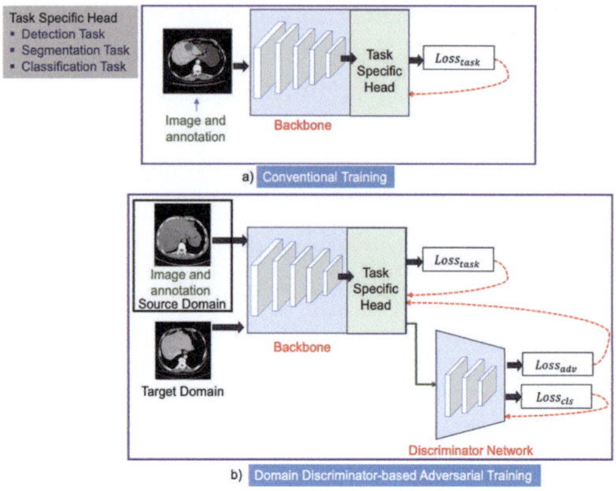

Fig. 8.2 Illustration of conventional and domain discriminator-based training approaches. **a** Conventional training approach. **b** Domain discriminator-based training, where a domain discriminator is added for adversarial learning during training

tasks and a detection head for object detection tasks. For more detailed information about task-specific networks, please refer to Chap. 2. Figure 8.2a illustrates the conventional training approach, which follows a standard process without incorporating domain adaptation techniques. In contrast, Fig. 8.2b demonstrates a domain discriminator-based training approach, where the domain discriminator is added to facilitate adversarial learning. This additional component enhances the alignment of feature distributions between the source and target domains, improving the model's adaptability to domain shifts. A detailed explanation of training method is provided in the subsequent sections.

Step 1: Training the Domain Discriminator—In the first step of training, the domain discriminator is trained to differentiate between features originating from the source and target domains. During this step, the deep learning framework (e.g., the feature extractor and task-specific head) is frozen and only the discriminator network is optimized.

Input samples from both the source and target domains are passed through the feature extractor to generate feature representations. These features are then fed into the domain discriminator, which predicts the domain label (source or target). This classification loss is computed to update the discriminator's weights, enhancing its ability to distinguish between the domains. This can be shown as:

$$L_{\text{cls}} = -\left[\mathbb{E}_{\mathbf{x}\sim\text{source}}[\log D(F(\mathbf{x});0)] + \mathbb{E}_{\mathbf{x}\sim\text{target}}[\log(1 - D(F(\mathbf{x});1))]\right] \tag{8.1}$$

where $D(\cdot)$ is the output of the domain discriminator, representing the probability that $\mathbf{x}$ (i.e., feature space) is from either the source or target domain.

Step 2: Training the Deep Learning Framework with Adversarial Learning—In the second step, the focus shifts to training the deep learning framework while freezing the domain discriminator. Adversarial training is employed to align the feature distributions of the source and target domains by tricking the domain discriminator.

Feature representations from both domains are provided to the discriminator, which classifies them as originating from either the source domain (label '0') or the target domain (label '1'). For adversarial learning, domain labels of the target domain features are flipped (i.e., target features are assigned label '0' instead of '1'). The adversarial binary classification loss, L_{adv}, is then computed as:

$$L_{adv} = -\mathbb{E}_{\mathbf{x}\sim\text{target}}[\log(1 - D(F(\mathbf{x}); 0))] \tag{8.2}$$

This adversarial loss encourages the feature extractor to generate representations that make it harder for the domain discriminator to distinguish between source and target domains. The overall objective function for the feature extractor combines the adversarial loss with the task-specific loss L_{task}. This can be given as:

$$L_{\text{total}} = L_{\text{task}} + \lambda \cdot L_{\text{adv}} \tag{8.3}$$

here λ is a hyperparameter to balance the two losses. L_{total} is the loss function used to train the primary framework. L_{task} are discussed in Chap. 2.

Inference Process: During inference, only the deep learning framework is used to make predictions, while the domain discriminator is excluded. The feature extractor and task-specific head work in tandem to process the input data and produce task-specific outputs (e.g., classification, detection, or segmentation results).

8.2.2.2 Gradient Reversal Layer (GRL)-Based Domain Adaptive Training Process

The Gradient Reversal Layer (GRL) [13] provides an alternative approach to adversarial loss-based methods for domain adaptation. Instead of explicitly calculating an additional adversarial loss, the GRL enables the training of a domain-adversarial neural network by inverting the gradients during backpropagation. This layer is placed between the feature extractor and the domain discriminator network to facilitate adversarial learning within the framework. The overall training process for the GRL-based approach is similar to the previously discussed adversarial loss-based method. It also follows a two-step training process:

Step 1: The domain discriminator is trained to classify features from the source and target domains accurately by minimizing the classification loss L_{cls}.

Step 2: The feature extractor and task-specific network are optimized to confuse the domain discriminator by using the GRL to invert the gradients during backpropagation. Unlike the adversarial loss-based approach, the GRL simplifies the process by directly using the classification loss of the domain discriminator L_{cls} to update both the feature extractor and the task-specific network. In this scenario, the GRL modifies the backpropagation process by flipping the sign of the gradient. The total loss function for the feature extractor network and task-specific head is given as:

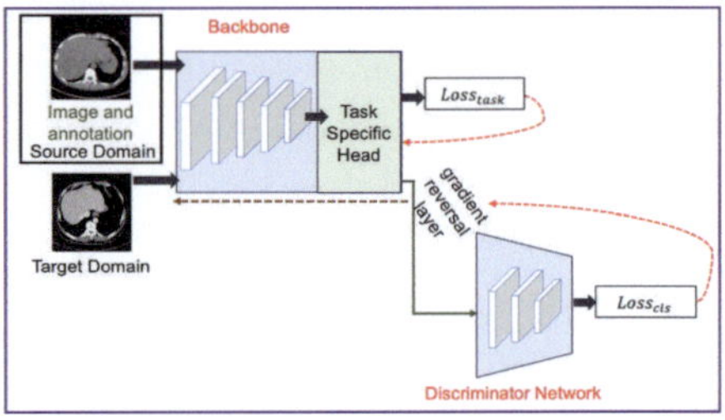

Fig. 8.3 Gradient reversal layer (GRL)-based domain adaptation. During backpropagation, the GRL inverts the gradient direction, enabling adversarial training. The feature extractor learns domain-invariant features

$$L_{\text{total}} = L_{\text{task}} + GRL\{\lambda \cdot L_{\text{cls}}\} \tag{8.4}$$

here, L_{task} is the task-specific loss (e.g., classification, detection, or segmentation loss). L_{cls} is the classification loss for distinguishing between source and target domains. λ is a hyperparameter to control the contribution of the domain adaptation term. Figure 8.3 illustrates the GRL-based framework.

These training process ensures that the model learns robust features through adversarial learning while maintaining high task-specific performance. The first stage optimizes the domain discriminator for accurate domain classification, and the second stage leverages adversarial learning to align source and target domain feature distributions. The inference process utilizes only the trained deep learning framework, making it efficient and streamlined for real-world applications.

8.2.3 Feature Alignment-Based Domain Adaptation

Alignment-based methods in domain adaptation focus on minimizing the discrepancy between source and target domain distributions by aligning their feature representations in a shared feature space. A feature extractor network extracts features from both source and target domain samples. The objective is to learn a feature representation that is useful for the primary task while being domain-invariant. This allows a model trained on the source domain to generalize well to the target domain. Figure 8.4 illustrates the process of aligning feature distributions between labeled source-domain images and unlabeled target-domain images to improve model performance. Alignment methods usually define a discrepancy metric or loss function to quantify the dissimilarity between the source and target distributions. This metric is minimized during training to promote domain-invariant feature learning. Various discrepancy metrics can be used, including Maximum Mean Discrepancy (MMD) [14, 15], Kullback–Leibler (KL) divergence [16], and Maximum Square Loss [17]. In this scenario, the total objective loss is formulated as:

$$L_{\text{total}} = L_{\text{task}} + \lambda \cdot L_{\text{align}} \tag{8.5}$$

Fig. 8.4 Illustration of alignment-based domain adaptation using labeled source-domain and unlabeled target-domain multiphase CT images. The figure demonstrates the process of aligning feature distributions to minimize domain discrepancy at output level (**a**), and feature level (**b**)

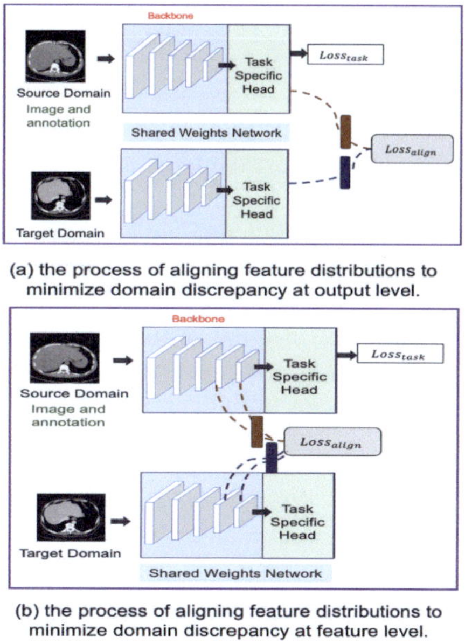

here, L_{align} represents a discrepancy metric or loss function used to quantify the dissimilarity between the source and target domain distributions. It is added to the task-specific loss (L_{task}) to encourage domain-invariant feature learning. The weighting factor λ controls the trade-off between the task-specific performance and the alignment of feature distributions.

These methods for aligning the distributions can be implemented at different stages within the frameworks. Specifically, alignment can be implemented at the output level, as illustrated in Fig. 8.4a, or at the feature level, as shown in Fig. 8.4b. Some strategies employing $Loss_{\text{align}}$ metrics are discussed as follows:

8.2.3.1 Domain Adaptation Using Kullback–Leibler (KL) Divergence

Kullback–Leibler (KL) divergence is widely used in domain adaptation tasks, particularly in Noisy Domain Adaptation [16], where it helps minimize the divergence between the output distributions of two classifiers. The objective is to align the feature distributions extracted from the source and target domains by measuring and minimizing their divergence.

It is a measure of the deviation of one probability distribution from another, i.e., reference probability distribution. For two probability distributions S and T defined over the same probability space, the KL divergence is defined as:

$$D_{KL}(S\|T) = \sum_{i=1}^{n} S(x)\log\frac{S(x)}{T(x)} \tag{8.6}$$

where $S(x)$ is the probability of x under the distribution S, $T(x)$ is the probability of x under the distribution T and n denotes the number of samples.

For domain adaptation, KL divergence can be used to minimize the difference between the source and target domain distributions, and it can be implemented at either the output level or the feature level. To apply KL divergence at the feature level, the softmax function is used to convert feature activations into probability distributions, then the softmax probabilities for features from both the source and target domains is computed. The KL divergence between the softmax probabilities of feature maps from the source $(feat_S)$ and target $(feat_T)$ domains is:

$$L_{KL_{DIV}} = D_{\mathrm{KL}}(feat_S \mid\mid feat_T) = \sum_{i=1}^{n} \mathrm{softmax}(feat_{x_i^s})\log\frac{\mathrm{softmax}(feat_{x_i^s})}{\mathrm{softmax}(feat_{x_i^t})} \tag{8.7}$$

The KL divergence loss can be combined with the objective optimization function (detection loss) to optimize the feature extractor. The updated loss function ensures that the feature maps from different domains are more aligned, which enhances the model's generalization capability across domains.

8.2.3.2 Domain Adaptation Using Maximum Squares Loss

The maximum squares loss is a domain adaptation technique introduced by Chen et al. [17] for semantic segmentation tasks. This method aims to mitigate domain shift by promoting confident predictions and avoiding overconfident incorrect predictions. The loss function operates on the output probability map of the segmentation model and encourages balanced class-wise predictions across the target domain. By maximizing the square of the softmax probabilities, the method implicitly aligns the target domain features with the source domain, enhancing the segmentation performance. This loss can be expressed as below:

$$L_{\mathrm{MSL}} = -\frac{1}{HW}\sum_{h=1}^{H}\sum_{w=1}^{W}\sum_{c=1}^{C} P_c(h,w)^2 \tag{8.8}$$

where $P_c(h, w)$ represents the predicted probability for class c at pixel location (h, w), H and W are the height and width of the image, and C is the number of classes. The objective optimization function, which incorporates the task-specific loss can be combined with maximum square loss to update the network's gradients. By maximizing the squared probabilities, this loss effectively encourages the model to make confident, discriminative predictions while maintaining robustness to domain shifts.

Furthermore, these different losses can be combined into a generalized framework that leverages Adversarial Loss, Maximum Square Loss, and KL Divergence to align feature distributions. Recent studies have introduced the implementation of loss combinations to enhance performance [3–5, 7]. This ensure effective object detection and robust performance across domains.

8.2.4 Ensemble Learning-Based Domain Adaptation

Ensemble learning-based methods use multiple models or classifiers to enhance domain adaptation performance. Ensemble approach involves training multiple models on different subsets of the source domain or using various weighting schemes to address domain shift [18]. The predictions from these models are then aggregated, typically through averaging or voting, to produce the final output. By combining the predictions from several models, these methods aim to achieve better generalization to the target domain. The underlying concept is that different models may capture various aspects of the data, and their collective knowledge can lead to improved overall performance.

These methods involves co-training of multiple frameworks on different views or subsets of the data. Each framework provides pseudo-labels for the target domain data, which are then used to train the other frameworks, iteratively improving the model's performance on the target domain. In scenarios with multiple source domains, ensemble methods can aggregate information from each source to better adapt to the target domain. For example, an ensemble of models trained on diverse source domains can collectively generalize more effectively to the target domain. Additionally, by incorporating different architectures or hyperparameters within the ensemble, these methods increase model diversity, which helps in capturing various domain-specific features and mitigating overfitting to any single domain.

In another ensemble learning framework for domain adaptation, a baseline Convolutional Neural Network (CNN), referred to as the student network, is trained using labeled samples from the source domain [19]. This network makes predictions based on a task-specific loss, such as segmentation loss. Concurrently, a second network, known as the teacher network, generates predictions using only the unlabeled samples from the target domain. The teacher network is updated using a temporal ensemble strategy, where its weights are adjusted by the exponential moving average of the student network's weights. This approach ensures that the teacher network accumulates knowledge over time, providing stable and robust predictions. During training, the domain shift between the source and target domains is addressed by a consistency loss. This loss function compares the predictions of the student and teacher networks, enforcing that the predictions remain consistent across domains. By minimizing this consistency loss, the model reduces the discrepancy between the source and target domains, enhancing its ability to generalize to the target domain.

In more advanced ensemble techniques, a meta-model (or second-level model) is trained using the predictions of the base learners as input features. This meta-learner combines these predictions to make a final decision, further refining the adaptation process. While adversarial ensemble learning approach combines adversarial learning with ensemble methods. Multiple domain discriminators are used to provide feedback to the feature extractor, ensuring that the feature representations are aligned across domains.

8.2.5 Application of Domain Adaptation to Lesion Detection in Multiphase CT Images

In recent study various domain adaptive methods for medical application is presented [9, 20]. In this section, we show application to mitigate domain shift from multiphase CT images. In real-world clinical settings, lesion detection often requires analyzing multiphase CT images. Multiphase CT imaging, also known as contrast-enhanced dynamic CT (CE-CT), is frequently used for diagnosing various malignant tumors. A typical multiphase CT scan consists of three or four distinct phases, each representing a different imaging domain. The initial phase is the non-contrast-enhanced (NC) phase, recorded before the administration of the contrast agent. Following the contrast injection, additional phases are acquired. The arterial (ART) phase is captured approximately 30–40 s post-injection. The portal venous (PV) phase is obtained around 70–80 s after injection. A delayed phase (DP), which occurs later. Other factors such as patient position, movement, respiration, and heartbeat can affect CT images obtained at different phases. Consequently, the texture and spatial characteristics, as well as the position of the liver and tumors, may vary significantly across these phases (i.e., different domains). Figure 8.5 illustrates corresponding slices of liver tumors across different phases. The texture and spatial attributes of liver tumors exhibit notable variations across these phases, influenced by the timing of contrast agent administration, as shown in the figure. These differences in contrast enhancement and vascular perfusion in phases result in different appearances of different tissues (cancers) in CT images, underscoring the importance of considering multiple phases for accurate diagnosis and analysis. The feature distributions between different phases of CT images contribute to domain shift issues, which pose significant challenges in various medical applications. As a result, deep learning models trained on images from one phase (source) often exhibit suboptimal performance when evaluated on another phase images (target domain).

Here, we aim to address this challenge by training a model using only PV phase images and applying it to ART and NC phases. By leveraging domain adaptation (DA) techniques, we seek to reduce phase-specific domain gaps and improve the model's generalization across different phases. This approach not only minimizes the reliance on multi-phase annotations but also ensures efficient and accurate lesion detection in practical clinical workflows. The subsequent sections outline the experimental setup, including datasets.

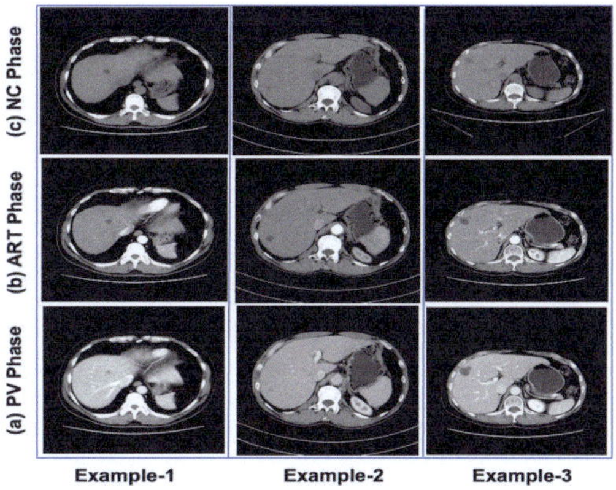

Fig. 8.5 Corresponding slices from NC, ART and PV phase of liver CT images

8.2.5.1 Dataset Description and Implementation Details

MPCT-FLLs Dataset: In this study, an in-house dataset MPCT-FLLs (Multi-Phase CT dataset of Focal Liver Lesions) is used. It is collected by Sir Run Run Shaw Hospital, Zhejiang University. This dataset has been used in several previous works [21]. Three CT phases from this dataset are used: Portal Venous (PV), Arterial (ART), and Non-Contrast (NC) phases. Expert radiologists annotated segmentation masks of the liver and lesions for each phase. The Canny edge detection algorithm [22] is applied to generate rectangular boxes around the lesion boundaries. The slice-plane resolution is set to 512 × 512 pixels, and the number of slices per phase ranged from 25 to 99. For image processing, pixel intensity values within the range of −70 to +180 HU were selected for each slice of each phase. This dataset includes five different types of liver lesions: Cyst, FNH, HCC, HEM, and METs. The MPCT-FLLs dataset comprises 121 sets of multi-phase abdominal CT images. A total of 121 patients were randomly divided into 80 for training and 41 for testing. Further details regarding the training and test slices for each phase are provided in Table 8.1. Only lesion images from each phase are used for model training and testing.

Table 8.1 Distribution of CT images across different phases in the MPCT-FLLs dataset

Phase	Training images (80 patients)	Test images (41 patients)	Total (121 patients)
PV phase	630	360	990
NC phase	635	367	1002
ART phase	632	377	1009

Table 8.2 Results of liver lesion detection: domain adaptation from MPCT-FLLs **PV** to MPCT-FLLs **ART** phase images

Network	Adversarial learning	Feature alignment		AP IoU-0.5	AP IoU-0.3
		Max. square loss	KL div. loss		
Baseline				61.25 ± 0.63	66.45 ± 1.34
Method A	✓			63.75 ± 0.21	68.8 ± 0.28
Method B			✓	64.5 ± 0.20	69.9 ± 0.70
Method C	✓	✓		64.55 ± 0.21	69.85 ± 0.49
Method D	✓		✓	64.8 ± 0.28	70.19 ± 0.84
Method E	✓	✓	✓	65.25 ± 0.21	70.55 ± 0.49

8.2.5.2 Results and Comprehensive Analysis

Performance Analysis: Domain Shift from MPCT-FLLs (PV) to MPCT-FLLs (ART): First, the evaluation is performed using PV phase as training data and ART phase as test data. Table 8.2 shows the results obtained through baseline method (traditional CenterNet without DA) [23], adversarial learning-based domain adaptation [3], maximum square loss using feature maps with adversarial learning [5] and KL divergence-based feature generalization [7] utilizing the DLA34 backbone. In normal training, in case of domain shift from the PV to ART phase of CT imaging (source-to-target), the network achieves an accuracy of 61.25% AP (IoU-0.5) and 66.45% AP (IoU-0.3). The incorporation of adversarial loss and maximum square loss enhances performance without requiring additional annotations. The DA-based approach, using feature generalization with adversarial loss, achieves an accuracy of 64.8% (IoU-0.5) and 70.19% (IoU-0.3). While the approach, incorporating feature generalization with adversarial and maximum square loss, achieves an accuracy of 65.25% (IoU-0.5) and 70.55% (IoU-0.3). These results demonstrate a significant enhancement in performance facilitated by our method.

Performance Analysis: Domain Shift from MPCT-FLLs (PV) to MPCT-FLLs (NC): For the domain shift from the PV phase to the NC phase, the DA-based approach demonstrates notable improvements in detection accuracy. DA-based methods enhance the performance by approximately 6% in terms of AP (IoU-0.5) and 7% in terms of AP (IoU-0.3). These results signify the efficacy of the proposed approach in mitigating the domain shift between the PV and NC phases, resulting in enhanced detection performance. The results are given in the Table 8.3.

Table 8.3 Results of liver lesion detection: domain adaptation from MPCT-FLLs **PV** to MPCT-FLLs **NC** phase images

Network	Adversarial learning	Feature alignment		AP IoU-0.5	AP IoU-0.3
		Max. square loss	KL div. loss		
Baseline				47.3 ± 0.56	52.95 ± 0.21
Method A	✓			51.35 ± 0.49	57.05 ± 0.35
Method B			✓	52.25 ± 0.25	57.75 ± 0.65
Method C	✓	✓		52.7 ± 0.98	58.5 ± 1.41
Method D	✓		✓	54.2 ± 0.00	59.25 ± 0.07
Method E	✓	✓	✓	54.65 ± 1.06	60.0 ± 0.28

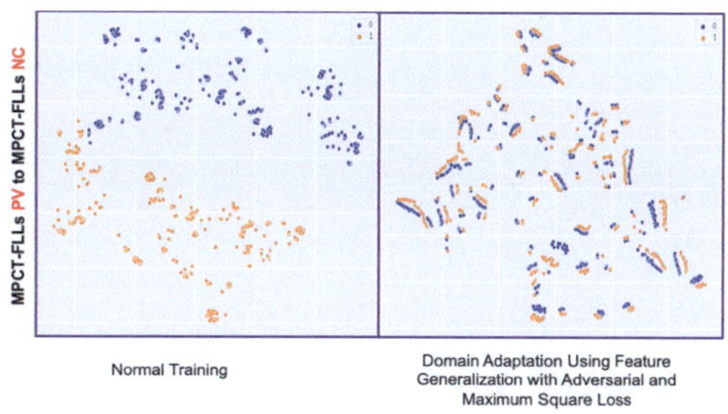

Fig. 8.6 Visualization of the t-SNE feature distributions of PV and NC liver lesion images [orange: PV (source); blue: NC (target)]

The t-SNE feature distributions [24] of training images from the PV and NC phases are illustrated in Fig. 8.6. The feature distributions for the PV and NC images exhibit significant differences in the standard training scheme. However, DA-based method effectively aligns these domains, as demonstrated by the t-SNE visualizations. This alignment is particularly evident in Fig. 8.7, which presents the detection results across various CT phases using MPCT-FLLs dataset.

8.2.6 Summary

This section discusses the challenges of domain shift across various medical centers and CT scan phases. Feature generalization is achieved through divergence-based optimization at the feature level and the integration of different domain adaptive losses, enabling robust and invariant feature extraction. The discussed methods effectively bridge the domain gap between different CT imaging phases using our private dataset. Notably, the approach requires annotations for only a single CT phase image during model training, minimizing annotation effort for other phases. Specifically, PV phase CT images with bounding box annotations were used as the source

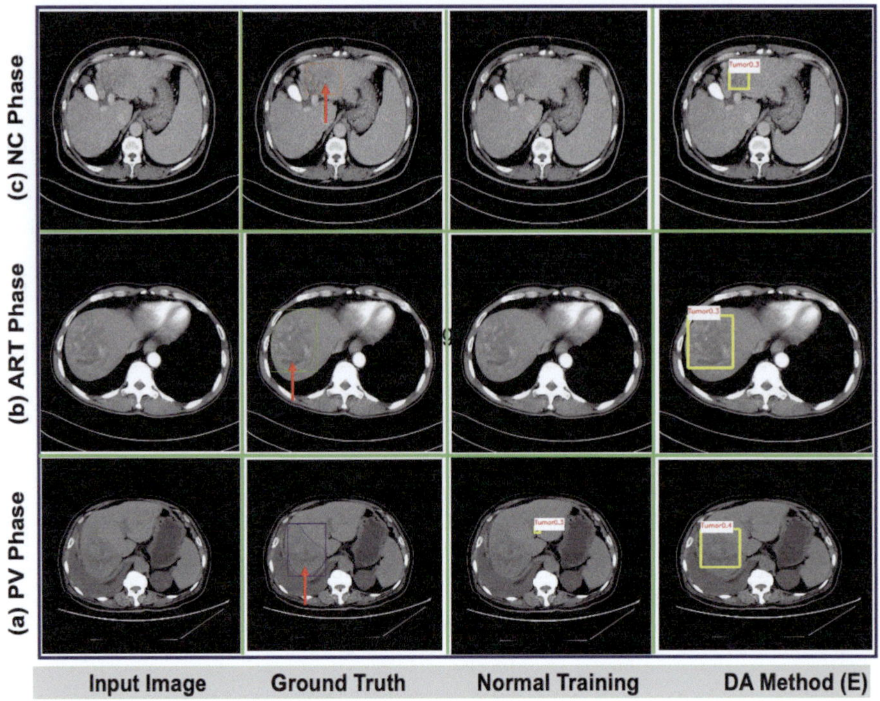

Fig. 8.7 Illustration of liver lesion detection across various CT phases

domain for training, while ART and NC phase images served as target (test) domains without annotations. Validation involved training models on datasets from various centers and evaluating them across different phases of our private dataset. The experiments demonstrated that domain adaptation approaches significantly enhanced performance compared to models applied directly without adaptation. In the next part we discuss some domain generalization techniques for medical imaging.

8.3 Domain Generalization Techniques for Medical Image Analysis

Domain Generalization (DG) aims to train a model on one or multiple source domains so that it performs well on unseen target domains during inference. An example illustrating the differences between images from various domains is shown in Fig. 8.8. Unlike Domain Adaptation, DG techniques do not utilize any labeled or unlabeled target data during training, making the task more challenging but also more applicable in realistic scenarios. Recently, the DG task has gained significant attention in computer vision, encompassing fields such as image recognition, scene segmen-

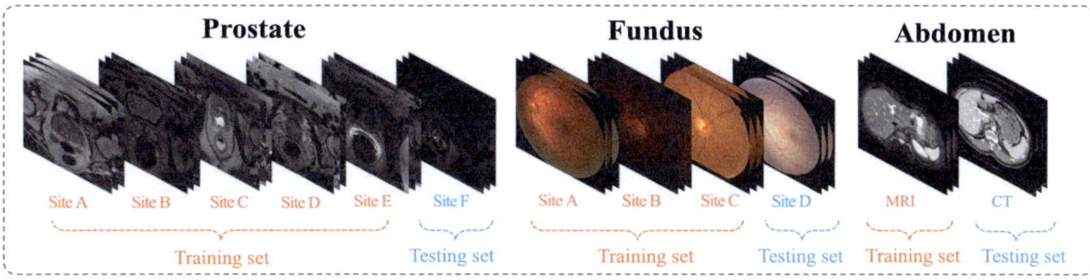

Fig. 8.8 Examples of domain generalization problems across three common datasets

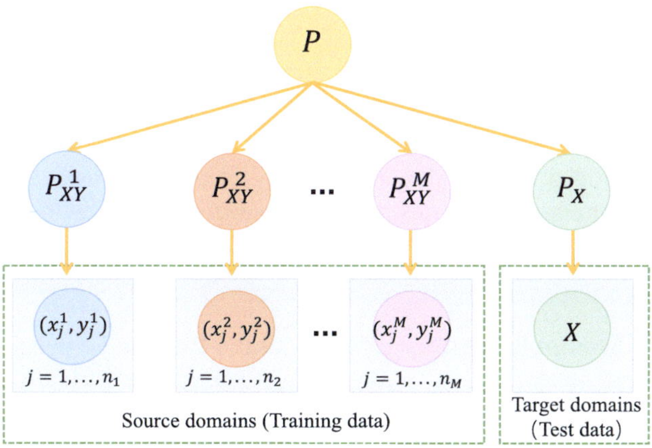

Fig. 8.9 An overview of the domain generalization technique. The labeled source data is used to generalize the network

tation, person identification, and medical image analysis. DG typically consists of two primary settings defined by the number of source datasets: multi-source DG and single-source DG. Additionally, alternative variations include federated DG, semi-supervised DG, and few-shot DG.

Formulaically, in domain generalization, $M(M \geq 1)$ domain(s) $S_{train} = \{S^i | i = 1, 2, \dots, M\}$ are provided for training, where $S^i = \{(\mathbf{x}^i_j, \mathbf{y}^i_j)\}_{j=1}^{n_i}$ represents the i-th domain, n_i represents the data size of source domain i. This setting is illustrated in Fig. 8.9. Each domain S^i is drawn from its corresponding specific distribution, and the distributions between any two domains are different: $P^i_{XY} \neq P^j_{XY}$. A model that performs well across all known source domains is a good choice to use when dealing with a previously unknown target domain. Therefore, the goal of domain generalization is to learn a robust and generalizable predictive function $h : X \rightarrow Y$ from the M training domain(s) to achieve a minimum prediction error on an unseen target domain and $P^{target}_{XY} \neq P^i_{XY}$.

$$\min_h \mathbb{E}_{(\mathbf{x},\mathbf{y})\in\mathcal{S}_{\text{target}}}[\ell(h(\mathbf{x}), \mathbf{y})] \tag{8.9}$$

where $\mathbb{E}$ is the expectation and $\ell(\cdot, \cdot)$ is the loss function.

In this section, a comprehensive overview of domain generalization is provided in the field of medical image analysis, which covers the theoretical foundations, research methodologies, representative works, benchmark datasets, and experimental results.

8.3.1 Multi-source Domain Generalization Setting

Multi-source domain generalization (MDG) aims to train a robust model using multiple labeled source domains, enabling it to generalize well to any unseen target domains. By leveraging data from diverse sources, multi-source domain generalization methods seek to learn general representations, making the model more generalizable to new distributions. Existing domain generalization methods can be categorized into three main groups: data manipulation-based, invariant representation-based, and learning strategy-based approaches. In this subsection, a detailed overview of each category and its corresponding representative works is provided.

8.3.1.1 Data Manipulation-Based Multi-source Domain Generalization

In traditional supervised learning, data manipulation is often employed to prevent model overfitting by augmenting the diversity of training samples. In MDG, This type of approach aims to enhance data diversity, enabling predictive models trained on manipulated datasets to achieve better generalization. Two widely used techniques for this purpose are data augmentation and data generation. Data augmentation involves modifying and transforming input data using techniques like randomization and augmentation. In contrast, data generation focuses on creating entirely new, diverse samples to enrich the training set. Typically, BigAug [25] proposes a deep stacked transformation approach for domain generalization. Specifically, a series of stacked transformations are applied to each image during network training. Consequently, a deep model trained on the augmented “big” data is able to generalize well on unseen domains. Differently, FedDG [26]considers exchanging the amplitude spectrum across domains to transmit distribution information, while keeping the phase spectrum intact to preserve core semantics for data augmentation. Based on this, it devises a continuous frequency space interpolation mechanism, which interpolates between the local and transferred distributions to enrich the established multi-domain distributions for each domain. Later on, AADG [27] proposes an automated augmentation method that effectively samples data augmentation policies, generating novel domains and diversifying the training set from an appropriate search space. Specifically, AADG introduces a novel proxy task that maximizes the diversity among multiple augmented novel domains, measured by the Sinkhorn distance in a unit sphere space, making automated augmentation tractable. Overall, data manipulation in domain generalization aims to enhance both the diversity and quantity of source

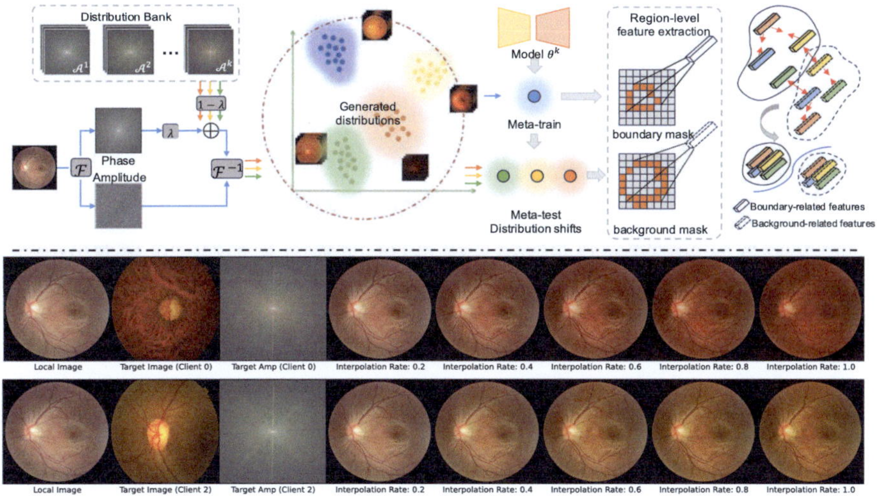

Fig. 8.10 A schematic view of FedDG method is presented. The method exchanges distribution information among clients using a continuous interpolation mechanism in frequency space, allowing each client to access multiple source distributions. It adopts an episodic training approach to expose local optimizations to domain shifts, employing explicit regularization to enhance feature cohesion and separation at ambiguous boundaries

data through sophisticated techniques, which simulate a wide range of domain shifts and improve model robustness to unseen distributions.

Among the methods discussed, FedDG [26] stands out for its innovative use of frequency domain augmentation, which addresses domain generalization challenges by enhancing data diversity. In the following, we provide a detailed introduction to this approach.

Frequency Domain Information Decoupling: Inspired by the concept from visual psychophysics that an image's low-level characteristics (style) and high-level semantics can be represented by its amplitude and phase spectrum in frequency space. FedDG proposes exchanging amplitude spectra across domains through a continuous frequency space interpolation mechanism to address federated multi-source domain generalization tasks. This interpolation enriches the multi-domain distributions accessible to each local client, as shown in Fig. 8.10 (upper part).

Specifically, given a sample $\mathbf{x}_i^k \in \mathbb{R}^{H\times W\times C}$ from the kth client, the frequency space signal obtained through Fast Fourier transform [28] as:

$$\mathcal{F}\left(\mathbf{x}_i^k\right)(u, v, c) = \sum_{h=0}^{H-1}\sum_{w=0}^{W-1} \mathbf{x}_i^k(h, w, c) e^{-j2\pi\left(\frac{h}{H}u+\frac{w}{W}v\right)} \tag{8.10}$$

This frequency space signal $\mathcal{F}_i^k$ can be further decomposed to an amplitude spectrum $\mathbf{A}_i^k \in \mathbb{R}^{H\times W\times C}$ and a phase spectrum $\mathbf{P}_i^k \in \mathbb{R}^{H\times W\times C}$, which respectively represent the low-level styles and high-level semantics of the image. To exchange the distribution information across client domains, a distribution bank $\mathbf{A} = [\mathbf{A}^1, \cdots, \mathbf{A}^K]$ is constructed to store the style information, where each $\mathbf{A}^K = \{\mathbf{A}_i^K\}_{i=1}^{N^k}$ contains all amplitude spectrum of images from the kth domain.

Continuous Frequency Space Interpolation: Based on the decomposed amplitude spectrum, which represents style information, and the phase spectrum, which encapsulates semantic information, a continuous interpolation mechanism within the frequency space is designed to transmit multi-source distribution information and generate out-of-domain distributions. For a local image $\mathbf{x}_i^k$ from client k, some low-frequency components of its amplitude spectrum are replaced with the ones in distribution bank $\mathbf{A}$, while its phase spectrum is unaffected to preserve the semantic content. Additionally, by continuously interpolating between amplitude spectrum of local data and the transferred amplitude spectrum of other domains, the method effectively enriches the diversity and quantity of source domains. Formally, this process generates new distributions of amplitude spectrum $\mathbf{A}_{i,\lambda}^{k\to n}$ for local client k and external client n:

$$\mathbf{A}_{i,\lambda}^{k\to n} = \left((1-\lambda)\mathbf{A}_i^k + \lambda\mathbf{A}_j^n\right) * \mathbf{M} + \mathbf{A}_i^k * (1-\mathbf{M}) \tag{8.11}$$

where $\mathbf{A}_j^n$ is amplitude spectrum randomly sampled from the distribution bank, $\mathbf{M} = \mathbb{I}_{(h,w)\in[-\alpha H:\alpha H, -\alpha W:\alpha W]}$ is a binary mask which controls the scale of low-frequency component within amplitude spectrum to be exchanged, λ is the interpolation ratio adjusting the amount of distribution information contributed by $\mathbf{A}_i^k$ and $\mathbf{A}_j^n$. Finally, after obtaining the interpolated amplitude spectrum $\mathbf{A}_{i,\lambda}^{k\to n}$, it can be combined with the original phase spectrum to generate the transformed image $\mathbf{x}_{i,\lambda}^{k\to n}$ via inverse Fourier transform $\mathcal{F}^{-1}$ as follows:

$$\mathbf{x}_{i,\lambda}^{k\to n} = \mathcal{F}^{-1}(\mathbf{A}_{i,\lambda}^{k\to n}, \mathbf{P}_i^k) \tag{8.12}$$

where the generated image $\mathbf{x}_{i,\lambda}^{k\to n}$ preserves the original semantics of $\mathbf{x}_i^k$ while incorporating a new distribution. As illustrated in Fig. 8.10 (lower part), the interpolation operation enables generated samples to span the intermediate space between different distributions across domains.

8.3.1.2 Invariant Representation-Based Multi-source Domain Generalization

Invariant representation-based MDG methods aim to minimize differences in feature representations across multiple source domains, thereby learning a domain-invariant feature representation that enhances the model's adaptability to unseen domains. Key techniques within this approach include explicit feature alignment, domain-adversarial learning, and feature disentanglement.

For instance, DoFE [29] introduces a domain-oriented feature embedding framework to enhance the generalization ability of CNNs to unseen target domains by leveraging knowledge from multiple source domains. This approach dynamically enriches image features with domain-specific knowledge from diverse sources, making semantic features more distinguishable. Similarly, ICSL [30] proposes a universal representation learning method by adjusting the inductive bias. It first diversifies the training distribution by blending the styles of training instances, enabling CNNs to learn from a broader range of domains or styles. Based on this diversified distribution, ICSL employs a dual-branch invariant content synergistic learning strategy to mitigate style-biased predictions while preserving invariant content. Furthermore, CDD [31] introduces a contrastive domain disentanglement network that decomposes medical images into anatomical and modality representation factors. To enhance domain-invariant representations, it uses a contrastive loss to encourage modality representations within the same domain to cluster closely while separating those from different domains. Finally, CDD incorporates a domain augmentation strategy to randomly generate new domains, further improving the model's generalization during training.

As a representative MDG method based on invariant representation, DoFE [29] aims to develop universal representations for multi-source domain data to tackle cross-center fundus image segmentation tasks. It introduces a domain-oriented feature embedding framework designed to dynamically enrich image features with domain-specific knowledge, incorporating discriminative information from diverse sources. This is achieved through a dynamic feature embedding mechanism, as illustrated in Fig. 8.11. This approach enhances network generalization in two ways:

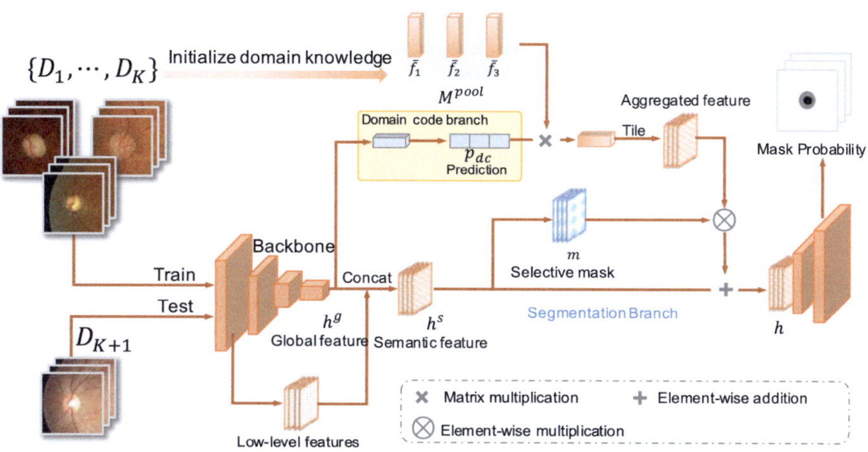

Fig. 8.11 A schematic view of DoFE framework is presented. DoFE adopt the domain knowledge pool ($\mathbf{M}^{pool}$) to learn and memorize the multi-source domain prior knowledge. The framework dynamically enriches the image semantic feature ($\mathbf{h}^s$) with domain-oriented aggregated feature ($\hat{\mathbf{h}}^{agg}$) extracted from ($\mathbf{M}^{pool}$) to improve the expressiveness of the semantic feature

(1) by integrating semantic features of test images with domain-oriented features, effectively augmenting features automatically; and (2) by incorporating domain prior knowledge and supplementary information to enrich features and improve their discriminative capability. The framework includes a domain knowledge pool $\mathbf{M}^{pool}$ to learn and retain domain-specific information from multi-source training datasets. It dynamically generates a domain-oriented aggregated feature $\mathbf{h}^{agg}$ from the domain knowledge pool based on the similarity between the input test image and images from multiple source domains. The framework is constructed in the following four main steps:

Domain Knowledge Pool Construction: The knowledge pool consists of multiple items, each representing domain-specific knowledge from a single training dataset in the source domains. Specifically, DoFE uses discriminative prototypes from each source domain as domain prior knowledge. For an input image $\mathbf{x}_i^k$ from the k-th source domain dataset $\mathcal{D}_k$, DoFE extracts high-level semantic features $\mathbf{f}_k^i$ from a pre-trained segmentation network. The average semantic feature across the spatial dimensions is then computed and used as the initialization for the k-th item in the domain knowledge pool.

$$\bar{\mathbf{f}}_k = \frac{1}{N_k \times H \times W} \sum_{j}^{H \times W} \sum_{i}^{N_k} \mathbf{f}_k^{i,j} \tag{8.13}$$

where N_k denotes the number of input images from the kth domain, and H, W represent the height and width of the feature map.

Domain Similarity Learning: The goal of this step is to generate the domain-oriented aggregated feature ($\mathbf{h}^{agg}$) from the domain knowledge pool ($\mathbf{M}^{pool}$) to enrich the original semantic features and enhance their discriminative capabilities. Since the input image may share varying degrees of similarity with images from different source domains, the domain prior knowledge contributes unequally during the calculation of $\mathbf{h}^{agg}$. To address this, DoFE employs a novel domain code prediction branch integrated with the segmentation network. This branch automatically learns the relationships between the input image and the multi-source domains. Specifically, it predicts a domain code ($\mathbf{p}_{dc}$) based on the domain attributes of the training dataset, which indicate the source domain of each training image. During training, the branch differentiates domains in the multi-source training dataset, and during testing, it estimates the similarity between the input image and the source domains to guide the aggregation process.

Domain-oriented Feature Aggregating: The domain-oriented aggregated feature ($\mathbf{h}^{agg}$) is formulated as a weight sum of the items in the domain knowledge pool according to the domain code ($\mathbf{p}_{dc}$):

$$h^{agg} = \mathbf{p}_{dc} \times \mathbf{M}^{pool} \tag{8.14}$$

where $\times$ represents the matrix multiplication, $\mathbf{p}_{dc} \in \mathbb{R}^{K}$, $\mathbf{M}^{pool} \in \mathbb{R}^{K\times H\times W\times C}$, and $\mathbf{h}^{agg} \in \mathbb{R}^{H\times W\times C}$. The feature $\mathbf{h}^{agg}$ is then tiled into $\hat{\mathbf{h}}^{agg}$ to match the shape of h^{s} for further calculation.

Dynamic Feature Embedding: In the final step, DoFE dynamically enhances the original semantic feature $\mathbf{h}^{s}$ by integrating the tiled domain-oriented aggregated feature $\mathbf{h}^{agg}$ using an attention mechanism to improve their discriminative capability. Specifically, it add a convolutional layer after $\mathbf{h}^{s}$ and use a tanh activation layer to generate a self-attention map $\mathbf{m}$:

$$\mathbf{m} = \tanh(\text{conv}(\mathbf{h}^{s})) \tag{8.15}$$

The final dynamic domain-oriented feature $\mathbf{h}$ can be formulated as:

$$\mathbf{h} = \mathbf{h}^{s} + m \otimes \mathbf{h}^{\hat{agg}} \tag{8.16}$$

where $\otimes$ is element-wise multiplication. The dynamic domain-oriented feature h is further fed into the following segmentation components to generate the fine-grained segmentation masks.

8.3.1.3 Learning Strategy-Based Multi-source Domain Generalization

Many learning strategies are widely A variety of learning strategies have been introduced to effectively address the challenges of domain generalization, each focusing on enhancing the model's adaptability to unseen domains. Methods in this category leverage advanced strategies, including ensemble learning, meta-learning, self-supervised learning, and other innovative strategies.

Ensemble learning methods play a crucial role in machine learning by enhancing generalization through specific network designs and training strategies that leverage relationships across multiple source domains. For example, MCN [32] proposed a mixture of calibrated networks for brain tumor segmentation, combining predictions from multiple base models, each with unique calibration features, to achieve more precise tumor boundaries and improved segmentation accuracy.

In contrast, meta-learning, often described as "learning to learn," focuses on improving the learning process by simulating domain shifts and enabling models to adapt quickly to new tasks or domains. This approach is particularly valuable in domain generalization (DG), where the model must generalize to unseen target domains. SAML [33] introduces a shape-aware meta-learning framework for prostate MRI segmentation, employing a gradient-based meta-learning strategy. It simulates domain shifts during training by creating virtual meta-train and meta-test tasks, thereby improving model generalization. Additionally, SAML incorporates two specialized loss functions designed to preserve object shapes and enhance segmentation boundary delineation.

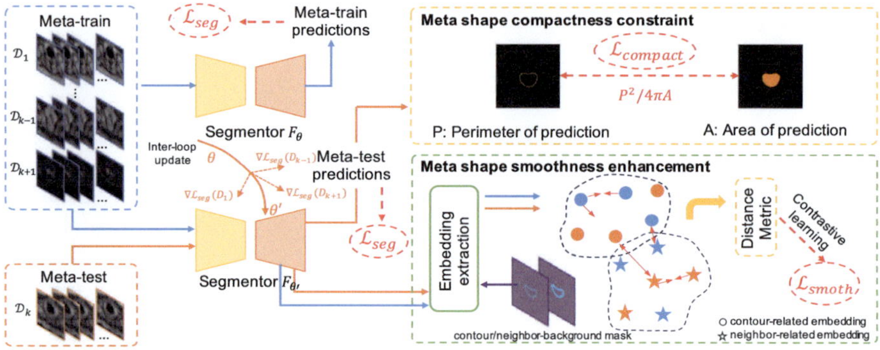

Fig. 8.12 A schematic view of SAML learning scheme is presented. The source domains are randomly split into meta-train and meta-test to simulate the domain shift. The shape compactness and smoothness in meta-test are constrained to preserve the complete shape and smooth boundary of the segmentation mask

Self-supervised learning (SSL) leverages large-scale unlabeled data to enhance feature learning and improve generalization. Inspired by this, FSS [34] proposed a superpixel-based SSL approach for few-shot semantic segmentation. By generating pseudo-labels to guide the learning process and integrating an adaptive local prototype pooling module into prototypical networks, FSS addresses challenges such as foreground-background imbalance and significantly improves segmentation performance.

Among these strategies, meta-learning techniques have shown exceptional promise in medical imaging, where annotated data is often scarce, and models must adapt rapidly to new data domains. In the following subsection, we delve into the details of the shape-aware meta-learning approach proposed by SAML [33], illustrated in Fig. 8.12. This method employs gradient-based meta-learning to explicitly simulate domain shifts and introduces tailored loss functions to preserve object shapes and refine segmentation boundaries.

Gradient-Based Meta-Learning Scheme: The learning scheme adopts a gradient-based meta-learning algorithm, which enhances the robustness of the optimization process by simulating real-world domain shifts during training. Specifically, in each iteration, the source domains D are randomly split into meta-train D_{tr} and meta-test D_{te} sets. Meta-learning consists of two steps. First, the model parameters $\boldsymbol{\theta}$ are updated on data from meta-train D_{tr} using segmentation loss L_{seg}:

$$\boldsymbol{\theta}' = \boldsymbol{\theta} - \alpha \nabla_{\boldsymbol{\theta}} L_{seg}(D_{tr}; \boldsymbol{\theta}) \tag{8.17}$$

where α is the learning-rate for this inner-loop update. Second, the method then aims at enforcing the learning on meta-train D_{tr} to exhibit certain properties on unseen meta-test D_{te} by applying a meta-learning step. Intuitively, this training scheme not only learns the segmentation task on meta-train D_{tr} but also learns how to generalize across the simulated domain shift between meta-train D_{tr} and meta-test D_{tr}.

Meta Shape Compactness Constraint and Smoothness Enhancement: In addition to meta-learning training strategies, SAML aims for the model to preserve the complete shape (compactness) and smooth boundary (smoothness) of the segmentation mask in unseen target domains. To achieve this, SAML designs meta shape compactness constraint and smoothness enhancement strategies. Firstly, based on the observation that the target region (prostate) typically exhibits a compact shape across different domains, SAML proposes to explicitly include a compact shape constraint in the meta-objective L_{Meta}. It uses the Iso-Perimetric Quotient (IPQ) measurement [35], a well-established method for quantifying shape compactness, defined as $C_{IPO} = 4\pi A/P^2$, where P and A represent the perimeter and area of the shape, respectively. The shape compactness loss is defined as the reciprocal of C_{IPO} and is applied pixel-wise as:

$$L_{\text{compact}} = \frac{P^2}{4\pi A} = \frac{\sum_{i\in\Omega}\sqrt{(\nabla \mathbf{p}_{u_i})^2 + (\nabla \mathbf{p}_{v_i})^2 + \epsilon}}{4\pi\left(\sum_{i\in\Omega}|\mathbf{p}_i| + \epsilon\right)} \tag{8.18}$$

where $\mathbf{p}$ is the prediction probability map, Ω is the set of all pixels in the map. ∇p_{u_i} and ∇p_{v_i} are the probability gradients for each pixel i in direction of horizontal and vertical. ε is a hyperparameter for computation stability.

In addition to focusing on the image shapes, SAML encourages smooth boundary delineation by enhancing intra-class cohesion and inter-class separation between contour-relevant and background-relevant embeddings across all domains D. For an image $\mathbf{x}_m \in \mathbb{R}^{H\times W\times 3}$ and its one-hot label $\mathbf{y}_m$, the activation map from layer l, $\mathbf{M}_m^l \in \mathbb{R}^{H_l\times W_l\times C_l}$, is interpolated to $\mathbf{T}_m^l \in \mathbb{R}^{H\times W\times C_l}$ using bilinear interpolation to keep consistency with the dimensions of $\mathbf{y}_m$. Contour-relevant embeddings $\mathbf{E}_m^{con} \in \mathbb{R}^{C_l}$ and background-relevant embeddings $\mathbf{E}_m^{bg} \in \mathbb{R}^{C_l}$ are extracted by first deriving binary contour mask $\mathbf{c}_m \in \mathbb{R}^{H\times W\times 1}$ and binary background mask $\mathbf{b}_m \in \mathbb{R}^{H\times W\times 1}$ from $\mathbf{y}_m$ using morphological operations. These embeddings $\mathbf{E}_m^{con}$ and $\mathbf{E}_m^{bg}$ are then obtained from $\mathbf{T}_m^l$ through a weighted average operation over $\mathbf{c}_m$ and $\mathbf{b}_m$:

$$\mathbf{E}_m^{con} = \frac{\sum_{i\in\Omega}(\mathbf{T}_m^l)_i \cdot (\mathbf{c}_m)_i}{\sum_{i\in\Omega}(\mathbf{c}_m)_i}, \quad \mathbf{E}_m^{bg} = \frac{\sum_{i\in\Omega}(\mathbf{T}_m^l)_i \cdot (\mathbf{b}_m)_i}{\sum_{i\in\Omega}(\mathbf{b}_m)_i} \tag{8.19}$$

where Ω denotes the set of all pixels in $\mathbf{T}_m^l$, $\mathbf{E}_m^{con}$ and $\mathbf{E}_m^{bg}$ are single vectors, representing the contour and background-relevant representations extracted from the whole image $\mathbf{x}_m$.

To enhance the domain-invariance of $\mathbf{E}^{con}$ and $\mathbf{E}^{bg}$ in latent space, SAML promotes intra-class cohesion and inter-class separation among samples from all source domains $\mathcal{D}$. SAML employs contrastive learning [36] for this purpose. Specifically, an embedding network H_ϕ is introduced to project the features $\mathbf{E} \in [\mathbf{E}^{con}, \mathbf{E}^{bg}]$ to a lower-dimensional space. The distance between feature vectors $\mathbf{E}_m$ and $\mathbf{E}_n$ from network H_ϕ is computed as $d_\phi(\mathbf{E}_m, \mathbf{E}_n) = \|H_\phi(\mathbf{E}_m) - H_\phi(\mathbf{E}_n)\|_2$, where the sample pair (m, n) is randomly selected from all domains $\mathcal{D}$. This approach aims to align the embedding spaces of $\mathcal{D}_{te}$ and $\mathcal{D}_{tr}$ to capture domain-invariant representations, particularly around the boundary region. The contrastive loss is defined as follows:

$$L_{contrastive}(m, n) = \begin{cases} d_\phi(\mathbf{E}_m, \mathbf{E}_n), & \text{if } \tau(\mathbf{E}_m) = \tau(\mathbf{E}_n) \\ (\max\{0, \zeta - d_\phi(\mathbf{E}_m, \mathbf{E}_n\})^2, & \text{if } \tau(\mathbf{E}_m) \neq \tau(\mathbf{E}_n) \end{cases} \tag{8.20}$$

where the function $\tau(\mathbf{E})$ indicates the class (1 for $\mathbf{E}$ being $\mathbf{E}^{con}$, and 0 for $\mathbf{E}^{bg}$), ζ is a pre-defined distance margin following the practice of metric learning (set as 10 in the experiments).

The overall objective L_{smooth} is computed within mini-batch of q samples. SAML randomly employs either E^{con} or E^{bg} for each sample, and the L_{smooth} is the average of $\ell_{contrastive}$ over all pairs of (m, n) embeddings:

$$L_{smooth} = \sum_{m=1}^{q} \sum_{n=m+1}^{q} \ell_{contrastive}(m, n)/C(q, 2) \tag{8.21}$$

where $C(q, 2)$ is the number of combinations. Overall, all training objectives including $L_{seg}(\mathcal{D}_{tr}; \theta)$ and $L_{meta}(\mathcal{D}_{tr}, D_{te}; \theta')$, are optimized together with respect to the original parameters θ. The L_{smooth} is also optimized with respect to H_ϕ.

8.3.2 Single Domain Generalization Setting

In multi-source domain generalization, models are trained using multiple labeled source domains to enhance robustness when faced with unseen domains. However, this approach requires extensive data collection and annotation, which is particularly costly in medical image segmentation due to the intensive labor and expertise involved. These challenges highlight the importance of single-source domain generalization (SDG), which seeks to overcome these limitations by enabling models to generalize effectively with only one source domain. In this section, we categorize SDG into two main approaches: domain augmentation-based and learning strategy-based. And then, we will review representative works in each category and further introduce a novel hybrid strategy-based SDG approach to address the limitations of current methods.

8.3.2.1 Domain Augmentation-Based Single-Source Domain Generalization

Similar to data manipulation methods in MDG, domain augmentation-based methods aim to use a single source domain to mimic unseen target domain distributions. Hence, enhancing the diversity of these simulated target domain distributions is crucial for improving generalization capability.

Recently, RST [37] proposes a random style transfer network designed for multi-vendor and multi-center cardiac image segmentation. This network randomly generates a modality vector during the style transfer process to simulate domain shifts for unseen domains. Additionally, it incorporates spatial information and shape regularization terms to improve segmentation accuracy before focusing on specific areas. ADS [38] introduces an adversarial domain generalization approach for organ segmentation using a single domain. ADS employs an adversarial domain synthesizer to generate synthetic domains, assuming these domains encompass a sufficiently diverse distribution to interpolate effectively for unseen domains. Furthermore, it introduces a mutual information regularizer to maintain semantic consistency across images from synthetic domains, which is estimated using patch-level contrastive learning. SLAug [39] developes a novel augmentation strategy, termed location-scale augmentation, to tackle challenges in single-source domain generalization. This approach applies constrained Bézier transformations to both global and local (class-level) regions, significantly increasing augmentation diversity while leveraging class-level information. SLAUG also proposes a Saliency-Balancing Fusion mechanism, which incorporates gradient information to guide augmentations with optimal orientation and magnitude, enhancing their informativeness. A detailed introduction to this method follows in the next section.

Previous augmentation methods, while effective, face several limitations in medical image segmentation. First, global-only augmentation, which transforms the entire image uniformly, restricts the diversity of augmented images. Second, random augmentation lacks constraints on sample distribution, potentially leading to over-generalization and performance decline. Third, general-task-specific augmentation, designed for natural images, often lacks diversity and informativeness crucial for covering diverse target domain distributions in medical image tasks. Their augmented samples are usually insufficient in diversity and informativeness, thus failing to cover the possible target domain distribution.

To overcome these challenges, SLAug introduces a novel augmentation strategy called location-scale augmentation, comprising two modules: global location-scale augmentation (GLA) and local location-scale augmentation (LLA). This dual approach significantly enhances the diversity of augmented samples. Additionally, it developes a saliency-balancing fusion (SBF) strategy to mitigate over-generalization in random augmentation. SBF uses gradient distribution to guide the augmentation process, ensuring it is both informative and directionally focused.

Location-scale Augmentation: SLAug explores two types of augmentation: Global Location scale Augmentation (GLA), which increases the source-like images through global distribution shifting. While Local Location-scale Augmentation (LLA) focuses on class-specific augmentation to capture diverse or extreme appearances from unseen domains. Specifically, it uses a monotonic non-linear transformation Cubic $B\acute{e}zier$ Curve, as the mapping function for augmentation. This function is constrained by the minimum and maximum values within the non-blank image regions.

$$Bézier(t) = \sum_{k=0}^{3}(1-t)^{3-k}t^k P_k,\ t \in [v_{\text{low}}, v_{\text{high}}] \tag{8.22}$$

where v_{low} and v_{high} are the lowest and highest value of input image, t is a fractional value along the length of the line. $P_0 = (v_{\text{low}}, v_{\text{low}})$ and $P_3 = (v_{\text{high}}, v_{\text{high}})$ are the start and end points to limit the range of value. P_1 and P_2 are the control points whose values are randomly generated from $[v_{\text{low}}, v_{\text{high}}]$. The inverse mapping function can be generated by swapping the start point and end point to $P_0 = (v_{\text{low}}, v_{\text{high}})$ and $P_3 = (v_{\text{high}}, v_{\text{low}})$. This non-linear transformation function is tagged as $\mathcal{F}_{\text{p}}(\cdot)$, where p is the probability to perform the inverse mapping.

For image **x** and corresponding mask **m**, before augmentation, the input image **x** is min–max normalized in the range of [0, 1]. Then, GLA performs global distribution shifting, it can be described as:

$$\text{GLA}(\mathbf{x}) = \alpha\mathcal{F}_0(\mathbf{x}) + \beta \tag{8.23}$$

where $\alpha \sim \mathcal{TN}(1, \sigma_1)$ and $\beta \sim \mathcal{TN}(0, \sigma_2)$ are location-scale factors. σ_1 and σ_2 are standard deviations of the two truncated Gaussian distributions, respectively. Using $\mathcal{F}_0$, where the probability of inversion is zero, ensures that GLA samples maintain a similar appearance to the original images, as depicted in Fig. 8.13 (right part A).

In contrast, Local Location-scale Augmentation (LLA) treats individual class-level regions as processing units and applies transformations independently. The augmented regions are then linearly combined. This approach can be illustrated as:

$$\text{LLA}(\mathbf{x}, \mathbf{m}) = \sum_{c=1}^{C} \alpha_c \mathcal{F}_{\text{p}_c}(\mathbf{x}^c) + \beta_c \tag{8.24}$$

where $\alpha_c \sim \mathcal{TN}(1, \sigma_1)$ and $\beta_c \sim \mathcal{TN}(0, \sigma_2)$ are location-scale factors. For all classes, $\text{p}_c = 0.5$ is set to apply random inversion, except $\text{p}_1 = 1$ is set to ensure the LLA augmented images are dissimilar to the GLA samples.

Saliency-balancing Fusion: SLAug introduces Saliency-balancing Fusion (SBF), a mechanism that uses gradient distribution to guide the augmentation process. The

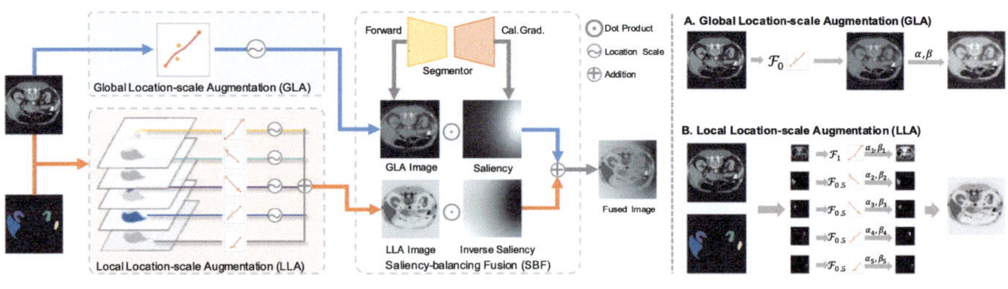

Fig. 8.13 Left: Workflow of the saliency-balancing location-scale augmentation (SLAug) proposed by Su et al. [39]. Right: Illustration of proposed global and local location-scale augmentation

goal is to preserve sensitive areas (high-gradient areas) while expanding the training data span. SBF computes a saliency map by taking the l_2 norm of gradient values across input channels, downsampling it to a $g \times g$ grid, and then interpolating it back to the original image size using quadratic B-spline kernels for smoothing. Algorithm 1 outlines the entire augmentation pipeline.

Algorithm 1 Saliency-balancing Location-scale Augmentation

Require: Training data $(\mathbf{x}, \mathbf{m})$, network f_θ, loss function L, global location-scale augmentation GLA, local location-scale augmentation LLA, common augmentation F.

1: $\mathbf{x}^g \leftarrow \text{GLA}(\mathbf{x})$ ▷ See Eq. (8.23)
2: $\mathbf{x}^l \leftarrow \text{LLA}(\mathbf{x}, \mathbf{m})$ ▷ See Eq. (8.24)
3: $\tilde{\mathbf{x}}^g, \tilde{\mathbf{x}}^l, \tilde{\mathbf{m}} \leftarrow F(\mathbf{x}^g, \mathbf{x}^l, \mathbf{m})$ ▷ Common augmentation
4: Calculate gradient $Grad = \nabla_{\tilde{\mathbf{x}}^g} L(f_\theta(\tilde{\mathbf{x}}^g), \tilde{\mathbf{m}})$
5: $\mathbf{s} \leftarrow normalize(smooth(|Grad|))$ ▷ Calculate saliency
6: $\tilde{\mathbf{x}}^{fused} \leftarrow \mathbf{s} \odot \tilde{\mathbf{x}}^g + (1 - \mathbf{s}) \odot \tilde{\mathbf{x}}^l$
7: **return** $\tilde{\mathbf{x}}^{fused}$

8.3.2.2 Learning Strategy-Based Single-Source Domain Generalization

As representative approaches under the category of learning strategy-based SDG methods, u-FUS [40] introduces a dynamic CNN for localization through ensemble learning, effectively combining image and optical flow data to provide the most reliable information for predictions. TASD [41] employs dictionary learning to extract shape patterns from a single source and devises a strategy for adapting these patterns during testing. It leverages dual-consistency regularization to enhance the model's capability to handle new domains, thereby improving generalizability. Meanwhile, GIN [42] presents a causal learning framework for CT image segmentation across different domains. It utilizes a structural causal model to represent the causal relationships among input data, domain shifts, and task-specific outputs, aiming to facilitate generalization from a single source to multiple domains.

Integrating Shape Dictionary: The semantic shapes of segmentation reflect the underlying anatomical structures in medical images, thus are unaffected by scanner variations across observed domains. TASD aims to leverage this general shape information, which remains consistent even within a single source domain, to improve model generalization in SDG. TASD starts by creating a dictionary $\mathcal{D}$ of K explicit shape templates, denoted as $\mathcal{D} = \{d_1, d_2, \ldots, d_K\}$, to represent the various segmentation masks $\{\mathbf{y}_i\}_{i=1}^N$ in the single source domain dataset $\mathcal{S} = \{(\mathbf{x}_i, \mathbf{y}_i)\}_{i=1}^N$, where N represents the number of labeled data pairs. To ensure the shape dictionary learned from one domain applies well to unseen domains, its construction follows two principles: (1) The number of shape templates (denoted as K) should be much smaller than the number of masks in the single source data (N), ensuring $K \ll N$. (2) Each mask y_i should be represented using the fewest possible elements.

For simplicity, consider a binary segmentation task (i.e., $\mathbf{y}_i \in \mathbb{R}^{H\times W}$). Extending this to multi-class segmentation involves creating a separate shape dictionary for each class in a similar manner. Then the creation of dictionary involves solving an optimization problem with respect to the shape dictionary $\mathcal{D} \in \mathbb{R}^{K\times H\times W}$ and the coefficients $\alpha = [\alpha_1, \alpha_2, \ldots, \alpha_N]$ ($\alpha_i \in \mathbb{R}^K$ denotes the learned coefficients of dictionary for each $\mathbf{y}_i$), defined as:

$$\arg\min_{\mathbf{D},\alpha} \sum_{i=1}^{N} \left(||\mathbf{y}_i - \sum_{j=1}^{K} \mathbf{d}_j \alpha_{ij}||_2^2 + \lambda ||\alpha_i||_1 \right) \tag{8.25}$$

where λ is a balancing parameter to regularize the learned coefficients.

After obtaining the learned shape dictionary, TASD further incorporates the shape dictionary into the deep model, aiming to enable the segmentation to be performed with references from these shape priors for boosting generalization. Specifically, the pre-extracted shape priors in $\mathcal{D}$ are linearly combined according to the shape coefficients $\hat{\alpha}$ to generate a reference mask:

$$M = \sum_{j=1}^{K} \mathbf{d}_j \hat{\alpha}_{ij} \tag{8.26}$$

Which is then concatenated with extracted features before the last convolutional block of decoder for further refinement and generate the final segmentation mask. In addiction, since α is regularized to be sparse during dictionary learning, the cosine similarity is adopted to minimize the distance between $\hat{\alpha}$ and α instead of the L1 or L2 loss, which is expressed as:

$$L_{regress}(\mathbf{x}_i) = 1 - \frac{\hat{\alpha}_i \cdot \alpha_i}{\max(||\hat{\alpha}_i||_2 \cdot ||\alpha_i||_2, \epsilon)} \tag{8.27}$$

where ϵ is a small value to avoid division by zero.

Dual-consistency Regularized Test-time Update: By adding a regression branch to the network, shape priors are incorporated to aid in segmentation. However, a concern arises: the model learns to use this prior information (including parameters from the regression branch and the original network) based on narrow data distributions within a single domain. When applied to new domains with different distributions, the reliability of the generated shape coefficients may be uncertain, potentially limiting the effective use of shape priors. Recognizing that test-time samples provide clues about their distribution, TASD proposes dynamically adjusting model parameters during inference. This approach aims to effectively utilize shape priors across diverse, unseen data distributions, thereby enhancing overall generalization performance. To achieve this goal, adaptation can be implemented as a consistency regularization mechanism. This approach can be applied flexibly to both the original network and the embedded regression branch.

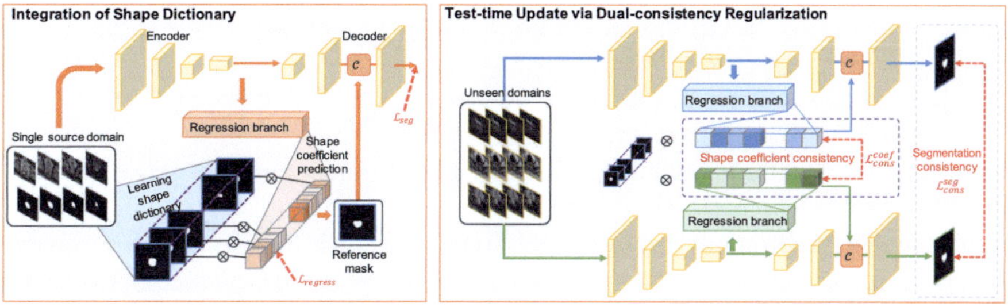

Fig. 8.14 Overview of the TASD. For training, TASD establishes dictionary learning to extract the explicit shape priors and integrate these prior information into the network. During testing on new domains, it dynamically adapts the model parameters especially on the regression branch to adaptively utilize the prior information to improve generalization ability

In the approach illustrated in the right part of Fig. 8.14, at test time with an inference data $\mathbf{x}^t$, different perturbations η and η' are applied (e.g., adding Gaussian noise to the input or dropout to network parameters). The model is then regularized to ensure consistent predictions for these perturbed samples, aiming to extract inherent information from the test data. Importantly, to fully adapt the parameters of the regression branch independently from the original network, consistency regularization is applied to both segmentation predictions and shape coefficient predictions. The intuition is that variations in input perturbations should not alter the semantic anatomy within a medical image. Therefore, the final segmentation mask of an inference sample and how the model utilizes pre-collected shape priors (i.e., shape coefficients) to generate the mask should remain unchanged. Formally, these consistency regularization terms can be expressed as:

$$
\begin{aligned}
L_{cons}^{seg}(\mathbf{x}^t) &= ||f(\mathbf{x}^t, \eta') - f(\mathbf{x}^t, \eta)||_2^2 \\
L_{cons}^{coef}(\mathbf{x}^t) &= 1 - \frac{\hat{\alpha}^t(\eta') \cdot \hat{\alpha}^t(\eta)}{\max(||\hat{\alpha}^t(\eta')||_2 \cdot ||\hat{\alpha}^t(\eta)||_2, \epsilon)}
\end{aligned}
\tag{8.28}
$$

where $L_{cons}^{seg}(\mathbf{x}^t)$ is the consistency regularization on segmentation results with mean square error loss, in which $f(\cdot)$ denotes the network, η and η' are different perturbations applied to the same testing data $\mathbf{x}^t$. $L_{cons}^{coef}(\mathbf{x}^t)$ denotes the consistency regularization on shape coefficients with cosine distance loss, in which $\hat{\alpha}^t(\cdot)$ is the shape coefficient predictions for $\mathbf{x}^t$ under perturbation $(\cdot)$. Our dual-consistency regularization objective can then be expressed as:

$$
L_{cons}^{dual}(\mathbf{x}^t) = L_{cons}^{seg}(\mathbf{x}^t) + L_{cons}^{coef}(\mathbf{x}^t) \tag{8.29}
$$

At test time, only one iteration of gradient descent is required to optimize Eq. (8.29) and adapt the parameters of the source model based on each inference data point.

8.3.2.3 Hybrid Strategy-Based Single-Source Domain Generalization

Although domain augmentation-based and learning strategy-based SDG approaches can improve model generalization performance, they each face limitations. Augmentation-based methods often struggle to accurately simulate target distributions, frequently leading to overfitting on the regular patterns of synthetic data and limiting generalization capabilities. Additionally, medical images contain intricate tissue and organ details, making it challenging for generation-based methods to create high-quality virtual data. Learning strategy-based methods, such as meta-learning, tend to be slow during inference, reducing their efficiency for practical applications.

To address these limitations, we propose a novel SDG method from a hybrid strategy perspective, named IRLSG [43], which incorporates three innovative features: Class-scale Photometric Augmentation simulates diverse and informative target domains by augmenting the original data, thereby enhancing the diversity and informativeness needed for robust generalization. Following this, a Dual-Consistency Framework is applied to enforce consistency between intermediate features and segmentation results of the original and augmented images, promoting the exploration of domain-invariant representations. Style Feature Whitening is a straightforward and effective method that removes domain-specific styles from higher-order covariance statistics, enhancing the network's ability to learn domain-invariant representations and improve generalization.

Class-scale augmentation (CSA): Different from previous methods that apply global-scale augmentation, IRLSG use class-specific information to perform class-scale augmentation for each class, enhancing the diversity and informativeness of the augmented samples. Inspired by ObjectAug [44], which separates an image into the background and objects using masks and then replaces the objects on an inpainted background, the image is decomposed into multiple categories based on the mask, and each part is enhanced separately. Specifically, given an image $\mathbf{x}$ and its mask $\mathbf{m} \in \{1, 2, \ldots, C\}^{H\times W}$, where C is the number of classes including the background, $\mathbf{m}^c$ is defined as the binary mask of class c. The decomposition of an image by semantic class is represented as:

$$\mathbf{x}^c = \sum_{c=1}^{C} \mathbf{m}^c \odot \mathbf{x} \tag{8.30}$$

where $\mathbf{x}^c$ are the image regions belong to class c, $\odot$ represents dot product. Furthermore, inspired by BigAug [25], photo-metric transformation is chosen as the augmentation strategy, including brightness, hue, saturation, contrast transformations, and Gaussian noise or blurring. The augmented data can be represented as:

$$\mathcal{A}(\mathbf{x}) = \sum_{c=1}^{C} \mathcal{PM}(\mathbf{x}^c) \tag{8.31}$$

where $\mathcal{PM}(\cdot)$ means photo-metric transformation. The original and augmented data with shared labels are used to update network weights. The combination of per-pixel cross-entropy and soft Dice loss is employed, formulated as:

$$L_{task} = \frac{1}{2}CE\left(f\left(\mathbf{X}_i\right), \mathbf{m}_i\right) + \frac{1}{2}Dice\left(f\left(\mathbf{X}_i\right), \mathbf{m}_i\right) \tag{8.32}$$

where i is the index of mini-batch, $\mathbf{X}_i$ represents original images $\mathbf{x}_i$ and the corresponding augmented images $\mathcal{A}\left(\mathbf{x}_i\right)$, $f(\cdot)$ denotes the network, $CE(\cdot)$ and $Dice(\cdot)$ are cross-entropy and soft Dice loss.

Dual-Consistency Framework (DCF): By applying instance augmentation to the source domain, the domain shift problem is reduced. However, the generalization benefit from data augmentation alone is limited. Observing that images with the same content but different styles can produce different segmentation results, the aim is to learn domain-invariant representations by using consistency constraints during training. This approach enhances the model's generalization performance, enabling it to handle unseen target data distributions more effectively.

As depicted in Fig. 8.15a, regularization is applied to the model to ensure consistent predictions between the original and augmented samples. This consistency regularization is enforced not only on the final predictions but also on intermediate features to improve the network's feature extraction and decision-making capabilities. The intuition is that varying input photo-metric augmentations should not alter the semantic anatomies within a medical image. Formally, the consistency regularization terms can be formulated as follows:

$$\begin{aligned} L_{cons}^{fea} &= \left\| E\left(\mathbf{x}^i\right) - E\left(\mathcal{A}\left(\mathbf{x}^i\right)\right)\right\|_2^2 \\ L_{cons}^{seg} &= \left\| f\left(\mathbf{x}^i\right) - f\left(\mathcal{A}\left(\mathbf{x}^i\right)\right)\right\|_2^2 \end{aligned} \tag{8.33}$$

where L_{cons}^{fea} and L_{cons}^{seg} are the consistency regularization on intermediate feature and segmentation results using mean square error loss. $E(\cdot)$ denotes the deepest encoder layer and $f(\cdot)$ denotes the network. The dual-consistency regularization objective can then be expressed as:

$$L_{cons}^{dual} = L_{cons}^{fea} + L_{cons}^{seg} \tag{8.34}$$

Style Feature Whitening (SFW): Providing images with the same content but different styles to the same network results in varied predictions, as certain features extracted from the network are sensitive to style changes [45]. Quantitatively identifying these style-sensitive features is crucial for selectively removing domain-specific style influences. To achieve this, IRLSG calculate the disparities between two feature maps derived from the original and augmented images, defined as $\mathbf{M}_{sub}$. Specifically, the difference between feature maps $\mathbf{M}_{sub}$ is represented as:

$$\mathbf{M}_{sub} = \left|\mathbf{M} - \mathbf{M}'\right| \tag{8.35}$$

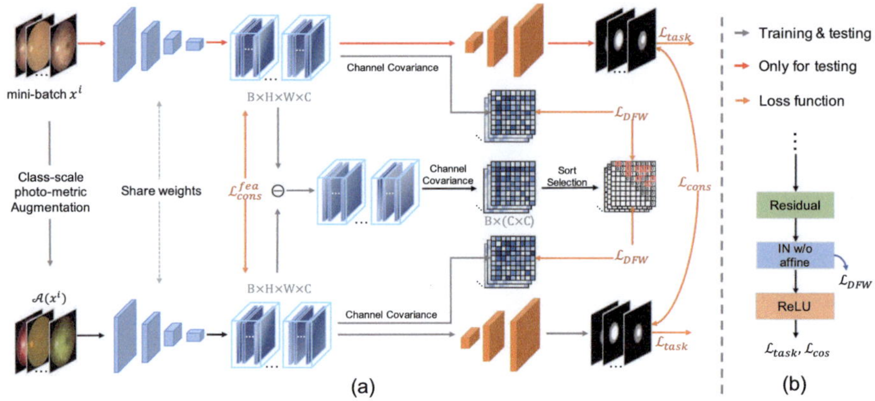

Fig. 8.15 An overview of the proposed method: IRLSG. **a** The training and testing process of IRLSG. **b** The proposed stage realizes the dual consistency and feature whitening

where $\mathbf{M}$ and $\mathbf{M}' \in \mathbb{R}^{B\times C\times H\times W}$ represent the feature maps extracted from the original and augmented images, respectively. The absolute value operation, denoted as $|\cdot|$, converts values into their absolute values. The style-sensitive feature vector $\mathbf{M}_s$ are obtained by standardizing $\mathbf{M}_{sub}$ channel-wise. This process can be formulated as:

$$\mathbf{M}_s = \frac{\mathbf{M}_{sub} - \text{Min}\,(\mathbf{M}_{sub})}{\text{Max}\,(\mathbf{M}_{sub}) - \text{Min}\,(\mathbf{M}_{sub})} \tag{8.36}$$

where Max$(\cdot)$ and Min$(\cdot)$ are the operation of calculating the maximum and minimum values along the channel dimension. A higher value in the style-sensitive feature vector $\mathbf{M}_s$ indicates greater sensitivity to changes in style. Therefore, selective removal of domain-specific styles from the feature map is achievable based on the sensitive locations identified by $\mathbf{M}_s$.

Besides enforcing consistency in intermediate features and segmentation results, removing domain-specific style information is crucial for encouraging the network to enhance its modeling ability and learn domain-invariant representations. The approach first separates features into domain-invariant and domain-specific components within the covariance space using style-sensitive vectors. Then, it selectively suppresses the domain-specific components. Importantly, the Style Feature Whitening (SFW) can be easily integrated into each stage of the network. Specifically, the elaboration on the three steps of SFW is as follows:

(1) *Covariance transformation*: The intermediate feature maps $\mathbf{M}$ and $\mathbf{M}'$, after instance normalization, along with the normalized style-sensitive vector $\mathbf{M}_s$, are used to compute the corresponding covariance matrix Σ_x, $\Sigma_{x'}$, $\Sigma_s \in \mathbb{R}^{B\times(C\times C)}$. For instance, considering the style-sensitive vector $\mathbf{M}_s$, the process can be formulated as:

$$\Sigma_s = \frac{1}{HW}\,(\mathbf{M}_s)\,(\mathbf{M}_s)^T \tag{8.37}$$

(2) *Sort selection*: The top largest positions in Σ_s are selected to create the Whitening Mask (WM). A higher value at a position indicates greater sensitivity to domain-specific style. The process can be formulated as:

$$WM = \Sigma_{s_{ij}} = \begin{cases} 1, & \text{if } \Sigma_{s_{ij}} \in \text{top } \gamma \\ 0, & \text{otherwise} \end{cases} \tag{8.38}$$

(3) *Directed whitening*: The values at corresponding positions in Σ_x and $\Sigma_{x'}$ are adjusted towards 0 based on the WM to reduce domain-specific style. Because the covariance matrix is symmetric, the loss is applied only to its upper triangular part. The directional feature whitening loss is formulated as:

$$L_{SFW} = \frac{1}{2}\mathbb{E}\left[\|WM \odot \Sigma_x\|_1\right] + \frac{1}{2}\mathbb{E}\left[\|WM \odot \Sigma_{x'}\|_1\right] \tag{8.39}$$

where $\odot$ is an element-wise multiplication and $\mathbb{E}$ denotes the arithmetic mean.

8.3.3 Other Domain Generalization Settings

In addition to MDG and SDG, several other DG settings have emerged to meet practical needs in medical image analysis.

Federated DG: Focuses on privacy protection and overcomes limitations related to data sharing and transmission.

Semi-supervised DG: Utilizes a combination of labeled and unlabeled data to reduce annotation costs, particularly beneficial for tasks like dense pixel segmentation.

Few-shot DG: Addresses scenarios where only a few labeled examples are available per target domain, aiming to generalize effectively with limited supervision.

For example, FedFA [46] FedFA tackles federated learning by augmenting local feature statistics, addressing local data bias. It uses global information to adjust each client's latent feature statistics, enhancing their understanding of the underlying distribution. The approach probabilistically augments the statistics with a determined variance, considering both local biases and global feature statistics. For semi-supervised DG, we design a IS2NET [47] framework, which integrates an inter-domain generalization branch with intra-domain semi-supervised learning branches. This framework utilizes a style and semantic memory mechanism, along with a confident pseudo-labeling strategy, to foster interaction between domains. For few-shot DG, Sun et al. [48] introduced a few-shot medical image segmentation approach that enables models to generalize quickly to unseen classes with minimal training data. They developed a deep convolutional network trained episodically, incorporating a global correlation module to model spatial consistency and regularity in medical images. Next, we will provide a detailed introduction to the proposed IS2NET.

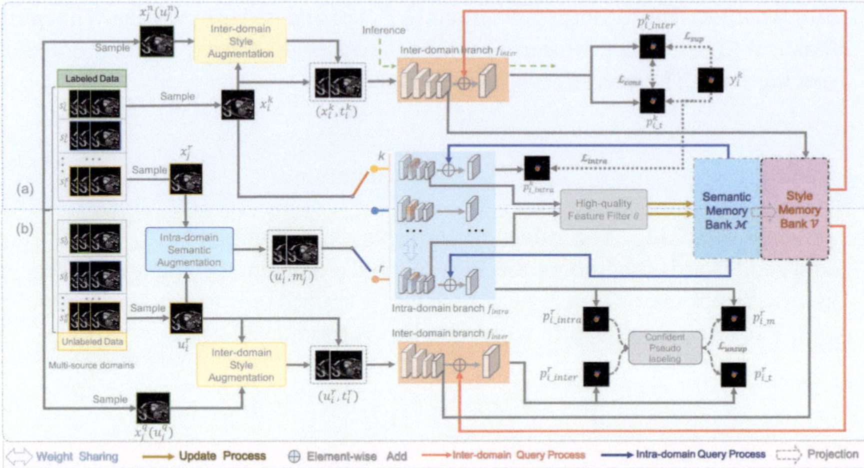

Fig. 8.16 The proposed IS2Net framework operates with multi-source domain data divided into labeled and unlabeled parts. **a** The supervised training process and **b** the unsupervised training process are illustrated. Specifically, during inference, only the inter-domain branch is utilized

Despite the success of DG methods in supervised settings, which require extensive effort for pixel-wise annotation, Semi-supervised Learning (SSL) aims to leverage a small number of annotated images alongside unlabeled data to realize more cost-effective learning. However, SSL is limited by the i.i.d. assumption and may not generalize effectively to unseen target domains. To address this, researchers have turned to a more practical approach called Semi-supervised Domain Generalization (SSDG) for medical image segmentation. In SSDG, only a small portion of data is labeled within each source domain, while a large amount remains unlabeled.

The mainstream approaches to SSDG for medical image segmentation can be divided into two categories: meta-learning-based and style-augmentation-based. Meta-learning-based methods primarily use domain-level information from unlabeled data, which may not fully exploit pixel-wise semantic features, leading to suboptimal segmentation performance. Style-augmented methods closely integrate domain generalization and pseudo-labeling, but the quality of pseudo-labels in the source domain can be affected by domain gaps.

To address these challenges, a novel Semi-supervised Domain Generalization (SSDG) scheme named IS2Net is proposed, as depicted in Fig. 8.16. Inspired by a parallel intra-inter domain architecture, the SSDG task is divided into a primary inter-domain DG task and several auxiliary intra-domain SSL tasks. Each intra-domain branch focuses independently on learning reliable label-related features without interference from other domains. A semantic and style memory bank is introduced to store high-quality label-related features. These features enhance intra-domain semantic features by establishing pixel-region relations, such as improving the quality of pseudo-labels. Additionally, IS2Net convert to domain representative information for inter-domain style feature enhancement, addressing domain distribution shifts. Given the differing learning capacities of inter and intra-domain branches, a confident pseudo-labeling strategy is implemented. This strategy forms an ensemble result by selecting more confident pixels from both branches, significantly enhancing the quality of pseudo-labels.

Mixing Augmentation Strategy: Style augmentation $\mathcal{A}ug_{inter}$ is applied during the training of the inter-domain branch to create diverse out-of-domain distributions. Currently, semantic augmentation $\mathcal{A}ug_{intra}$ is performed during the training of each intra-domain branch to thoroughly explore the semantic features. Specifically, for a labeled sample $(\mathbf{x}_i^k, \mathbf{y}_i^k)$ from the k-th domain S_L^k, a fast Fourier transform $\mathcal{F}$ is performed to obtain its frequency space signal $\mathcal{F}\left(\mathbf{x}_i^k\right)$. This signal is decomposed into an amplitude spectrum $\mathbf{A}_i^k$ and a phase spectrum $\mathbf{P}_i^k$. Another labeled or unlabeled image $\mathbf{x}_j^n(\mathbf{u}_j^n)(n \neq k)$ is randomly selected from a different domain n, and its amplitude spectrum $\mathbf{A}_j^n$ is obtained. A new amplitude spectrum $\mathbf{A}_{i,\lambda}^{k\to n}$ is synthesized by interpolating between $\mathbf{A}_i^k$ and $\mathbf{A}_j^n$. Subsequently, a transformed image $\mathbf{x}_{i,\lambda}^{k\to n}$ is generated to preserve the original semantics of $\mathbf{x}_i^k$ while incorporating a new distribution formed between $\mathbf{x}_i^k$ and $\mathbf{x}_j^n$, achieved by the inverse Fourier transform $\mathcal{F}^{-1}$. The $\mathcal{A}ug_{inter}$ process can be further formulated as:

$$\begin{aligned} \mathbf{A}_{i,\lambda}^{k\to n} &= \left((1-\lambda)\mathbf{A}_i^k + \lambda\mathbf{A}_j^n\right) * \mathbf{M} + \mathbf{A}_i^k * (1-\mathbf{M}) \\ \mathbf{x}_{i,\lambda}^{k\to n} &= \mathcal{F}^{-1}\left(\mathbf{A}_{i,\lambda}^{k\to n}, \mathbf{P}_i^k\right), \mathcal{A}ug_{inter}(\mathbf{x}_i^k, \mathbf{x}_j^n) \to \mathbf{x}_{i,\lambda}^{k\to n} \end{aligned} \tag{8.40}$$

where λ is a parameter that adjusts the ratio between the phase information of $\mathbf{A}_i^k$ and $\mathbf{A}_j^n$. $\mathbf{M}$ is a binary mask which controls the scale of low-frequency component to be exchanged.

During the training process of intra-domain branch, semantic augmentation $\mathcal{A}ug_{intra}$ is achieved by exchanging the semantic information within each domain. For the unlabeled image u_i^r from the r-th domain, another labeled pair $(\mathbf{x}_j^r, \mathbf{y}_j^r)$ is randomly sampled within the same domain. The phase spectrums $\mathbf{P}_i^r$, $\mathbf{P}_j^r$ and amplitude spectrums $\mathbf{A}_i^r$, $\mathbf{A}_j^r$ can be calculated similarly by decomposing the frequency space signal of these two samples. A new phase spectrum $\mathbf{P}_{i,\lambda}^r$ is synthesized by interpolating between $\mathbf{P}_i^r$ and $\mathbf{P}_j^r$, and it is then combined with the original amplitude spectrum $\mathbf{A}_i^r$ to generate the transformed image $\mathbf{u}_{i,\lambda}^r$ via $\mathcal{F}^{-1}$. Similar to the CutMix operation [49], a transformed pseudo-label $\hat{\mathbf{y}}_{i,\lambda}^r$ is also obtained by interpolating between the original pseudo-label $\hat{\mathbf{y}}_i^r$ and $\mathbf{y}_j^r$. The $\mathcal{A}ug_{intra}$ process can be further formulated as:

$$\begin{aligned} \mathbf{P}_{i,\lambda}^r &= \left((1-\lambda)\mathbf{P}_i^r + \lambda\mathbf{P}_j^r\right) * \mathbf{M} + \mathbf{P}_i^r * (1-\mathbf{M}) \\ \mathbf{u}_{i,\lambda}^r &= \mathcal{F}^{-1}\left(\mathbf{A}_i^r, \mathbf{P}_{i,\lambda}^r\right) \hat{\mathbf{y}}_{i,\lambda}^r = (1-\lambda)\hat{\mathbf{y}}_i^r + \lambda\mathbf{y}_j^r \\ &\mathcal{A}ug_{intra}((\mathbf{u}_i^r, \hat{\mathbf{y}}_i^r), (\mathbf{x}_j^r, \mathbf{y}_j^r)) \to (\mathbf{u}_{i,\lambda}^r, \hat{\mathbf{y}}_{i,\lambda}^r) \end{aligned} \tag{8.41}$$

For clarity, the transformed image after $\mathcal{A}ug_{inter}$ as $\mathbf{t}_i^k$ is denoted as $\mathbf{t}_i^k$, where $\mathbf{t}_i^k = \mathbf{x}_{i,\lambda}^{k\to n}$. The generated image after performing $\mathcal{A}ug_{intra}$ is denoted as $\mathbf{m}_i^r$, where $\mathbf{m}_i^r = \mathbf{u}_{i,\lambda}^r$.

Inter-domain Branch f_{inter}: To address the domain generalization task, a labeled pair $(\mathbf{x}_i^k, \mathbf{y}_i^k)$ is first sampled, and inter-domain style augmentation $\mathcal{A}ug_{inter}$ is performed to obtain the transformed image $\mathbf{t}_i^k$, which integrates style information from another domain while maintaining its semantic information. The augmented $\mathbf{t}_i^k$ and the original input $\mathbf{x}_i^k$ are then fed into the inter-domain branch f_{inter}, yielding the predictions $\mathbf{p}_{i\_t}^k$ and $\mathbf{p}_{i\_inter}^k$. To ensure that $\mathbf{t}_i^k$ retains the same semantic information as $\mathbf{x}_i^k$, Mean Squared Error (MSE) is used as the consistency loss L_{cons} to minimize the discrepancy, and Dice loss is used as the supervised segmentation loss L_{inter}. This process can be formulated as follows:

$$\begin{aligned} L_{cons} &= MSE(\mathbf{p}_{i\_inter}^k, \mathbf{p}_{i\_t}^k) \\ L_{inter} &= \frac{1}{2}Dice(\mathbf{p}_{i\_inter}^k, \mathbf{y}_i^k) + \frac{1}{2}Dice(\mathbf{p}_{i\_t}^k, \mathbf{y}_i^k) \end{aligned} \tag{8.42}$$

Intra-domain Branch f_{intra}: Training in each domain is treated as an independent task to minimize disturbances from other domains during pseudo-labeling. A straightforward approach would be to train an ensemble of models, but this would result in a large number of parameters. As demonstrated in [50], domain-specific knowledge is encapsulated in the statistics of the Batch Normalization (BN) layer. Therefore, in our proposed design, all intra-domain branches f_{intra} share a common feature extractor, while each branch has its own BN layer and classifier. For a labeled sample $(\mathbf{x}_i^k, \mathbf{y}_i^k)$, it is first fed into the k-th intra-domain branch f_{intra}^k based on its domain label k to obtain the corresponding prediction $\mathbf{p}_{i\_intra}^k$. The supervised loss is then applied to constrain this prediction $\mathbf{p}_{i\_intra}^k$:

$$L_{intra} = Dice(\mathbf{p}_{i\_intra}^k, \mathbf{y}_i^k) \tag{8.43}$$

Semantic and Style Memory Bank: Because of the parallel structure design, each intra-domain branch creates its own label-related features without being affected by other domains. To make the most of these features, inspired by [51], a memory bank is used to store high-level representations of classes based on these features in each source domain. Every training iteration updates this memory bank by choosing high-quality class representations from each source domain. Since category-wise dependency is crucial for context modeling, an *Intra-domain Query Process* is employed to enhance pixel representations. Additionally, these high-quality class-wise representations can reflect the distribution of a domain, providing reliable domain-specific knowledge for the inter-domain branch. The *Inter-domain Query Process* in the inter-domain branch utilizes this knowledge to establish relationships within source domains and between source and target domains. This process aids in training with source domain data and bridging the gap between different domains.

Confidence-based Pseudo-labeling Strategy: As previously mentioned, the intra-domain branch and inter-domain branch have different learning capacities, resulting in distinct prediction maps. Instead of directly averaging predictions from both branches to generate the pseudo-label, the uncertainty in these predictions is considered. This approach effectively eliminates unreliable and noisy pseudo-labels. Specifically, given the unlabeled image $\mathbf{u}_i^r$ from the r-th domain, the pixel-wise probability maps $\mathbf{p}_{i\_intra}^r$ and $\mathbf{p}_{i\_inter}^r$ from the branches of intra-domain and inter-domain can be obtained. To obtain higher quality pseudo-labels, the entropy of the pixel-wise probability distribution is calculated in both $\mathbf{p}_{i\_intra}^r$ and $\mathbf{p}_{i\_inter}^r$, formulated as:

$$\mathcal{H}(\mathbf{p}^{(h,w)}) = -\sum_{c=0}^{C-1} \mathbf{p}^{(h,w)}(c) \log \mathbf{p}^{(h,w)}(c) \tag{8.44}$$

where $\mathbf{p}^{(h,w)}$ represents the prediction $\mathbf{p}_{i\_inter}^r$ ($\mathbf{p}_{i\_intra}^r$) at position (h, w). Predictions with lower entropy (less uncertainty) are typically more accurate. The more confident parts from either the intra-domain or inter-domain branch are selected to create a reliable pseudo-label $\hat{\mathbf{y}}_i^r$ for the unlabeled image $\mathbf{u}_i^r$.

$$\begin{aligned} \hat{\mathbf{y}}_i^{r(h,w)} &= R^{(h,w)} \cdot \hat{\mathbf{y}}_{i\_\text{inter}}^{(h,w)} + (1 - R^{(h,w)}) \cdot \hat{\mathbf{y}}_{i\_\text{intra}}^{(h,w)} \\ R^{(h,w)} &= \mathbf{1}\{\mathcal{H}(\mathbf{p}_{i\_inter}^{r(h,w)} < \mathcal{H}(\mathbf{p}_{i\_intra}^{r(h,w)})\} \end{aligned} \tag{8.45}$$

where $R^{(h,w)}$ records the position of pixels with lower uncertainty from inter-domain branch. While $(1 - R^{(h,w)})$ can select the confident pixels from intra-domain branch.

8.3.4 Application of Domain Generalization to Mitigate Domain Shift in Segmentation Tasks

8.3.4.1 Dataset Description and Implementation Details

Multi-source Domain Generalization Datasets: The datasets used in these experiments include a cross-center retinal fundus segmentation dataset [26] and a cross-center prostate segmentation dataset [33]. The retinal fundus dataset comprises data collected from four medical institutions across three public datasets: REFUGE [52], DrishtiGS [53], and RIM-ONE-r3 [54]. The data are uniformly center-cropped to an 800 × 800 disc region and then resized to 384 × 384 as the network input. The cross-center prostate dataset includes prostate T2-weighted MRI scans from six different clinical centers across three public datasets: NCI-ISBI13 [55], I2CVB [56], and PROMISE12 [57]. All the data are pre-processed to ensure a similar field of view for the prostate region, resized to 384 × 384 in the axial plane, and individually normalized to zero mean and unit variance in intensity values.

Single-source Domain Generalization Dataset: The datasets used in these experiments include a cross-modality abdominal segmentation dataset [58, 59] and cross-sequence cardiac segmentation dataset [60]. The cross-modality abdominal dataset consists of 30 Computed Tomography (CT) scans and 20 T2 spectral presaturation with inversion recovery (SPIR) MRI scans with four organs, namely liver, left kidney, right kidney, and spleen. The cross-sequence cardiac dataset contains 45 balanced steady-state free precession (bSSFP) MRI scans and 45 late gadolinium enhanced (LGE) MRI scans with three organs, namely left ventricle (LVC), right ventricle (RVC), and myocardium (MYO). All datasets were originally in 3D format and were reformatted to 2D, then resized to 192×192 and padded along the channel dimension to fit into the network. For the abdominal CT dataset, a window of $[-275, 125]$ in Hounsfield units was applied. For all MRI images, the top 0.5% of the histograms was clipped. All 3D scans were normalized to have zero mean and unit variance.

Multi-source Domain Generalization Implementation Details: For all the methods, a DeepLabV3+ backbone is utilized as the segmentation network. All methods are implemented using PyTorch with one NVIDIA TITAN Xp GPU. All experiments were optimized with batch size 16, and the learning rate started at 1×10^{-4} for 80 epochs before dropping to 2×10^{-4} after 40 epochs.

Single-source Domain Generalization Implementation Details: For all the methods, a U-Net with an EfficientNet-b2 backbone is utilized as the segmentation network. For all the methods, the network was trained from scratch. All methods are implemented using PyTorch with one NVIDIA GeForce RTX 3090 GPU (24G memory). For all experiments, a batch size of 32 was set, and the methods were evaluated at the 2000th epoch. Optimization is performed using the Adam optimizer with an initial learning rate of 3×10^{-4} and a weight decay of 3×10^{-5}.

8.3.4.2 Results and Comprehensive Analysis

Multi-source Domain Generalization Experiments and Analysis: The experimental results of the MDG methods mentioned above in the optic disc/cup segmentation task and the prostate segmentation task are summarized in Tables 8.4 and 8.5, respectively. In these experiments, one domain is designated as the target domain, while all remaining domains are used as source domains for training. To evaluate the segmentation performance of each method, the Dice score is employed as the primary metric, quantifying the overlap between predictions and ground truth. The tables include comparisons between the baseline empirical risk minimization (Baseline) approach and several state-of-the-art methods, such as BigAug [25], SAML [33], DoFEDoFE [29], FedDG [26], ICSL [30], DADA [61] and RFA [62]. As observed, the empirical risk minimization (Baseline) model achieves moderate performance on both tasks, with average Dice scores of 85.54 and 83.79, respectively. However, all MDG methods exhibit notable improvements over the baseline, demonstrating their ability to effectively address performance degradation in unseen target domains. These results highlight the strengths of advanced techniques like domain augmenta-

Table 8.4 Comparison of MDG methods for cross-center optic disc and cup segmentation on fundus images

Task	Optic disc segmentation					Optic cup segmentation					Overall
Unseen site	A	B	C	D	Avg.	A	B	C	D	Avg.	
Baseline	94.96	89.69	89.33	90.09	91.02	77.03	78.21	80.28	84.74	80.07	85.54
BigAug [25]	93.49	86.18	92.09	93.67	91.36	81.62	69.46	82.64	84.51	79.56	85.46
SAML [33]	94.30	91.17	92.28	87.95	91.43	83.16	75.68	82.00	82.88	80.93	86.18
DoFE [29]	92.88	85.73	92.07	93.21	90.97	80.84	69.71	82.28	83.35	79.05	85.01
FedDG [26]	95.37	87.52	93.37	94.50	92.69	84.13	71.88	83.94	85.51	81.37	87.03
ICSL [30]	95.39	93.49	93.94	94.88	94.43	86.23	81.95	87.47	87.14	85.70	90.06
DADA [61]	96.33	92.15	92.58	94.01	93.77	85.68	71.88	80.07	87.06	87.85	85.17
RFA [62]	95.64	90.97	95.12	95.60	94.33	86.10	83.37	87.33	83.88	85.17	89.72

Table 8.5 Comparison of MDG methods for cross-center prostate segmentation on prostate MRI images

Unseen site	A	B	C	D	E	F	Average
Baseline	83.51	81.94	82.29	85.44	84.66	84.89	83.79
BigAug [25]	89.63	84.62	83.86	87.66	81.20	88.96	85.99
SAML [33]	89.66	87.53	84.43	88.67	87.37	88.34	87.67
DoFE [29]	89.02	84.48	84.11	86.30	80.38	89.15	85.57
FedDG [26]	90.19	87.17	85.26	88.23	83.02	90.47	87.39
ICSL [30]	91.26	90.22	87.60	88.88	85.15	89.52	88.77
DADA [61]	91.07	89.45	85.88	90.82	90.78	90.86	89.81
RFA [62]	91.01	90.92	88.52	88.00	85.19	89.79	88.91

tion, invariant representation learning, and learning strategy optimization in enhancing the generalization capability of segmentation models. Moreover, the consistent improvement across two datasets underscores the robustness and versatility of the MDG approaches in medical image analysis.

Single-Source Domain Generalization Experiments and Analysis: The experimental results of the aforementioned SDG methods on the cross-modality abdominal segmentation task and the cross-sequence cardiac segmentation task are summarized in Tables 8.6 and 8.7, respectively. These tables include comparisons between the baseline empirical risk minimization (Baseline) approach and several state-of-the-art methods, such as BigAug [25], CSDG [42], ADS [38], TASD [41], SLAug [39], and IRLSG [43]. The Baseline model achieves average Dice scores of 69.46 and 80.63 for the cross-modality and cross-sequence tasks, respectively, indicating lower performance compared to cross-center tasks. This demonstrates that that cross-modallity and cross-sequence tasks are more challenging than cross-center tasks due to the more significant domain shifts. But fortunately, the SDG methods significantly outperform the Baseline, demonstrating a substantial improvement in the model's generalization capability. The SDG methods not only reduce the labeling cost of the data but also

Table 8.6 Comparison of single-source domain generalization results on cross-modality abdominal segmentation

Method	Abdominal cross-modality										
	CT to MRI					MRI to CT					Overall
	Liver	R-Kidney	L-Kidney	Spleen	Avg.	Liver	R-Kidney	L-Kidney	Spleen	Avg.	
Baseline	78.03	78.11	78.45	74.65	77.31	84.90	40.44	65.17	55.90	61.60	69.46
BigAug [25]	85.51	84.76	82.09	81.16	83.38	85.17	73.35	77.40	73.92	77.46	80.51
CSDG [42]	86.62	87.48	86.88	84.27	86.31	85.70	79.91	80.18	74.73	80.13	83.22
ADS [38]	85.24	89.87	86.92	81.69	85.93	83.06	76.85	79.44	73.05	78.10	82.02
TASD [41]	84.76	86.40	82.85	80.04	83.51	82.74	75.09	77.50	71.36	76.67	80.09
SLAug [39]	90.08	89.23	87.54	87.67	88.63	89.26	80.98	82.05	79.93	83.05	85.71
IRLSG [43]	87.80	88.97	87.22	85.45	87.36	87.11	79.66	81.58	75.56	80.98	84.17

Table 8.7 Comparison of single-source domain generalization results on cross-sequence cardiac segmentation

Method	Cardiac cross-sequence								
	bSSFP to LGE				LGE to bSSFP				Overall
	LVC	MYO	RVC	Avg.	LVC	MYO	RVC	Avg.	
Baseline	86.06	66.98	74.94	75.99	90.16	78.59	87.04	85.26	80.63
BigAug [25]	87.22	69.78	79.35	78.78	90.96	79.37	87.70	86.01	82.40
CSDG [42]	90.35	77.82	86.87	85.01	91.37	80.43	89.16	86.99	86.00
ADS [38]	89.66	78.12	86.94	84.90	91.54	79.21	89.86	86.87	85.89
TASD [41]	88.50	76.86	85.35	83.57	90.83	79.39	88.10	86.11	84.84
SLAug [39]	91.53	80.65	87.90	86.69	91.92	81.49	89.61	87.67	87.18
IRLSG [43]	89.88	77.64	87.02	84.85	91.17	79.91	89.20	86.76	85.81

address the privacy concerns associated with multi-center data sharing, showcasing their potential for wide application across various domains and further emphasizing their practical value in real-world medical scenarios.

8.3.5 Summary

In summary, this section discusses various approaches, methods, and datasets used to tackle domain generalization challenges in medical image segmentation. Solutions are discussed to address different practical scenarios, including multi-source domain generalization, single-source domain generalization, and specialized settings such as decentralized federated domain generalization for privacy protection, as well as semi-supervised and few-shot learning to reduce labeling costs. These applications hold significant practical value. Additionally, several commonly used datasets are highlighted in medical domain generalization and review recent methods applied to these datasets. These methods are versatile and can be adapted beyond medical imaging to address generalization issues in other fields.

References

1. Jain, R. K., Watasue, T., Nakagawa, T., Sato, T., Iwamoto, Y., Ruan, X., & Chen, Y. W. (2021). LogoNet: A robust layer-aggregated dual-attention anchorfree logo detection framework with an adversarial domain adaptation approach. *Applied Science*, *11*(20), 9622.
2. Jain, R. K., Sato, T., Watasue, T., Nakagawa, T., Iwamoto, Y., Ruan, X., & Chen, Y. W. (2023). Unsupervised logo detection using adversarial learning from synthetic to real images. *IEEE Transactions on Emerging Topics in Computational Intelligence* (pp. 1–14).
3. Jain, R. K., et al. (2023). Unsupervised domain adaptation approach for liver tumor detection in multi-phase CT images. In *Artificial intelligence and machine learning for healthcare. Intelligent systems reference library* (Vol. 228). Springer.

4. Jain, R. K., et al. (2022). Unsupervised domain adaptation with adversarial learning for liver tumors detection in multi-phase CT images. In *Innovation in medicine and healthcare. Smart innovation, systems and technologies* (Vol. 308). Springer.
5. Jain, R. K., et al. (2022). Unsupervised domain adaptation using adversarial learning and maximum square loss for liver tumors detection in multi-phase CT images. In *International Conference of the IEEE Engineering in Medicine & Biology Society (EMBC).*
6. Jain, R. K., et al. (2023). Unsupervised domain adaptation using Fourier phase enhanced training images for liver tumors detection in multi-phase CT images. In *International Conference of the IEEE Engineering in Medicine & Biology Society (EMBC).*
7. Jain, R. K., et al. (2024). Enhanced liver lesion detection in multi-phase CT images using unsupervised domain adaptation and feature generalization. In *Advances in intelligent disease diagnosis and treatment*. Springer.
8. Guan, H., & Liu, M. (2022). Domain adaptation for medical image analysis: A survey. *IEEE Transactions on Biomedical Engineering*, *69*(3), 1173–1185.
9. Rongbo, S., Jianhua, Y., Yan, K., Tan, K., Jiang, C., & Zhou, K. (2020). Unsupervised domain adaptation with adversarial learning for mass detection in mammogram. *Neurocomputing*, *393*, 27–37.
10. Panfilov, E., Tiulpin, A., Klein, S., Nieminen, M. T., & Saarakkala, S. (2019). Improving robustness of deep learning based knee MRI segmentation mixup and adversarial domain adaptation. In *International Conference on Computer Vision Workshops (ICCVW)* (pp. 450–459).
11. Ganin, Y., et al. (2016). Domain-adversarial training of neural networks. *Journal of Machine Learning Research*, 2096–2030.
12. Bousmalis, K., Silberman, N., Dohan, D., Erhan, D., Krishnan, D. (2017). Unsupervised pixel-level domain adaptation with generative adversarial networks. In *The IEEE Conference on Computer Vision and Pattern Recognition (CVPR).*
13. Gannin, Y., & Lempitsky, V. (2015). Unsupervised domain adaptation by backpropagation. In *Proceedings. International Conference on Machine Learning and Cybernetics* (pp. 1180–1189).
14. Gretton, A., Borgwardt, K., Rasch, M., Schölkopf, B., Smola, A. J. (2007). A kernel method for the two-sample-problem. In *Advances in neural information processing systems* (pp. 513–520).
15. Gretton, A., Borgwardt, K. M., Rasch, M. J., Scholkopf, B., & Smola, A. (2012). A kernel two-sample test. *Journal of Machine Learning Research*, 723–773.
16. Yu, Q., Hashimoto, A., & Ushiku, Y. (2021). Divergence optimization for noisy universal domain adaptation. In *Proceedings of the IEEE/CVF Conference on Computer Vision and Pattern Recognition* (pp. 2515–2524).
17. Chen, M., Xue, H., & Cai, D. (2019). Domain adaptation for semantic segmentation with maximum squares loss. In *International Conference on Computer Vision. Seoul, South Korea* (pp. 2090–2099).
18. Perone, C. S., et al. (2019). Unsupervised domain adaptation for medical imaging segmentation with self-ensembling. *NeuroImage*, 194, 1–11.
19. Liu, P., Kong, B., Li, Z., Zhang, S., Fang, R. (2019). CFEA: Collaborative feature ensembling adaptation for domain adaptation in unsupervised optic disc and cup segmentation. In *Medical Image Computing and Computer Assisted Intervention–MICCAI.*
20. Ananda, S., Iwamoto, Y., Han, X., Lin, L., Hu, H., & Chen, Y.-W. (2022). Dual discriminator-based unsupervised domain adaptation using adversarial learning for liver segmentation on multiphase CT images. In *IEEE Engineering in Medicine and Biology Society (EMBC). Scotland, UK* (pp. 1552–1555).
21. Xu, Y., et al. (2021). PA-ResSeg: A phase attention residual network for liver tumor segmentation from multi-phase CT images. *Medical Physics*, *48*(7), 3752–3766.
22. Canny, J. (1986). A computational approach to edge detection. *IEEE Transactions on Pattern Analysis and Machine Intelligence (PAMI-8)*, *6*, 679–698.
23. Zhou, X., Wang, D., & Krahenbuhl, P. (2019). *Objects as points*. arXiv:1904.07850

24. Maaten, L. V. D., & Hinton, G. (2008). Visualizing data using t-SNE. *Journal of Machine Learning Research*, *9*(86), 2579–2605.
25. Zhang, L., Wang, X., Yang, D., Sanford, T., Harmon, S., Turkbey, B., & Xu, Z. (2020). Generalizing deep learning for medical image segmentation to unseen domains via deep stacked transformation. *IEEE Transactions on Medical Imaging*, *39*(7), 2531–2540.
26. Liu, Q., Chen, C., Qin, J., Dou, Q., & Heng, P. A. (2021). FEDDG: Federated domain generalization on medical image segmentation via episodic learning in continuous frequency space. In *Proceedings of the IEEE/CVF Conference on Computer Vision and Pattern Recognition. CVPR* (pp. 1013–1023).
27. Lyu, J., Zhang, Y., Huang, Y., Lin, L., Cheng, P., & Tang, X. (2022). AADG: Automatic augmentation for domain generalization on retinal image segmentation. *IEEE Transactions on Medical Imaging*, *41*(12), 3699–3711.
28. Nussbaumer, H. J., & Nussbaumer, H. J. (1982). *The fast Fourier transform* (pp. 80–111). Springer.
29. Wang, S., Yu, L., Li, K., Yang, X., Fu, C. W., & Heng, P. A. (2020). DOFE: Domain-oriented feature embedding for generalizable fundus image segmentation on unseen datasets. *IEEE Transactions on Medical Imaging*, *39*(12), 4237–4248.
30. Kang, Y., Li, H., Zhao, X., Shi, X., Liu, F., Yan, Q., Yang, L., et al. (2022) Invariant content synergistic learning for domain generalization on medical image segmentation. In *2022 IEEE International Conference on Bioinformatics and Biomedicine (BIBM)* (pp. 450–456).
31. Gu, R., Lu, J., Zhang, J., Lei, W., Zhang, X., Wang, G., & Zhang, S. (2022). *Contrastive domain disentanglement for generalizable medical image segmentation*. arXiv preprint arXiv:2205.06551
32. Kamraoui, R. A., Ta, V. T., Tourdias, T., Mansencal, B., Manjon, J. V., & Coup, P. (2022). DeepLesionBrain: Towards a broader deep-learning generalization for multiple sclerosis lesion segmentation. *Medical Image Analysis*, *76*. Article 102312.
33. Liu, Q., Dou, Q., & Heng, P. A. (2020). Shape-aware meta-learning for generalizing prostate MRI segmentation to unseen domains. In *Medical Image Computing and Computer Assisted Intervention—MICCAI 2020: 23rd International Conference, Lima, Peru, October 4–8, 2020, Proceedings, Part II* (Vol. 23, pp. 475-485). Springer International Publishing.
34. Ouyang, C., Biffi, C., Chen, C., Kart, T., Qiu, H., & Rueckert, D. (2020). Self-supervision with superpixels: Training few-shot medical image segmentation without annotation. In *Computer Vision–ECCV 2020: 16th European Conference, Glasgow, UK, August 23–28, 2020, Proceedings, Part XXIX* (Vol. 16, pp. 762–780). Springer International Publishing.
35. Li, W., Goodchild, M. F., & Church, R. (2013). An efficient measure of compactness for two-dimensional shapes and its application in regionalization problems. *International Journal of Geographical Information Science*, *27*(6), 1227–1250.
36. Chen, T., Kornblith, S., Norouzi, M., & Hinton, G. (2020, November). A simple framework for contrastive learning of visual representations. In *International Conference on Machine Learning* (pp. 1597–1607). PMLR.
37. Li, L., Zimmer, V. A., Ding, W., Wu, F., Huang, L., Schnabel, J. A., & Zhuang, X. (2021). Random style transfer based domain generalization networks integrating shape and spatial information. In *Statistical atlases and computational models of the heart. M&Ms and EMIDEC Challenges: 11th International Workshop, STACOM 2020, Held in Conjunction with MICCAI 2020, Lima, Peru, October 4, 2020, Revised Selected Papers* (Vol. 11, pp. 208–218). Springer International Publishing.
38. Xu, Y., Xie, S., Reynolds, M., Ragoza, M., Gong, M., & Batmanghelich, K. (2022). Adversarial consistency for single domain generalization in medical image segmentation. In: *International Conference on Medical Image Computing and Computer-Assisted Intervention* (pp. 671–681). Springer Nature Switzerland.
39. Su, Z., Yao, K., Yang, X., Huang, K., Wang, Q., & Sun, J. (2023). Rethinking data augmentation for single-source domain generalization in medical image segmentation. In *Proceedings of the AAAI Conference on Artificial Intelligence* (Vol. 37(2), 2366–2374).

40. Philipp, M., Alperovich, A., Gutt-Will, M., Mathis, A., Saur, S., Raabe, A., & Mathis-Ullrich, F. (2022). Dynamic CNNs using uncertainty to overcome domain generalization for surgical instrument localization. In *Proceedings of the IEEE/CVF Winter Conference on Applications of Computer Vision* (pp. 3612–3621).
41. Liu, Q., Chen, C., Dou, Q., & Heng, P. A. (2022). Single-domain generalization in medical image segmentation via test-time adaptation from shape dictionary. In *Proceedings of the AAAI Conference on Artificial Intelligence* (Vol. 36(2), pp. 1756–1764).
42. Ouyang, C., Chen, C., Li, S., Li, Z., Qin, C., Bai, W., & Rueckert, D. (2022). Causality-inspired single-source domain generalization for medical image segmentation. *IEEE Transactions on Medical Imaging*, *42*(4), 1095–1106.
43. Niu, Z., Sun, H., Ouyang, S., Xie, S., Chen, Y., Tong, R., & Lin, L. (2024) IRLSG: Invariant representation learning for single-domain generalization in medical image segmentation. In *IEEE International Conference on Acoustics, Speech and Signal Processing (ICASSP)* (pp. 5585–5589).
44. Zhang, J., Zhang, Y., & Xu, X. (2021). OBJECTAUG: Object-level data augmentation for semantic image segmentation. In *2021 International Joint Conference on Neural Networks (IJCNN)* (pp. 1–8). IEEE.
45. Choi, S., Jung, S., Yun, H., Kim, J. T., Kim, S., & Choo, J. (2021). ROBUSTNET: Improving domain generalization in urban-scene segmentation via instance selective whitening. In *Proceedings of the IEEE/CVF Conference on Computer Vision and Pattern Recognition* (pp. 11580–11590).
46. Zhou, T., & Konukoglu, E. (2023). *FEDFA: Federated feature augmentation*. arXiv preprint arXiv:2301.12995
47. Xie, S., Niu, Z., Huang, H., Sun, H., Qin, R., Chen, Y. W., & Lin, L. (2023). IS2Net: Intra-domain semantic and inter-domain style enhancement for semi-supervised medical domain generalization. In *Proceedings of the 31st ACM International Conference on Multimedia* (pp. 8285–8293).
48. Sun, L., Li, C., Ding, X., Huang, Y., Chen, Z., Wang, G., & Paisley, J. (2022). Few-shot medical image segmentation using a global correlation network with discriminative embedding. *Computers in Biology and Medicine*, *140*. Article 105067.
49. Yun, S., Han, D., Oh, S. J., Chun, S., Choe, J., & Yoo, Y. (2019). CutMix: Regularization strategy to train strong classifiers with localizable features. In *Proceedings of the IEEE/CVF International Conference on Computer Vision* (pp. 6023–6032).
50. Chang, W. G., You, T., Seo, S., Kwak, S., & Han, B. (2019). Domain-specific batch normalization for unsupervised domain adaptation. In *Proceedings of the IEEE/CVF Conference on Computer Vision and Pattern Recognition* (pp. 7354–7362).
51. Alonso, I., Sabater, A., Ferstl, D., Montesano, L., & Murillo, A. C. (2021). Semi-supervised semantic segmentation with pixel-level contrastive learning from a class-wise memory bank. In *Proceedings of the IEEE/CVF International Conference on Computer Vision* (pp. 8219–8228).
52. Sivaswamy, J., Krishnadas, S., Chakravarty, A., Joshi, G., & Tabish, A. S. (2015). A comprehensive retinal image dataset for the assessment of glaucoma from the optic nerve head analysis. *JSM Biomedical Imaging Data Papers*, *2*(1), 1004.
53. Fumero, F., Alayón, S., Sanchez, J. L., Sigut, J., & Gonzalez-Hernandez, M. (2011). RIM-ONE: An open retinal image database for optic nerve evaluation. In *2011 24th International Symposium on Computer-based Medical Systems (CBMS)* (pp. 1–6). IEEE.
54. Orlando, J. I., Fu, H., Breda, J. B., Van Keer, K., Bathula, D. R., Diaz-Pinto, A., & Bogunović, H. (2020). Refuge challenge: A unified framework for evaluating automated methods for glaucoma assessment from fundus photographs. *Medical Image Analysis*, *59*, Article 101570.
55. Bloch, N. , Madabhushi, A., Huisman, H., Freymann, J., Kirby, J., Grauer, M., Enquobahrie, A., Jaffe, C., Clarke, L., & Farahani, K. (2015) NCI-ISBI 2013 challenge: Automated segmentation of prostate structures. *The Cancer Imaging Archive*, 370.
56. Lemaître, G., Martí, R., Freixenet, J., Vilanova, J. C., Walker, P. M., & Meriaudeau, F. (2015). Computer-aided detection and diagnosis for prostate cancer based on mono and multiparametric MRI: A review. *Computers in Biology and Medicine*, *60*, 8–31.

57. Litjens, G., Toth, R., Van De Ven, W., Hoeks, C., Kerkstra, S., Van Ginneken, B., & Madabhushi, A. (2014). Evaluation of prostate segmentation algorithms for MRI: The PROMISE12 challenge. *Medical image analysis*, *18*(2), 359–373.
58. Landman, B., Xu, Z., Igelsias, J., Styner, M., Langerak, T., & Klein, A. (2015). MICCAI multi-atlas labeling beyond the cranial vault–workshop and challenge. In *Proceedings of the MICCAI Multi-Atlas Labeling Beyond Cranial Vault—Workshop Challenge* (Vol. 5, p. 12).
59. Kavur, A. E., Gezer, N. S., & BarısÌ§, M., Aslan, S., Conze, P. H., Groza, V., Selver, M. A., et al. (2021). CHAOS challenge-combined (CT-MR) healthy abdominal organ segmentation. *Medical Image Analysis*, *69*, Article 101950.
60. Zhuang, X., Xu, J., Luo, X., Chen, C., Ouyang, C., Rueckert, D., & Li, L. (2022). Cardiac segmentation on late gadolinium enhancement MRI: A benchmark study from multi-sequence cardiac MR segmentation challenge. *Medical Image Analysis*, *81*, Article 102528.
61. Lai, H., Luo, Y., Li, B., Zhang, G., & Lu, J. (2023) Domain-aware dual attention for generalized medical image segmentation on unseen domains. *IEEE Journal of Biomedical and Health Informatics*.
62. Kang, Y., Zhao, X., Zhang, Y., Li, H., Wang, G., Cui, L., & Yang, L. (2023). Improving domain generalization performance for medical image segmentation via random feature augmentation. *Methods*, *218*, 149–157.

Chapter 9
Deep Learning Models for Medical Image Translation

9.1 Introduction

Medical imaging plays a crucial role in clinical practice, offering invaluable insights into internal structures and pathologies. Various imaging modalities, such as X-rays, CT scans, MRI, and PET are routinely used for patient examination. However, some of these modalities have notable drawbacks, including high costs, prolonged examination times, and exposure to ionizing radiation. For instance, contrast-enhanced computed tomography (CE-CT) scans often require contrast agents and prolonged X-ray exposure, which can impose a significant burden on patients. Addressing these limitations is essential for improving diagnostic efficiency and patient safety.

Medical image translation has emerged as a promising solution to these challenges. This task involves generating one imaging modality from another, enabling the acquisition of difficult-to-obtainmedical images using more readily available ones. For example, converting non-contrast computed tomography (NC-CT) images into CE-CT images can reduce radiation exposure and eliminate the need for contrast agents. Other applications include generating CT images from radiation-free MRI for pediatric radiotherapy planning or synthesizing PET images from MRI to avoid high cost of radiotracers. Despite its potential, medical image translation faces significant technical challenges, such as maintaining fine-grained local details, ensuring anatomical consistency, and addressing domain-specific variations in medical imaging.

In the field of medical image synthesis, two dominant paradigms have been widely explored: Generative Adversarial Networks (GANs) [1] and Diffusion Models (DMs) [2]. GAN-based models consist of a generator G and a discriminator D, which are trained adversarially to produce synthetic images that closely resemble real ones. While GANs excel in generating high-quality images, they often suffer from issues such as mode collapse and lack of stability during training. On the other hand, Diffusion Models, a class of probabilistic generative models, adopt a unique approach by incorporating both forward and reverse processes for image synthesis.

Y. Chen et al., *Recent Advances in Deep Learning for Medical Image Analysis*, Intelligent Systems Reference Library 278, https://doi.org/10.1007/978-3-031-94791-9_9

In the forward process, random noise is progressively added to the original image, while the reverse process systematically removes noise to reconstruct the image. Although DMs offer better controllability and interpretability compared to GANs, they are computationally expensive and typically slower in generating images.

The following subsections will focus on the translation tasks of CE-CT images from non-CE-CT images as an example and explore the principles, design, and mathematical foundations of GAN-based and DM-based methods.

9.2 Generative Adversarial Networks (GANs)-Based Medical Image Translation

GANs [1] have significantly advanced medical image synthesis by enabling the generation of high-fidelity images from random noise or conditioned inputs. In this section, we explore a variety of GAN-based methods [3–5] that have been widely adopted and refined for medical imaging tasks. To establish a solid foundation, Sect. 9.2.1 will introduce the basic principles of GANs, including the adversarial training mechanism involving a generator and a discriminator. Following this, Sects. 9.2.2–9.2.4 will focus on specific GAN variants, such as CycleGANs [3], Pix2pix GANs [5], OA-GAN [6], and CDD-GAN [7], discussing their unique architectures, applications, and the advantages they offer in overcoming challenges unique to medical imaging. This progression ensures a thorough understanding of both the theoretical underpinnings and practical implementations of GANs in medical image synthesis.

9.2.1 GANs and Conditional GANs (Paired Image-to-Image Translation Methods)

9.2.1.1 Fundamentals of GANs

GANs [1] are a class of deep learning models that consist of two neural networks: a generator G and a discriminator D, which compete against each other in a process called adversarial training. The generator's goal is to create realistic data x, from random noise z, while the discriminator's task is to distinguish between the real data and the generated (fake) data $\widetilde{x}$. As the generator improves, it tries to fool the discriminator by producing more convincing samples, while the discriminator becomes better at detecting the fakes. This process can be formulated as:

$$\min_{G} \max_{D} L(D, G) = \mathbb{E}_{x \sim p_x}[\log(D(x))] + \mathbb{E}_{z \sim p_z}[\log(1 - D(G(z))] \tag{9.1}$$

This dynamic push and pull leads to the generator producing increasingly realistic outputs. GANs have gained wide applications in image generation, super-resolution, data augmentation, and creative content creation. However, they can be challenging to train due to issues like instability and mode collapse [8].

9.2.1.2 Conditional GANs: Paired Image-to-Image Translation Methods

Pix2Pix GAN [5] shown in Fig. 9.1 includes a pair of adversarial networks: generator G and discriminator D. The generator G takes image y as input and outputs synthetic image $\widetilde{x}$. The goal of the generator G is to produce a synthesized image $\widetilde{x}$ that will be judged as true by the discriminator D. While the discriminator D attempts to differentiate the types of input samples as much as possible: real or synthetic images. This process can be formulated as:

$$L_{adv}^{G} = \mathbb{E}_{x \sim p_x}[\log(D(x))] + \mathbb{E}_{y \sim p_y}[\log(1 - D(G(y))] \tag{9.2}$$

where p_y and p_x are the distribution of real images in the source domain and target domain, respectively. $D(x)$ is a probability that the discriminator D assumes the input image x comes from the target domain.

The above adversarial loss is inspired by the original GANs [1] framework and aims to train the generator to generate images that are indistinguishable from real images in the target domain. In addition to above adversarial loss, there is a reconstruction loss to ensure that the generated images preserve the essential information of the input images. The reconstruction loss is defined as follows:

$$L_{rec}^{G} = -E_{x \in X, y \in Y}[|x - G(y)|_1] \tag{9.3}$$

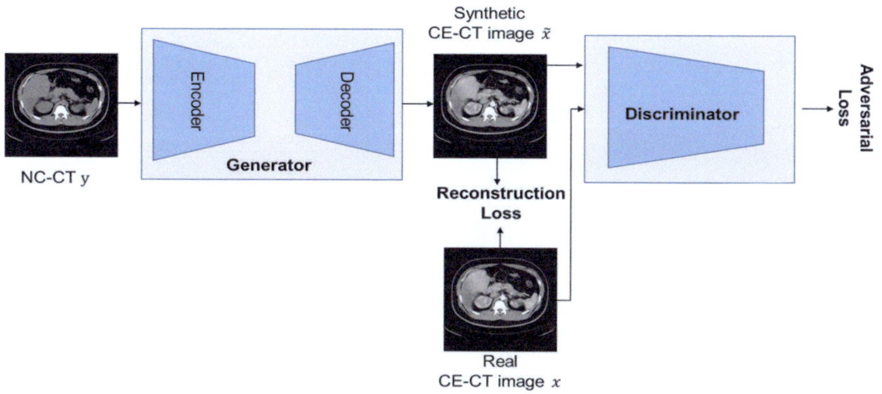

Fig. 9.1 The schematic diagram of paired image-to-image translation method called Pix2Pix GAN

The adversarial loss L_{adv}^{G} and reconstruction loss L_{rec}^{G} together form the overall loss function of the generator in Pix2Pix GAN [5], guiding the generator to produce high-quality images. During training, the generator aims to minimize both functions to produce high-quality images.

9.2.2 Unpaired Image-to-Image Translation Method

Unpaired Image-to-Image Translation is technique that transform images from one domain to another without requiring paired data during training. Zhu et al. [3] proposed a CycleGAN, which can learn mapping between two domains in an unsupervised manner, enabling transformations between diverse image styles, such as converting photos to paintings or horses to zebras. The overview of CycleGAN is shown in Fig. 9.2, which consists of three neural networks: two generators and a discriminator D. CycleGAN eliminates this requirement by leveraging the principle of cycle consistency. This principle states that if an image is translated from NC-CT image to CE-CT image and then back to NC-CT image, it should ideally remain unchanged. This cycle-consistent loss can be defined as follows:

$$L_{cyc}^{G} = -E_{y \in Y}[|y - GB(GA(y))|_1] \tag{9.4}$$

where GA and GB are the generator A and B, respectively. The cycle-consistent loss L_{cyc}^{G} is used to ensure that the structure of the synthesized image remains unchanged.

$$L_{adv}^{G} = \mathbb{E}_{x \sim p_x}[\log(D(x))] + \mathbb{E}_{y \sim p_y}[\log(1 - D(GA(y))] \tag{9.5}$$

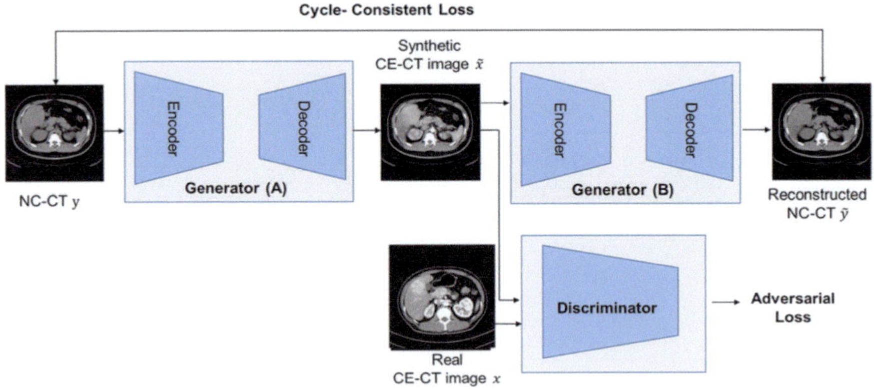

Fig. 9.2 The schematic diagram of unpaired image-to-image translation method called cycle-GAN

The adversarial loss L_G^{adv} is used to ensure that the generated image $GA(y)$ are realistic and resemble real images in the target domain. The adversarial loss L_{adv}^G and cycle-consistent loss L_{cyc}^G guide the generator to produce high-quality images by minimizing both during training.

9.2.3 Improved Image-to-Image Translation Methods with Attention

This section will take NC-CT to CE-CT image synthesis as an example to introduce some attention-based image-to-image translation methods: OA-GAN [5], CDD-GAN [7].

9.2.3.1 Supervised Attention GAN: Organ-Aware GAN

Organ-Aware Generative Adversarial Network (OA-GAN) [6] is a paired image-to-image translation technique specifically designed to synthesize high-quality medical images with enhanced focus on organ-level details. Its architecture, as shown in Fig. 9.3, consists of two interconnected components: an organ-aware (OA) network and a dual-decoder generator. The OA network incorporates a supervised attention mechanism, enabling precise localization of liver regions during synthesis, which significantly improves the diagnostic relevance of the generated images. Complementing this, the Dual-Decoder Generator plays a pivotal role in balancing global structure generation with localized organ refinement. By integrating two specialized decoders—a Global Decoder (GD) for synthesizing the entire image and a Local Decoder (LD) for focusing on liver-specific regions—the generator ensures both

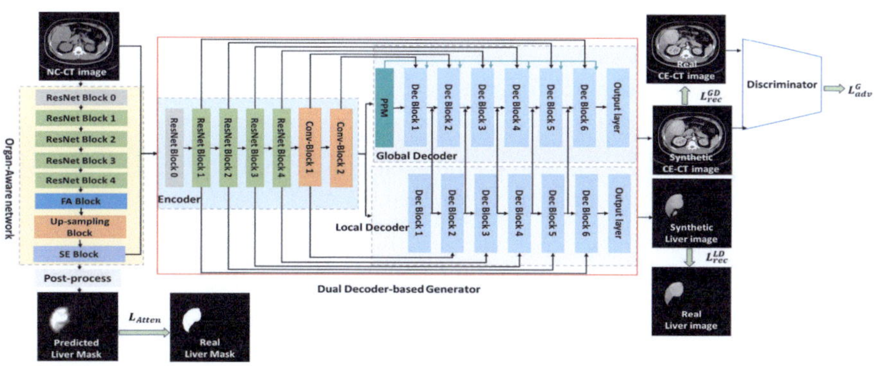

Fig. 9.3 The architecture of OA-GAN consists of two associated sub-networks. The organ-aware network extracts the spatial feature from the input NC-CT image. The dual-decoder based generator reconstructs the liver and the entire CE-CT image using a local decoder and global decoder

contextual coherence and detailed organ representation, enabling more accurate and realistic image synthesis.

Organ-Aware Network. The organ-aware network includes the first four blocks of the ResNet-34 [9] framework as the feature extraction backbone. This network is initialized with pre-trained weights on the ImageNet dataset except for the first block. The FA block [10] is introduced to capture long-range dependencies for making a rough estimate of the region of interest in the input image. Semantic features extracted from the backbone and the FA block are resized to 1/8 of the input image. The up-sampling block is used to translate the feature to the original input image size for obtaining the spatial attention map x_{atten}, which consists of three consecutive Conv-BN-ReLU layers (i.e., a convolutional layer, a batch normalization layer, and a ReLU activation layer) followed by a bilinear up-sampling layer with upsampling rate $r = 8$. The SE block [11] is to highlight more representative features by modeling the dependency across channels. At the top of the OA network, we use a post-process that averages the weighted channels of the spatial attention x_{atten} to obtain a predicted mask. L2 loss is calculated to guide the OA network using the real liver mask M_{liver}, emphasizing the target disease-related organ in the image.

Dual Decoder-based Generator. The dual decoder-based generator, as shown in Fig. 9.3, consists of a shared encoder network, a Global Decoder (GD) network, and a Local Decoder (LD) network. The encoder-GD networks combination can be viewed as a global image generator, responsible for synthesizing the entire CE-CT image, which is consistent with the generator architecture proposed in Improved cGAN [12]. On the other hand, the encoder-LD network combination serves as a local image generator, focusing on synthesizing the liver region within the CE-CT image. The encoder extracts high-level semantic features from the input, which are then processed by the GD and LD networks to synthesize the complete CE-CT image and the liver-specific region, respectively. While the GD network captures global structural information with components like a pyramid pooling module and attention mechanisms, the LD network focuses on refining organ-specific details using residual and up-sampling layers. Together, these components ensure a complementary balance, with the GD providing context-aware synthesis and the LD delivering enhanced organ-level precision, ultimately improving the quality of medical image generation. The detailed architecture of GD and LD networks can be seen in OA-GAN [6].

Adversarial Loss. Adversarial loss is employed to make the distribution of synthetic image close to real image, which is formulated as

$$L_{adv}^{G} = \mathbb{E}_{\widetilde{x} \sim p_{\widetilde{X}}} - [\log(1 - D(\widetilde{x}))] \tag{9.6}$$

where $D(\widetilde{x})$ represents the probability assigned by the discriminator D to the synthetic image $\widetilde{x}$ being real data. We use the discriminator proposed in PatchGAN [5] in our framework.

Global Reconstruction Loss. L1 loss is used for ensuring that the content of synthetic image $\widetilde{x}$ generated from GD network is consistent with ground truth image x, which can be defined as

$$L_{rec}^{GD} = \mathbb{E}_{\widetilde{x} \sim p_{\widetilde{X}}, x \sim p_X}[||x - \widetilde{x}||_1] \tag{9.7}$$

Local Reconstruction Loss. x_{liver} is liver region of CE-CT image, which is obtained by $x \times M_{liver}$. To guarantee that synthetic image $\widetilde{x}_{liver}$ generated from the LD network is consistent with real x_{liver}, The used L1 loss can be defined as:

$$L_{rec}^{LD} = \mathbb{E}_{\widetilde{x}_{liver} \sim p_{\widetilde{X}_{liver}}, x_{liver} \sim p_{X_{liver}}}[||x_{liver} - \widetilde{x}_{liver}||_1] \tag{9.8}$$

Attention Loss. To make the OA network pay more attention to liver region, we use MSE loss to restrict predicted and real liver mask M_{liver}, which can be defined as

$$L_{Atten} = MSE(Avg(OA(y)), M_{liver}) \tag{9.9}$$

OA and $Avg(\cdot)$ are the OA network and the average of the weighted channels. Finally, the overall loss function of the generator is given as follows:

$$L_G = \lambda_1 \times L_{adv}^{G} + \lambda_2 \times L_{rec}^{GD} + \lambda_3 \times L_{rec}^{LD} + \lambda_4 \times L_{Atten} \tag{9.10}$$

where $\lambda_1, \lambda_2, \lambda_3$, and λ_4 are hyper-parameters, which are set to $1, 25, 5, 2$, respectively.

9.2.3.2 Unsupervised Attention GAN: Crossing Dual Decoding GAN

While supervised attention mechanisms can enhance image translation quality by focusing on target organ regions, the training of OA-GAN [6] relies on additional organ masks, increasing annotation costs and training complexity. To address this, Crossing Dual Decoding Generative Adversarial Network (CDD-GAN) [7] is proposed to introduce an unsupervised attention mechanism that uses self-supervision to focus on critical organ regions without extra masks, reducing dependency while maintaining high-quality results. The overall architecture of CDD-GAN [7] is illustrated in Fig. 9.4 , which comprises of two networks: a crossing dual decoding generator G and a discriminator D.

The detailed structure of the generator G is illustrated in Fig. 9.4. It comprises four main components: a shared encoder, an attention decoder (AD) network, an improved transformed decoder (ITD) network, and a Pixel-Level Fusion Module (PLFM). The shared encoder extracts high-level semantic features from the NC-CT image y, which are then leveraged by both the ITD and AD networks. The ITD network generates a transformed image F_{TD}, while the AD network produces a spatial attention map F_{AD} to emphasize important regions for synthesis. These two networks interact through attention gates to enhance their outputs. The PLFM integrates these outputs to produce the final synthetic CE-CT image, offering a more robust and accurate result.

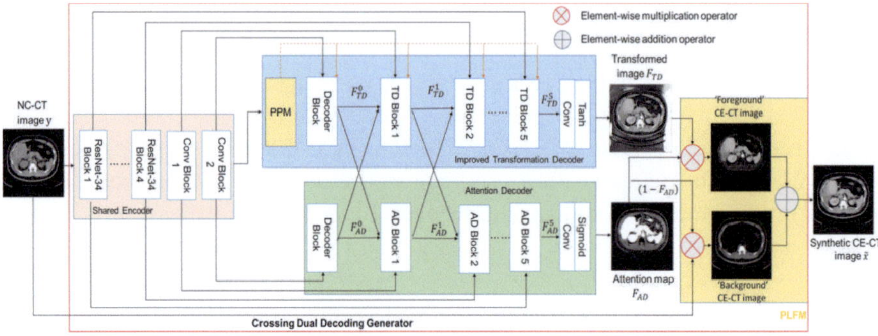

Fig. 9.4 The structure of crossing dual decoding generator G. The attention decoder aims to predict attention map that highlight task-related areas and suppress unrelated areas in input NC-CT image. The improved transformation decoder produces a transformed image. At the top of the aboved decoders, the Pixel-Level Fusion Module (PLFM) takes NC-CT image, attention map and transformed image as input to synthetize the final CE-CT image

Shared Encoder Network The shared encoder is based on the ResNet-34 architecture [9], adapted to handle our dataset. By incorporating additional Con-BN-ReLU blocks [9], it tailors the extracted semantic features for the synthesis task. This modification enables the shared encoder to efficiently capture high-level representations while ensuring compatibility with our specific input characteristics.

AD Network The AD network focuses on learning a spatial attention map F_{AD} to prioritize regions critical for synthesis. Its design emphasizes iterative refinement through attention mechanisms. Attention gates [13] replace conventional skip connections, enabling seamless integration of hierarchical features while suppressing irrelevant information. This approach ensures that the generated attention map highlights essential areas for accurate synthesis. The detailed architecture of the AD network can be seen in the CDD-GAN [7].

ITD Network The ITD network transforms semantic features into a preliminary synthetic image. A key innovation is the use of a PPM module [14] to capture spatial contextual information, which is further refined by attention mechanisms. Unlike traditional skip connections, attention gates facilitate smoother feature fusion, addressing disparities between hierarchical representations. The ITD network ensures high-quality initial transformations, which are crucial for generating the final output. A detailed explanation of the ITD network can be found in the description of CDD-GAN [7].

PLFMAs previously mentioned, the AD network focuses on highlighting important organs or regions that exhibit significant enhancement after CAs injection, while suppressing regions that show no significant changes. Ideally, important organs or regions predicted by the AD network should be 1 (or as close as possible), while those unimportant regions should be 0. Therefore, we introduce a PLFM to obtain the final result by utilizing a linear combination of F_{AD}, F_{TD} and y, which can be described as

$$\widetilde{x} = F_{AD} \times F_{TD} + (\mathbf{1} - F_{AD}) \times y, \tag{9.11}$$

where $\mathbf{1}$ is a matrix with all values of 1, which has the same size as the input image y. According to the above equation, we can see that the final synthetic CE-CT image $\widetilde{x}$ is divided into two parts: $F_{AD} \times F_{TD}$ and $(\mathbf{1} - F_{AD}) \times y$. In the final synthetic image $\widetilde{x}$, important regions predicted by AD network will be consistent with the output of transformation decoder by using $F_{AD} \times F_{TD}$. We called it Foreground CE-CT image. Meanwhile, unimportant regions do not change before and after proposed crossing dual decoding generator G by using $(\mathbf{1} - F_{AD}) \times y$. We call it Background CE-CT image. In summary, while the transformation decoder synthesizes CE-CT images from NC-CT images, the AD network, aided by the PLFM, can restrict the extent of alteration in the synthetic CE-CT images.

We combine adversarial learning to optimize the generator G for synthesizing higher quality and more accurate CE-CT images. Specifically, we use adversarial learning to improve the overall quality of synthetic CE-CT images. Smarter discriminator D would be more effective for this task in theory. Therefore, we use the PatchGAN discriminator proposed in [3, 5], which can evaluate the authenticity of 70×70 overlapping patches of the synthetic image, rather than the entire image. During the training phase, the parameters of the other two networks are fixed while optimizing one of the three networks. Next, we will present a detailed introduction about the objective function of each network. The objective function of generator G is formulated as

$$L_G = \lambda_1 \times L^G_{adv} + \lambda_2 \times L^G_{rec} + \lambda_3 \times L^G_{atten} \tag{9.12}$$

where λ_1, λ_2, and λ_3 are hyper-parameters to maintain the weight of generator, discriminator and tumor segmentation network, which are set to 2, 25, and 1, respectively. L^G_{adv}, L^G_{rec}, and L^G_{atten} stand for Adversarial Loss, Reconstruction Loss, and Attention Loss. Full and detailed explanation of above loss functions are shown in subsequent paragraphs. The generator G tries to minimize L_G during training.

Adversarial Loss To make the distribution of synthetic image as close as possible real image, we employ an adversarial loss for the generator G, which is formulated as

$$L^G_{adv} = \mathbb{E}_{\widetilde{x} \sim p_{\widetilde{x}}} - [log(1 - D(\widetilde{x}))] \tag{9.13}$$

where $D(\widetilde{x})$ refers to a probability that the discriminator D considers that the synthetic image $\widetilde{x}$ comes from real data. The generator G tries to deceive the discriminator D, which is equivalent to the value of $D(\widetilde{x})$ is as close as possible to 1.

Reconstruction Loss Although the distribution of synthetic images is almost indistinguishable from real image by using the adversarial loss in theory, it cannot guarantee that the content of synthetic image $\widetilde{x}$ is consistent with the ground truth image x. To solve this limitation, we optimize the generator G by minimizing the L1 distance between the ground truth image x and the generated image $\widetilde{x}$, defined as

$$L^G_{rec} = \mathbb{E}_{\widetilde{x}\sim p_{\widetilde{X}}, x\sim p_X}[||x - \widetilde{x}||_1] \tag{9.14}$$

Attention Loss The AD network is to produce a spatial attention map F_{AD} with same size as the input image x, which aims to highlight important organs and crucial regions in the input image while restraining irrelevant areas. Theoretically, the spatial attention map of the NC-CT image predicted by AD network should be consistent with that of the synthetic CE-CT image. A convincing explanation is that the liver region should be emphasized while the background should be neglected whether in the task of synthesizing CE-CT images from NC-CT images or complete reversal. To this end, we propose an AD network loss to encourage the AD network pay attention to task-related areas rather than whole image, which is defined as

$$L^G_{atten} = \mathbb{E}_{\widetilde{x}\sim p_{\widetilde{X}}, x\sim p_X}[||AD(Enc(x)) - AD(Enc(\widetilde{x}))||_1)] \tag{9.15}$$

where Enc and AD are the shared encoder network and AD network, respectively.

The vanilla GANs is notorious for unstable training. To tackle this limitation, we replace original adversarial loss with WGAN-GP for optimizing the discriminator D. The objective function for the discriminator D is formulated as

$$\begin{aligned} L^D_{adv} = &-\lambda_5 \times E_{\widetilde{x}\sim\widetilde{X}}[log(1 - D(\widetilde{x}))] \\ &+\lambda_6 \times E_{x\sim X}[log(1 - D(x)] \\ &-\lambda_7 \times E_{\widetilde{x}\sim\widetilde{X}}[log(1 - D(\widetilde{x}))] \end{aligned} \tag{9.16}$$

where $\widetilde{x}$ is generated by linear interpolation between synthetic and real CE-CT images. λ_5, λ_6 and λ_7 are 2, 10 and 5, respectively. The discriminator D aims to distinguish the type of all input samples as correctly as possible: is it from real or synthetic images, which is equivalent to minimizing the above L^D_{adv} during training. Compared to traditional discriminator, the discriminator D needs to distinguish the type of interpolated samples $\widetilde{x}$ in addition to the real and synthetic image.

9.2.4 *Performance Validation*

9.2.4.1 Contrast-Enhanced CT Dataset

We utilize an in-house CE-CT dataset [12] collected from Sir Run Run Shaw Hospital between 2015 and 2017 to evaluate the performance of the proposed method. The resolution of the CT images is 512 × 512 pixels, and the thickness of each slice is 7 mm. The dataset comprises 86 patients with four different types of focal liver lesions (FLLs): focal nodular hyperplasia (FNH), hepatocellular carcinoma (HCC), cysts, and metastases (METS). Each patient has CT images acquired across three distinct phases: Non-Contrast (NC), Arterial (ART), and Portal Venous (PV). The dataset includes real liver-tumor masks and corresponding FLL types labeled by

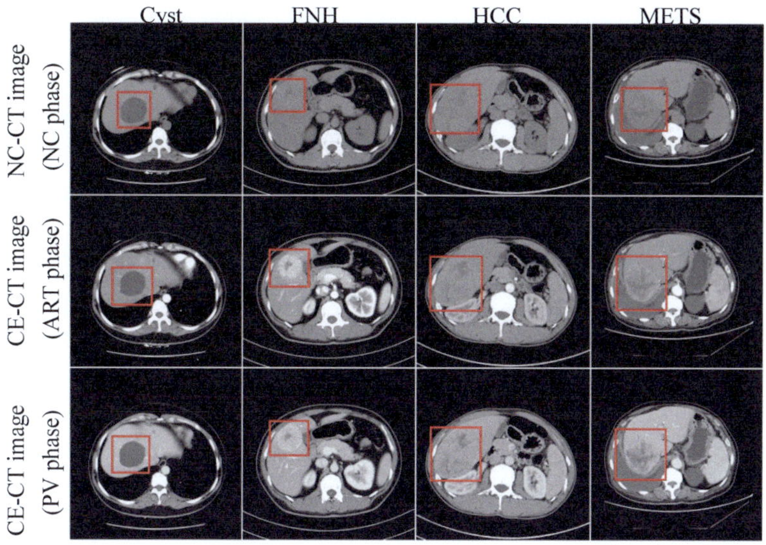

Fig. 9.5 Representative CT images of different types of focal liver lesions. The lesion regions are highlighted by the red box. From left to right: Cyst, Focal nodular hyperplasia (FNH), Hepatocellular carcinoma (HCC), and Metastases (METS), respectively

experienced radiologists. For training and testing, the subjects were randomly split with an 80% training and 20% testing ratio. The training set consisted of 16 FNH, 18 HCC, 26 cyst, and 9 METS patients, while the test set included 4 FNH, 4 HCC, 6 cyst, and 3 METS patients.

As shown in Fig. 9.5, the CE-CT scan encompasses three distinct phases, each providing unique insights into the vascular properties of liver tissues and lesions. The Non-Contrast (NC) phase, acquired before the injection of contrast agents (CAs), serves as a baseline image of the liver's anatomy. After contrast administration, the Arterial (ART) phase (captured within 20–30 s) emphasizes the arterial supply to the liver and hyper-vascular lesions, such as HCC. Subsequently, the Portal Venous (PV) phase (captured within 70–80 s post-injection) offers detailed visualization of the liver parenchyma and portal vein system, aiding in assessing the extent of liver diseases and tumors. These CE-CT phases are critical for differentiating various liver tumor types, as each is characterized by distinct patterns of contrast enhancement.

9.2.4.2 Implementation Details

The OA-GAN [6] and CDD-GAN [7] models are implemented on the PyTorch platform, utilizing four Nvidia 2080Ti GPU cards, each with 9 GB of memory. The training process lasted approximately 30 h with a batch size of 4 and a total of 400 epochs. We adopt the Two-Timescale Update Rule (TTUR) [15] to optimize both the generator G and the discriminator D. The Adam optimizer is employed for both networks with momentum parameters $\beta_1 = 0.5$ and $\beta_2 = 0.999$, and learning rates are set at 2×10^{-4} for the generator and 4×10^{-4} for the discriminator. The optimizer configuration remains fixed for both the generator G and the discriminator D throughout the training. In the second stage, we alternately train the two associated networks: the generator G and the discriminator D.

9.2.4.3 Image Fidelity Evaluation Metrics

PSNR, SSIM and MSE are used to evaluate the quality of reconstructed images and corresponding liver regions. The MSE is defined as follows:

$$MSE(x, \widetilde{x}) = \frac{1}{H \times W} \left(\sum_{i=0}^{h-1} \sum_{j=0}^{w-1} [x(i, j) - \widetilde{x}(i, j)]^2 \right) \tag{9.17}$$

where H and W represent the height and width of the image, both set to 256 pixels. $x(i, j)$ and $\widetilde{x}(i, j)$ are pixel values in (i, j)-th position of real and synthetic images. Smaller MSE score indicates that synthetic CE-CT image $\widetilde{x}$ is closer to real CE-CT image x.

PSNR is a commonly employed metric to evaluate the quality disparity between two images, which is computed by comparing the peak signal power of the real and synthetic images. PSNR is calculated using the following formula:

$$PSNR(x, \widetilde{x}) = 10 \times \log_{10} \frac{MAX_I^2}{MSE(x, \widetilde{x})} \tag{9.18}$$

where MAX_I is maximum pixel value, which is set to 255. A higher PSNR score indicates a high quality reconstructed image.

SSIM is a metric that measures the similarity between the real and synthetic images based on three aspects: luminance, contrast, and structure. It provides a comprehensive assessment of image similarity by considering both local and global characteristics. SSIM is calculated using the following formula:

$$\begin{aligned} SSIM(x, \widetilde{x}) &= [l(x, \widetilde{x})^{\alpha} c(x, \widetilde{x})^{\beta} s(x, \widetilde{x})^{\gamma}] \\ &= \frac{2\mu_x \mu_{\widetilde{x}} + c_1}{\mu_x^2 + \mu_{\widetilde{x}}^2 + c_1} \frac{2\sigma_x \sigma_{\widetilde{x}} + c_2}{\sigma_x^2 + \sigma_{\widetilde{x}}^2 + c_2} \frac{2\sigma_{x\widetilde{x}} + c_3}{\sigma_x \sigma_{\widetilde{x}} + c_3} \end{aligned} \tag{9.19}$$

where $l(x, \widetilde{x})$, $c(x, \widetilde{x})$ and $s(x, \widetilde{x})$ stand for luminance, contrast and structure. α, β and γ are weight parameters used to adjust the relative importance of each component. μ_x and σ_x are mean and variance. $\sigma_{x\widetilde{x}}$ is the covariance between real and synthetic images. c_1, c_2 and c_3 are three constants avoiding division by zero, which are 0.01^2, 0.03^2 and 0.015^2. In addition to assess the entire image quality, we also evaluate the quality of the liver region using the aforementioned three metrics. We substitute $(x, \widetilde{x})$ with $(x \times M_{\text{liver}}, \widetilde{x} \times M_{\text{liver}})$ to specifically focus on the liver region.

To evaluate the quality of tumor region, we adopt Pearson Correlation Coefficients (PCC) to calculate the correlation between density feature [16] of synthetic CE-CT image and real image. The density feature indicates the ratio of the average pixel density inside tumor region and that of liver parenchyma, which is calculated using the following formula:

$$PCC(\mathrm{d}f(x), \mathrm{d}f(\widetilde{x})) = \frac{\mathbb{E}[(\mathrm{d}f(x) - \mu_{\mathrm{d}f(x)})(\mathrm{d}f(\widetilde{x}) - \mu_{\mathrm{d}f(\widetilde{x})})]}{\sigma_{\mathrm{d}f(x)}\sigma_{\mathrm{d}f(\widetilde{x})}} \tag{9.20}$$

where $df(x)$ is density feature as calculated in [16].

9.2.4.4 Diagnostic Efficacy Evaluation Metrics

We compare the performance in the task of FLLs recognition from the perspectives of both computer-aided diagnosis and clinical diagnosis. Specifically, a DL-based FLLs classification experiment and a pilot trial are carried out to assess their effectiveness in these diagnostic contexts. During the pilot trial, two experienced radiologists will diagnose FLLs based on real and synthetic CE-CT images. Note that the real CE-CT images consist of CT images from real NC, ART and PV phases and the synthetic CE-CT images consists of CT images from real NC phase, synthetic ART and synthetic PV phases.

DL-based FLLs classification. In the DL-based FLLs classification experiments, we train a three-phase model based on the classification method proposed in [17]. The ResNet-50 with frozen convolutional layers is utilized as our classification model. Tumor region of interest (ROI) images were extracted from real and synthetic CE-CT images using labeled tumor masks. The tumor ROI images are resized to a uniform size of 224 × 224. Subsequently, these tumor images from NC, ART and PV phases are combined into a three-channel tumor image. The three-phase model takes the three-channel tumor images as input and predicts the types of FLLs: Cyst, FNH, HCC, or METS. The schematic representation of our three-phase classification model is shown in Fig. 9.6, which provides a detailed overview of above steps. The accuracy (ACC) metric is used to evaluate the performance of the three-phase model in FLLs classification task.

We divide real CE-CT images into training and test sets using the same partitioning approach as described the above section. The three-phase model is trained on the training set, which is implemented using PyTorch on a Nvidia 2080Ti GPU with batch size 10. We evaluated the performance of trained model on the test set of real CE-CT images. To ensure the fair comparison, we also evaluated the trained model on synthetic CE-CT images produced from different generative models, while NC phase is consistent with that in the test set of real CE-CT images and synthetic CE-CT images. Additionally, to evaluate the additional benefits of synthetic CE-CT images

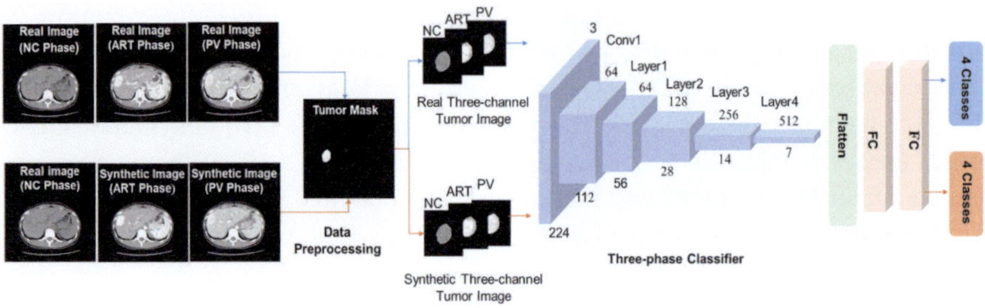

Fig. 9.6 Schematic representation of three-phase classification model. The tumor mask is utilized to extract a three-channel tumor image from real images, covering the NC, ART, and PV phases. In the synthetic three-channel tumor image, ART and PV are derived from synthesized images. Both real and synthetic three-channel tumor images are input into the three-phase classifier to assess their performance in tumor classification tasks, respectively

in FLLs classification, we trained a single-phase classification model. The single-phase model has same network structure and dataset partitioning as the three-phase model. However, only NC phase is utilized during training and testing phases. Note that, we performed classification for real and synthetic CE-CT images in all test sets. Specifically, the three-phase and single-phase models predict the type of FLLs of each sample in their respective dataset. Note that all the slices from the same patient should have the same FLLs category.

Pilot Trial. In addition to the FLLs classification experiment described above, we conducted a pilot trial to assess the effectiveness of synthetic CE-CT images in clinical diagnosis. For this purpose, we invited two experienced radiologists to diagnose samples in the test set from the aforementioned classification experiment. However, in this pilot trial, the radiologists diagnose FLLs by considering the entire CT images of a patient, rather than relying on a single slice image. The experimental procedure is as follows: we first randomly composed a new test set with both real and synthetic CE-CT images. The radiologists will independently diagnose each sample including the entire CT images of a patient in the test set and provide diagnostic reasoning. Importantly, the radiologists are unaware whether the sample is from real or synthetic CE-CT images to eliminate subjective bias. And NC-CT images alone are irrigorous for clinical diagnosis, which are not provided in this study. The example of diagnostic reasoning is as follows: the patient was identified as FNH due to tumor region exhibiting enhancement in the ART phase and fading in the PV phase. Similarly, the patient was identified as METs, since the edge region of tumor is enhanced in ART phase and slightly decreases in the PV phase. Another patient was identified as HCC because the interior of the tumor region displayed enhancement in the ART phase and subsequent decrease in the PV phase. We record the diagnostic results of each radiologist for each sample in our test set. We compare their diagnoses result with the ground truth labels to calculate the classification ACC for real and synthetic CE-CT images generated from different models.

9.2.4.5 Qualitative and Quantitative Assessment in CE-CT Image Synthesis

We compare OA-GAN [6] with other synthesis models such as Pix2pix GAN [5], MedGAN [2], CDD-GAN [7], and Improved cGAN [12]. We select Improved cGAN [12] as our baseline model, all synthesis models are trained using our paired data samples. Table 9.1 provides the quantitative results of the OA-GAN in comparison with other approaches for synthesizing CE-CT images of the ART phase and PV phase. OA-GAN achieves the lowest MSE score and the highest SSIM and PSNR scores for both synthesis tasks in the entire image, surpassing the comparison models. Specifically, for the liver region, OA-GAN outperforms the state-of-the-art models. These results demonstrate that OA-GAN can generate more realistic CE-CT images in both the entire image and liver region. In addition to the quantitative metrics discussed earlier, we conducted T-tests to evaluate whether OA-GAN brings statistically significant improvements to image synthesis performance compared to the comparative models. A p-value below 0.01 is generally considered indicative of a highly significant difference. Our experimental results consistently reveal p-values below 0.01 across all statistical tests, affirming that the enhancements introduced by proposed methods are statistically significant when compared to the comparison models.

Figure 9.7 presents qualitative results of synthesizing ART phase and PV phase from NC-CT images including four different types of liver tumors: Cyst, FNH, HCC, and METS. Qualitative results show that CE-CT images generated from OA-GAN, both in ART phase and PV phase, appear to be more sharper than those generated by comparison models in entire images. Most importantly, tumor area synthesized from OA-GAN are very close to real tumor images. In the case of FNH, the tumor area synthesized by OA-GAN exhibits substantial enhancement during the arterial (ART) phase and a slight reduction during the portal venous (PV) phase, aligning with the characteristic enhancement pattern of real tumors. Furthermore, in other cases, the tumor areas synthesized from OA-GAN display enhancement patterns that closely resemble those observed in real tumors. These results indicate that OA-GAN exhibits a greater capacity for capturing the variations in enhancement patterns among various liver tumors. Despite the quality of the synthesized liver images in both synthesis tasks may be considered subpar, the overall image generated by our proposed OA-GAN surpasses that of state-of-the-art methods by a significant margin. We are of the opinion that refining the quality of local images, potentially through the incorporation of a local GAN loss, may result in additional enhancements in the synthesized images. However, striking a balance between the global GAN loss and the local GAN loss in this approach can be computationally intensive, requiring substantial resources.

In above section, we utilized SSIM, PSNR, MSE, and PCC scores to assess the quality of synthesized CE-CT images. In this section, we carry out a detailed comparative analysis to assess the effectiveness of real and synthetic CE-CT images for identifying FLLs. This comparative analysis includes both computer-aided diagnosis and clinical diagnosis contexts. The real CE-CT images comprise CT scans from the

Table 9.1 Quantitative results of synthesized CE-CT images in ART and PV phases using different methods

Model	NC → ART							NC → PV						
	PSNR ↑		SSIM ↑		MSE ↓		PCC ↑	PSNR ↑		SSIM ↑		MSE ↓		PCC ↑
	Image	Liver	Image	Liver	Image	Liver		Image	Liver	Image	Liver	Image	Liver	
Pix2pix GAN [5]	32.86*	36.34*	0.7974*	0.9293*	34.30*	18.15*	0.9486	32.80*	35.65*	0.7961*	0.9301*	34.77*	20.31*	0.9602
MedGAN [2]	32.62*	35.78*	0.7504*	0.9273*	36.29*	21.25*	0.9251	32.60*	35.03*	0.7565*	0.9288*	36.52*	24.04*	0.9431
Improved cGAN [12]	33.18*	36.53*	0.8186*	0.9305*	31.92*	16.51*	0.9595	33.02*	36.04*	0.8185*	0.9319*	33.05*	19.31*	0.9609
CDD-GAN [7]	**33.30**	37.28	0.8371*	0.9333	**31.07**	14.22*	0.9770	33.14*	36.66*	0.8247*	0.9338*	32.16*	17.16*	**0.9788**
OA-GAN [6]	33.28	**37.42**	**0.8389**	**0.9335**	31.14	**13.88**	**0.9776**	**33.19**	**36.89**	**0.8329**	**0.9348**	**31.85**	**16.60**	**0.9788**

Images: results for entire image; Liver: results for liver region. The significance of the OA-GAN in terms of SSIM, PSNR, and MSE scores has been conformed. * represents that *p*-value less than 0.01 based on T-test results. Note: Bold represents the highest performance

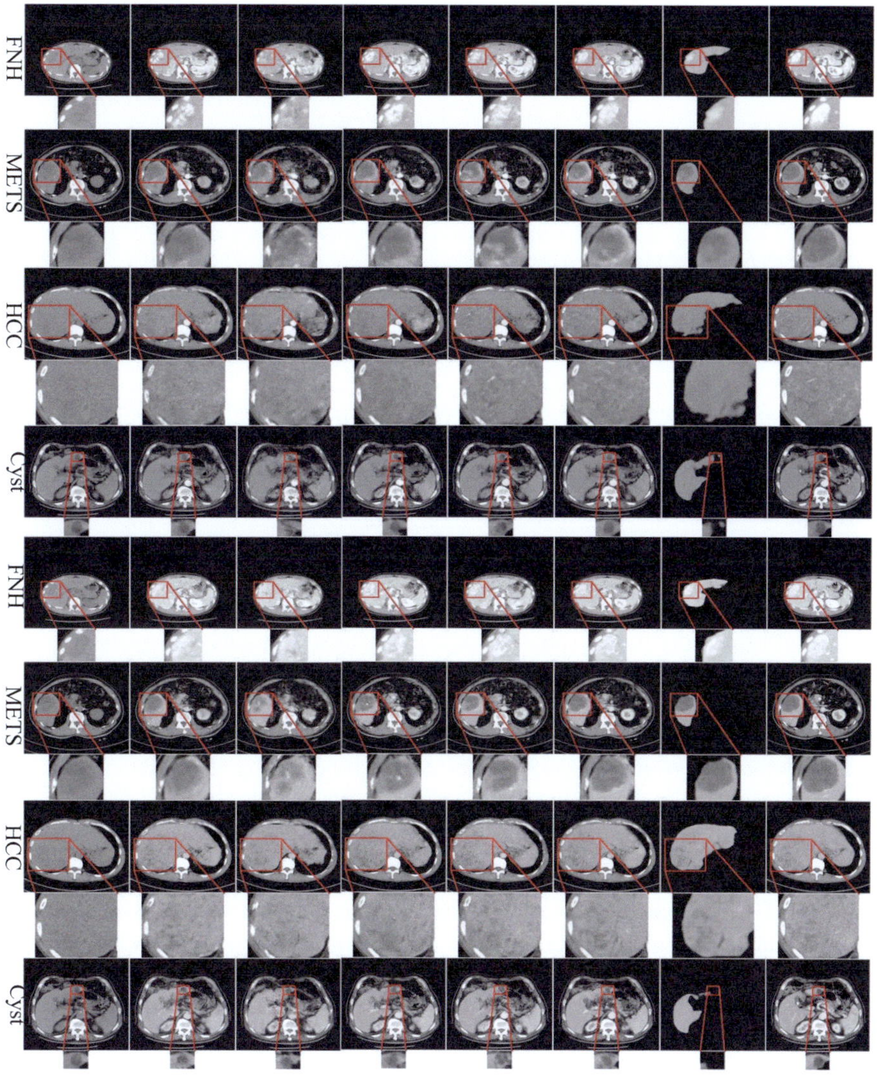

Fig. 9.7 The qualitative results of synthetic CE-CT images of ART phase (the first four lines) and PV phase (the last four lines). NC-CT (input) images, synthetic liver images produced by the LD network and real CE-CT images are presented in the first and last two columns, respectively. The remaining columns are synthetic CE-CT images produced from Pix2Pix GAN [5], MedGAN [2], Improved cGAN [12], CDD-GAN [12], and OA-GAN [6], respectively. The red box indicates the FLLs in each image

Table 9.2 Qualitative results of the DL-based FLLs classification

		NC-CT image	Pix2pix GAN [5]	MedGAN [2]	CDD-GAN [7]	Improved cGAN [12]	OA-GAN [6]	Real image
DL	ACC	0.6349	0.7874	0.7795	0.8268	0.8031	**0.8346**	0.8976

Evaluation of NC-CT images is performed by the trained single-phase model. Other columns are evaluated by the trained three-phase model. CE-CT image includes real NC, ART, and PV phases. Other GAN-based models include real NC phase, synthetic ART, and synthetic PV phases. Note: Bold represents the highest performance

real NC, ART, and PV phases, whereas synthetic CE-CT images consist of CT scan from the real NC phase, synthetic ART, and PV phases.

The accuracy (ACC) results are summarized in Table 9.2, indicating that when using only NC-CT images, the ACC is 0.6349 while the three-phase model achieves the lowest ACC of 0.7638 and the highest ACC of 0.8346. These findings suggest the potential benefit of incorporating synthetic images in FLLs classification. Notably, our proposed OA-GAN achieves an impressive ACC of 0.8346, surpassing existing GAN-based methods. This ACC result is approximately 6% lower than that of real CE-CT images with ACC of 0.8976. However, it represents a 20% improvement over using only the NC phase. These experimental results clearly demonstrate the effectiveness of our OA-GAN approach. Furthermore, we have presented evidence demonstrating significant improvements in the quality of synthesized images as shown in the Sub-section. The success of our approach can be attributed to the unique combination of two modules in OA-GAN, enabling it to extract semantic features from local images during the test phase. We also noticed that real and synthetic CE-CT images may represent a domain shift, which will negatively impacts the classification results.

Pilot Trial. We further conducted a pilot trial to evaluate the effectiveness of synthetic CE-CT images in clinical diagnosis. Our results, presented in Table 9.3 , show that synthetic CE-CT images generated by OA-GAN have achieved an ACC with 100% and 94.11%, which is close to that of real CE-CT images (ACC = 100.00%). The accuracy results for different FLL types are outlined in Table 5.4 [24], showcasing that the proposed OA-GAN demonstrates the highest accuracy in each specific category when contrasted with state-of-the-art models. Notably, the synthetic CE-CT images from OA-GAN exhibit a remarkable similarity to the real CE-CT images across all categories, indicating the effectiveness of our approach in clinical diagnosis. Neither of the two radiologists mistakenly identified patients with tumors as healthy individuals. This observation underlines the reliability of our proposed OA-GAN, as it did not generate any false negatives.

Table 9.3 Qualitative results of radiologist's diagnosis conducted by two radiologists: RD.1 and RD.2

Model	Pix2pix GAN [5]		MedGAN [2]		CDD-GAN [7]		Improved cGAN [12]		OA-GAN [6]		Real image	
	RD.1	RD.2	RD.1	RD.2	RD.1	RD.2	RD.1	RD.2	RD.1	RD.2	RD.1	RD.2
Cyst	1.0000	1.0000	0.8333	0.8333	1.0000	1.0000	0.8333	0.8333	**1.0000**	**1.0000**	1.0000	1.0000
FNH	0.7500	0.7500	0.5000	0.2500	1.0000	1.0000	1.0000	1.0000	**1.0000**	**1.0000**	1.0000	1.0000
HCC	0.5000	0.5000	0.5000	0.7500	0.7500	0.7500	0.5000	0.5000	**1.0000**	**0.7500**	1.0000	1.0000
METS	1.0000	1.0000	1.0000	1.0000	1.0000	1.0000	0.6667	0.6667	**1.0000**	**1.0000**	1.0000	1.0000
Overall	0.8235	0.8235	0.7058	0.7058	0.9411	0.9411	0.8235	0.8235	**1.0000**	**0.9411**	1.0000	1.0000

The CE-CT image is composed of real NC, ART, and PV phases of a patient, whereas the other columns consist of the real NC phase, synthetic ART, and synthetic PV phases produced from GAN-based models. Cyst, FNH, HCC, and METS indicate category-specific accuracy, whereas "Overall" reflects the accuracy across all categories. Note: Bold represents the highest performance

9.2.5 *Summary*

In this section, we introduce two types of image translation models: unpaired image translation models [3, 4] and paired image translation models [5–7]. Paired models typically provide better results but require paired datasets, where each image in the source domain corresponds to one in the target domain. For paired image translation, we also discuss two attention-based models: supervised attention [6] and unsupervised attention [7]. Attention mechanisms improve performance by allowing the model to focus on key regions of the image, but they require additional labels or supervision, especially for supervised attention.

9.3 Diffusion Model-Based Medical Image Translation

9.3.1 *Diffusion Model and Conditional Diffusion Model (Paired Diffusion Model)*

9.3.1.1 Diffusion Model

Diffusion Models [18] are a class of generative models that operate by iteratively diffusing noise to generate images. The diffusion process gradually refines the noise samples until they converge to realistic images. The training objective of diffusion models involves maximizing the likelihood of generating the training data under the diffusion process.

Figure 9.8 provides an overview of unsupervised diffusion models, comprising two fundamental stages: the forward process and the reverse process. In the forward process, gaussian noise ϵ is continuously infused into the input sample (CE-CT image), aiming to transform the input image into gaussian noise x_T. Conversely, the reverse process attempts to reconstruct input sample by utilizing a noise sample x_T.

Forward Process. The diffusion process is a T-step Markov Chain that starts from a clean image distribution $q(x)$ and repeatedly injects gaussian noise ϵ to the image

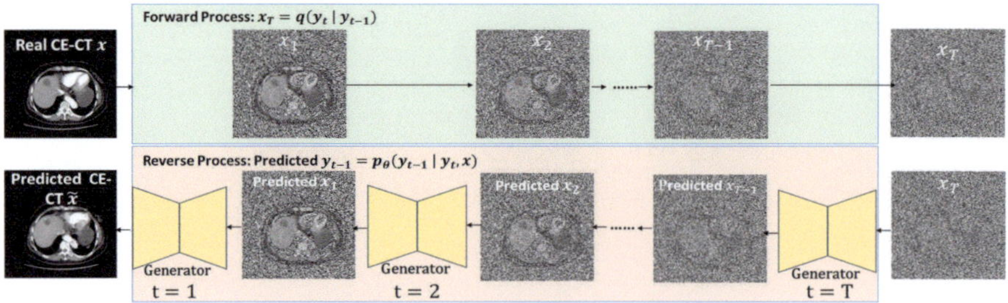

Fig. 9.8 The schematic diagram of DM-based synthesis models

according to a variance schedule $\beta_1, \ldots, \beta_T$. This process can be formalized as following.

$$q(x_t \mid x_{t-1}) = \mathcal{N}(x_t : \sqrt{1-\beta_t}x_{t-1}, \beta_t \mathcal{I}) \tag{9.21}$$

where $\beta_t \in (0, 1)$ is a constant hyperparameter that controls the variance of noise added corresponding to the diffusion step $1, \ldots, T$. $x_0 = x$. x_t represents that input sample after adding noise at time T in the forward process.

$$x_t = \sqrt{\bar{a}_t}x_0 + (1-\bar{a}_t)\epsilon_t, \epsilon_t \in \mathcal{N}(0, 1) \tag{9.22}$$

where $a_t = 1 - \beta_t, \bar{a}_t = \prod_i^T a_i$, the latent variables $x_{1:T}$ have identical shapes to the original image x_0.

Reverse Process. The reverse process in the original DDPM [18] can be considered as an unconditional image synthesis algorithm, which can be used to synthesize arbitrary images that obey the distribution of the target images, but cannot constrain the detailed information of the synthetic image. The noise-prediction loss function at any moment t in the reverse process can be formalized as following:

$$loss = \| \epsilon_t - \varepsilon_{\theta,t}(x_t, t) \|_2^2 \tag{9.23}$$

where ϵ_t is the noise adding at time t during the forward process. While the $\varepsilon_{\theta,t}$ is a learnable network in the reverse process, which attempts to predict the corresponding noise adding at the forward process.

9.3.1.2 Conditional Diffusion Model (Paired Diffusion Model)

Figure 9.9 provides an overview of Conditional Diffusion Models (CDMs) [19], comprising two fundamental stages: the forward process and the reverse process. The forward process in CDMs is same as the Unsupervised Diffusion Models. The reverse process in the original DDPM [18] cannot constrain the detailed information of the synthetic image. However, for our task, synthesizing CE-CT image from

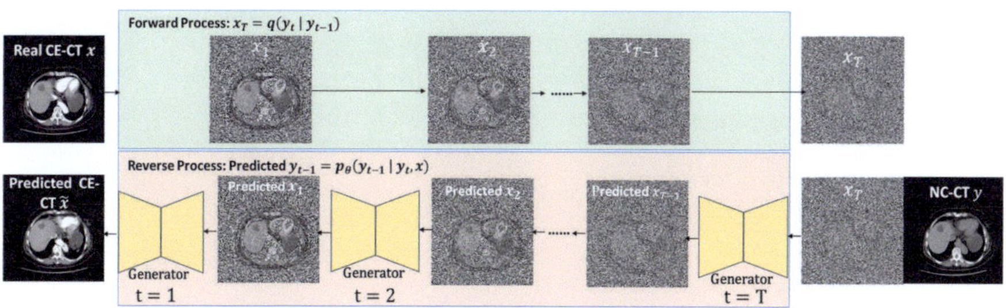

Fig. 9.9 The schematic diagram of CDM-based synthesis models

NC-CT image, generated CE-CT image must be the counterpart of input NC-CT image. Several approaches have been proposed to enforce the condition on diffusion models to control the sample generation. The input NC-CT image serves as an extra conditional information, which aims to guide the DMs model to synthesize target images that are highly correlated with the input samples as much as possible. With the two extra conditional information introduced in the reverse process, the noise estimator $\epsilon_{\theta,t}(x_t, t)$ in DDPM becomes $\epsilon_{\theta,t}(x_t, t, y)$ in our CDMs. Given the paired NC-CT and CT-CT samples, the noise-prediction loss function at any moment t in the reverse process can be formalized as follows:

$$loss = \parallel \epsilon_t - \varepsilon_{\theta,t}(x_t, y, t) \parallel_2^2 \tag{9.24}$$

where ϵ_t is the noise added at time t during the forward process. While the $\varepsilon_{\theta,t}$ is a learnable network in the reverse process, which attempts to predict the corresponding noise added at the forward process.

9.3.2 Improved Conditional Diffusion Model

Figure 9.10 provides an overview of Improved CDMs [20], including two fundamental process: the forward process and the reverse process. In the forward process, Gaussian noise ϵ is continuously infused into the input sample (CE-CT image), aiming to transform the input image into gaussian noise x_T. Conversely, the reverse process attempts to reconstruct the input samples by utilizing a noise sample x_T

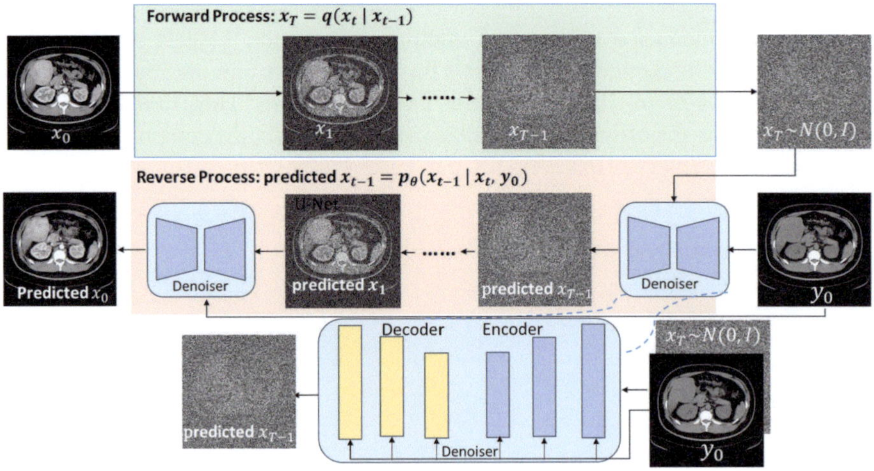

Fig. 9.10 The overview of proposed improved CDMs. The noisy image x_T, sampled from a Gaussian distribution, is fed into the encoder network of denoiser. While the NC-CT image y, acting as conditional information, is fed into the decoder network of denoiser

and conditional sample (NC-CT image y). Subsequently, we will offer an in-depth exploration of these distinct processes.

Forward Process. The diffusion process is a T-step Markov Chain that starts from a clean image distribution $q(x)$ and repeatedly injects Gaussian noise ϵ to the image using a variance schedule $\beta_1, \ldots, \beta_T$. The forward process can be formalized as following.

$$q(x_t \mid x_{t-1}) = \mathcal{N}(x_t : \sqrt{1-\beta_t} x_{t-1}, \beta_t \mathcal{I}) \tag{9.25}$$

where $\beta_t \in (0, 1)$ is a constant hyperparameter that controls the variance of noise added corresponding to the diffusion step $1, \ldots, T$. And the sample x_0 is x. Generally, when the T value exceeds 2000, x_T can be regarded as a pure Gaussian distribution. Equation (6.25) can be further expressed as Eq. (6.26) [25], x_t represents that input sample after adding noise at time T in the forward process.

$$x_t = \sqrt{\bar{a}_t} x_0 + (1 - \bar{a}_t)\epsilon_t, \epsilon_t \in \mathcal{N}(0, 1) \tag{9.26}$$

where $a_t = 1 - \beta_t, \bar{a}_t = \prod_i^T a_i$, the latent variables $x_{1:T}$ have identical shapes to the original image x_0.

Reverse Process. The reverse process in the original DDPM [2] can be considered as an unconditional image synthesis algorithm, which can be used to synthesize arbitrary images that obey the distribution of the target images, but cannot constrain the detailed information of the synthetic image. However, for our task, synthetizing CE-CT image from NC-CT image, generated CE-CT image must be the counterpart of input NC-CT image. Several approaches have been proposed to enforce the condition on diffusion models to control the sample generation. Specifically, the extra conditional information y and noise sample x_T are fed into a learnable neural networks also called denoiser (i.e., U-Net) to synthesize target images that are highly correlated with the conditional information y. However, the encoder network within the U-Net architecture is tasked with extracting features from input samples, while the decoder network is responsible for synthesizing target images. We hypothesize that directly integrating conditional information with noisy images into the encoder network may result in a gradual loss of semantic information regarding the conditional data.

Therefore, we introduce a novel U-Net architecture depicted in SDM [21], maintaining the conventional structure for the decoder component found in the traditional diffusion model. However, the encoder component is repurposed to accommodate the conditional information. Through this approach, the decoder can fully utilize conditional information when synthesizing target images. With an extra conditional informations introduced in the reverse process, the noise estimator of our improved CDMs can be simplified as $\epsilon_{\theta,t}(x_t, t, y)$. Given the paired NC-CT and CT-CT samples, the noise-prediction loss function at any moment t in the reverse process can be formalized as following:

$$loss = \parallel \epsilon_t - \varepsilon_{\theta,t}(x_t, y, t) \parallel_2^2 \tag{9.27}$$

where ϵ_t is the noise adding at time t during the forward process. While the $\varepsilon_{\theta,t}$ is the U-Net network in the reverse process, which attempts to predict the corresponding noise adding at the forward process.

9.3.3 GANs-CDM: GAN-Guided Conditional Diffusion Model

The overview of GANs-CDM [22] is depicted in Fig. 9.11, which comprises two components: a generator is to synthesize an initial CE-CT image $\tilde{x}_{init}$, while a CDM including forward and reverse processes that generates the final CE-CT image.

Compared to DM-based methods, GAN-based models can produce target image in a single step during the testing phase. However, the quality of synthesized images is limited due to mode collapse and training instability. Consequently, we firstly employ a GANs-based image synthesis model to synthesize a coarse CE-CT image $\tilde{x}_{init}$ from input NC-CT image y. To explore the robustness of our GANs-CDM, we validate the impact of using generators (Pix2Pix GAN [5] and Improved cGAN [12]) with different performances on the synthesized images. In this process, the loss function is formulated as

$$L_G = -\log(1 - D(G(y)) + ||x - G(y)||_1 \tag{9.28}$$

where $D(G(y))$ is the probability that synthetic sample $\tilde{x}_{init} = G(y)$ come from CE-CT images.

The crude CE-CT image $\tilde{x}_{init}$ produced from the previous generator G is used as conditional information to guide the CDM to generate final CE-CT image $\tilde{x}$. The forward and reverse processes are shown in the following two sections.

Forward Process. The diffusion process is a T-step Markov Chain that starts from a clean image distribution $q(x_0)$ and repeatedly injects Gaussian noise ϵ_t to the image according to a variance schedule $\beta_1, \ldots, \beta_T$. This process can be formalized as following.

$$q(x_t \mid x_{t-1}) = \mathcal{N}(x_t : \sqrt{1 - \beta_t} x_{t-1}, \beta_t \mathcal{I}) \tag{9.29}$$

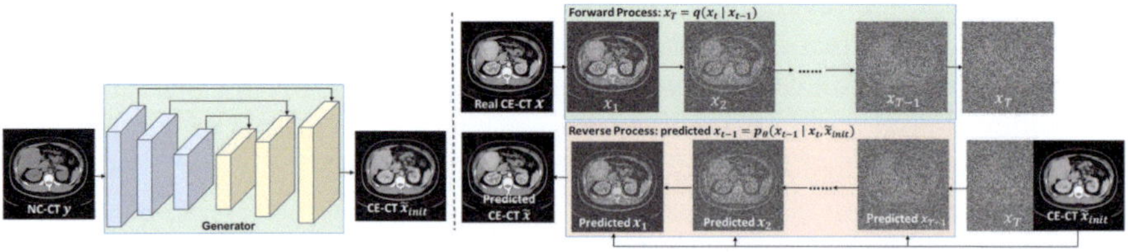

Fig. 9.11 The overview of proposed GANs-CDMs. The generator is to generate initial CE-CT image $\tilde{x}_{init}$. The forward process continuously adds noise into CE-CT image x, while the reverse process aims to synthesize final CE-CT image $\tilde{x}$

where $\beta_t \in (0, 1)$ is a constant hyperparameter that controls the variance of noise added corresponding to the diffusion step $1, \ldots, T$. The above equation can be further expressed as following.

$$x_t = \sqrt{\bar{a}_t} \times x + (1 - \bar{a}_t) \times \varepsilon_t, \varepsilon_t \in \mathcal{N}(0, 1) \tag{9.30}$$

where $a_t = 1 - \beta_t, \bar{a}_t = \prod_i^T a_i$, the latent variables $x_{1:T}$ have identical shapes to the original image x.

Reverse Process. The crude CE-CT image $\tilde{x}_{init}$ serves as the extra conditional information to guide the CDM model to synthesize target images that are highly correlated with the target image x as much as possible. Namely, the CDM attempts to predict the noise added at each step of the forward process instead of predicting the target image directly, the loss function of our CDMs can be formalized as following:

$$L = \| \varepsilon_t - \varepsilon_{\theta,t}(x_t, t, \tilde{x}_{init}) \|_2^2 \tag{9.31}$$

9.3.4 Performance Validation

In this section, we evaluate the performance of GANs-CDM with other state-of-the-art models in two tasks: synthesizing CE-CT images of ART phase and PV phase. The data partitioning and detailed description are provided in Sect. 9.2.4.1. Qualitative and quantitative results are shown in Fig. 9.12 and Table 9.4. We use the Palette [23] as our baseline model. Figure 9.12 shows that CE-CT images synthesized from GANs-CDM are more realistic than other methods, especially in tumor area.

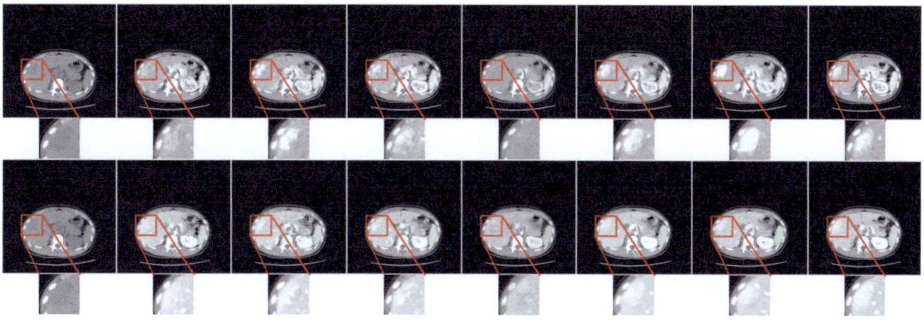

Fig. 9.12 The qualitative results of different models in synthesizing CE-CT images in ART phase and PV phase from NC-CT images. The first and last columns are input images and ground-truth images, respectively. The rest columns are synthetic CE-CT images generated from different models. From left to right are Pix2pix GAN [5], Improved CDM [20], Improved CGAN [12], GANs-CDM 1 [22] and GANs-CDM 2 [22], respectively. The red box represents pathological regions in NC-CT images

Table 9.4 Quantitative results of different models in synthesizing CE-CT images in ART phase and PV phase from NC-CT images

Model	NC → ART							NC → PV						
	PSNR ↑		SSIM ↑		MSE ↓		PCC ↑	PSNR ↑		SSIM ↑		MSE ↓		PCC ↑
	Image	Liver	Image	Liver	Image	Liver		Image	Liver	Image	Liver	Image	Liver	
Pix2pix GAN [5]	32.86	36.34	0.7974	0.9293	34.30	18.15	0.9486	32.80	35.65	0.7961	0.9301	34.77	20.31	0.9602
Improved CDM [20]	33.38	37.18	0.8299	0.9313	30.48	15.49	0.9620	33.24	36.66	0.8257	0.9337	31.67	18.42	0.9678
Improved cGAN [12]	33.18	36.53	0.8186	0.9305	31.92	16.51	0.9595	33.02	36.04	0.8185	0.9319	33.05	19.31	0.9609
Palette [23]	33.02	36.21	0.8099	0.9283	33.08	18.96	0.9418	32.82	35.23	0.8006	0.9292	34.80	24.43	0.9558
GANs-CDM 1 [22]	33.37	37.15	0.8296	0.9313	30.57	16.74	0.9560	33.31	36.04	0.8272	0.9319	31.10	19.31	0.9680
GANs-CDM 2 [22]	**33.44**	**37.62**	**0.8330**	**0.9339**	**30.08**	**14.26**	**0.9684**	**33.30**	**36.84**	**0.8305**	**0.9324**	**31.19**	**16.67**	**0.9686**

Image and Liver represent that we calculate PSNR, SSIM and MSE scores of synthetic and real CE-CT images as well as their liver regions. GANs-CDM 1 [22] utilizes Pix2pix GAN [5] to produce coarse CE-CT image while and GANs-CDM 2 [22] utilizes improved cGAN [12]. Note: Bold represents the highest performance

Table 9.4 shows that GANs-CDM 2 achieves higher SSIM and PSNR scores and lower MSE score in whole CE-CT image and liver region. In addition, higher PCC score also demonstrates that proposed GANs-CDM is very competitive in tumor region. Furthermore, our quantitative and qualitative results indicate that both GANs-CDM 2 [22] and GANs-CDM 1 [22] outperforms Pix2pix GAN [5], Improved cGAN [12] and Palette [23], suggesting that the strategy that combines GANs and CDM is effective for synthesizing superior CT images.

9.3.5 Summary

In this section, we introduce two key approaches for medical image synthesis: unsupervised Diffusion Models (DMs) [18] and supervised Diffusion Models [18, 20, 23]. We also explore the combination of GANs and Diffusion Models [22], a hybrid approach aimed at leveraging the strengths of both methods for improved image synthesis. Our results indicate that while Diffusion Models alone achieve remarkable performance in medical image synthesis, the fusion of GANs and Diffusion Models yields a significantly more effective approach for generating high-quality CT images.

References

1. Goodfellow, I., Pouget-Abadie, J., Mirza, M., Xu, B., Warde-Farley, D., Ozair, S., & Bengio, Y. (2020). Generative adversarial networks. *Communications of the ACM*, *63*(11), 139–144.
2. Armanious, K., Jiang, C., Fischer, M., Küstner, T., Hepp, T., Nikolaou, K., Gatidis, S., & Yang, B. (2020). MedGAN: Medical image translation using GANs. In *Computerized medical imaging and graphics* (p. 101684).
3. Zhu, J. Y., Park, T., Isola, P., & Efros, A. A. (2017). Unpaired image-to-image translation using cycleconsistent adversarial networks. In *Proceedings of IEEE/CVF Conference on Computer Vision and Pattern Recognition*.
4. Kaneko, T., Kameoka, H., Tanaka, K., & Hojo, N. (2019). Cyclegan-VC2: Improved Cyclegan-based non-parallel voice conversion. In *IEEE International Conference on Acoustics, Speech and Signal Processing* (pp. 6820–6824).
5. Isola, P., Zhu, J. Y., Zhou, T., & Efros, A. A. (2017). Image-to-image translation with conditional adversarial networks. In *Proceedings of the IEEE Conference on Computer Vision and Pattern Recognition* (pp. 1125–1134).
6. Yang, Y., Liu, J., Zhan, G., Chen, Q., Wang, F., Li, Y., Jain, R. K., Lin, L., Hu, H., Chen, Y. W. (2024). OA-GAN: organ-aware generative adversarial network for synthesizing contrast-enhanced medical images. *Biomedical Physics and Engineering Express*, *10*(3), 269.
7. Yang, Y., Chen, Q., Li, Y., Wang, F., Han, X. H., Iwamoto, Y., Liu, J., Lin, L., Hu, H., & Chen, Y. W. (2024). Segmentation guided crossing dual decoding generative adversarial network for synthesizing contrast-enhanced computed tomography images. *IEEE Journal of Biomedical and Health Informatics*.
8. Gulrajani, I., Ahmed, F., Arjovsky, M., Dumoulin, V., & Courville, A. C. (2017) Improved training of Wasserstein GANs. In *Advances in neural information processing systems* (p. 30).

9. He, K., Zhang, X., Ren, S., & Sun, J. (2016). Deep residual learning for image recognition. In *Proceedings of the IEEE Conference on Computer Vision and Pattern Recognition* (pp. 770–778).
10. Ouyang, X., Karanam, S., Wu, Z., Chen, T., Huo, J., Zhou, X. S., & Cheng, J. Z. (2020). Learning hierarchical attention for weakly-supervised chest X-ray abnormality localization and diagnosis. *IEEE Transactions on Medical Imaging*, *40*(10), 2698–2710.
11. Hu, J., Shen, L., & Sun, G. (2018). Squeeze-and-excitation networks. In *Proceedings of the IEEE Conference on Computer Vision and Pattern Recognition* (pp. 7132–7141).
12. Yang, Y., Xu, C., Chen, Q., Yutaro, I., & Han X. H. (2022). Synthesizing contrast-enhanced computed tomography images with an improved conditional generative adversarial network. In *Annual International Conference of the IEEE Engineering in Medicine Biology Society* (pp. 2097–2100).
13. Oktay, O., Schlemper, J., Folgoc, L. L., Lee, M., Heinrich, M., Misawa, K., & Rueckert, D. (2018). *Attention U-Net: Learning where to look for the pancreas*. arXiv preprint arXiv:1804.03999.
14. Zhao, H., Shi, J., Qi, X., Wang, X., & Jia, J. (2017). Pyramid scene parsing network. In *Proceedings of the IEEE Conference on Computer Vision and Pattern Recognition* (pp. 2881–2890).
15. Heusel, M., Ramsauer, H., Unterthiner, T., Nessler, B., & Hochreiter, S. (2023). GANs trained by a two time-scale update rule converge to a local nash equilibrium. In *Advances in neural information processing systems* (pp. 30–35).
16. Saito, N., Coifman, R. R., Geshwind, F. B., & Warner, F. (2002). Discriminant feature extraction using empirical probability density estimation and a local basis library. *Pattern Recognition*, 2841–2852.
17. Weibin, W., Qingqing, C., Iwamoto, Y., Xianhua, H., Zhang, Q., Hongjie, H., & Yen-Wei, C. (2019). Deep learning-based radiomics models for early recurrence prediction of hepatocellular carcinoma with multi-phase CT images and clinical data. In *41st Annual International Conference of the IEEE Engineering in Medicine and Biology Society* (pp. 4881–4884).
18. Ho, J., Jain, A., & Abbeel, P. (2020). Denoising diffusion probabilistic models. In *Advances in neural information processing systems* (pp. 6840–6851).
19. Dhariwal, P., & Nichol, A. (2021). Diffusion models beat GANs on image synthesis. In *Advances in neural information processing systems* (pp. 8780–8794).
20. Yang, Y., Xu, C., Chen, Q., Yutaro, I., Han X. H., et al. (2024). An improved conditional diffusion model for synthesizing contrast-enhanced computed tomography images. In *The KES International Conference on Innovation in Medicine and Healthcare* (pp. 2–10).
21. Wang, W., Bao, J., Zhou, W., Chen, D., Chen, D., Yuan, L., & Li, H. (2022). *Semantic image synthesis via diffusion models*. arXiv preprint arXiv:2207.00050
22. Yang, Y., Xu, C., Chen, Q., Yutaro, I., Han X. H., et al. (2024). GANs-guided conditional diffusion model for synthesizing contrast-enhanced computed tomography images. In *Annual International Conference of the IEEE Engineering in Medicine Biology Society* (pp. 2097–2100).
23. Saharia, C., Chan, W., Chang, H., Lee, C., Ho, J., Salimans, T., Fleet, D., & Norouzi, M. (2022). Palette: Image-to-image diffusion models. In *ACMSIGGRAPH 2022 Conference Proceedings* (pp. 1–10).
24. Yang, Y., Jing, L., Qingqing, Chen., Yinhao, Li., Xian-Hua, H., Hongjie, H., Lanfen, Lin., and Yen-Wei, Chen (2024). GANs-guided Conditional Diffusion Model for Synthesizing Contrast-enhanced Computed Tomography Images. In *2024 46th Annual International Conference of the IEEE Engineering in Medicine and Biology Society (EMBC)* (pp. 1–4).
25. Song, J., Chenlin, M., and Stefano, Ermon. (2020). Denoising diffusion implicit models. arXiv preprint arXiv:2010.02502.

Chapter 10
Foundation Models for Medical Image Analysis

10.1 Introduction

Over the past decade, deep learning models have revolutionized image processing by autonomously extracting task-specific features from training data, outperforming traditional machine learning methods that relied on manually crafted features. This shift has significantly advanced medical image analysis. However, most techniques depend on fully supervised training with task-specific labeled datasets, limiting their generalizability across diverse medical imaging modalities, datasets collected from different imaging devices, and task-specific function requirements. Creating high-quality object-level annotaions medical datasets requires substantial time and effort, placing a significant burden on radiologists. Medical image analysis often faces the challenge of data scarcity, which limits the applicability of deep learning models to narrow domains.

In recent years, the concept of foundation models has gained prominence in deep learning research. A foundation model is a deep learning model trained on large datasets, enabling its application across a wide range of tasks. In natural language processing, models like BERT [1] and GPT-4 [2] have demonstrated the feasibility of using general-purpose models to solve diverse tasks. Inspired by this success, similar approaches have been explored in computer vision (CV), with models such as SegGPT [3] and SEEM [4] serving as foundational models for image segmentation. In 2023, Meta introduced the Segment Anything (SAM) model [5], which represents one of the most prominent and widely adopted foundation models in this domain. However, the model, while demonstrating remarkable versatility and generalization, is predominantly trained on natural images. Numerous studies [6, 7] have shown that applying these models directly to medical images results in a significant performance drop. This decline can be attributed to the substantial differences in feature distributions between natural and medical images. As medical images like CT, MRI, and X-rays are generated based on tissue properties, unlike natural images captured

Y. Chen et al., *Recent Advances in Deep Learning for Medical Image Analysis*, Intelligent Systems Reference Library 278, https://doi.org/10.1007/978-3-031-94791-9_10

through visible light. Additionally, natural images feature diverse textures and complex backgrounds, while medical images have uniform backgrounds and focus on anatomical structures with low contrast and subtle texture variations.

While SAM's pre-training on large-scale natural image datasets equips it with strong generalization capabilities, it provides a solid foundation for improving the model's generalization performance on medical image tasks through targeted fine-tuning. The most conventional adaptation strategy is full fine-tuning, a traditional approach where all the parameters of the pre-trained model are updated for the downstream task. However, full fine-tuning has significant drawbacks, particularly when applied to foundation models. Given the typically enormous parameter size of foundation models, full fine-tuning demands extensive computational resources and storage. For each specific task, a fully fine-tuned, independent model needs to be stored, further increasing storage requirements.

In the medical domain, these challenges are further compounded by the relatively small size of task-specific datasets compared to the large-scale datasets used for pre-training foundation models. This discrepancy makes full fine-tuning susceptible to overfitting, resulting in suboptimal performance on medical imaging tasks. These limitations highlight the need for parameter-efficient and task-specific adaptation strategies that address computational efficiency and data scarcity constraints in medical image analysis.

The following subsections will use the SAM model as an example to introduce several parameter-efficient fine-tuning methods for foundation models.

10.2 Segment Anything Model (SAM)

SAM [5] is trained on the SA-1B dataset, which includes over 1 billion segmentation masks from 11 million images, making it 400 times larger than any existing segmentation dataset. Leveraging this vast training data, SAM adopts a modular and hierarchical architecture specifically designed for interactive segmentation tasks. Its architecture consists of three key components: the Image Encoder, the Prompt Encoder, and the Mask Decoder, each contributing to its flexibility and efficiency in handling diverse segmentation scenarios, as shown in Fig. 10.1.

The Image Encoder of SAM utilizes a Vision Transformer (ViT) [8], previously introduced in Chap. 3, pre-trained with a Masked Autoencoder (MAE) [9] approach and specifically adapted for high-resolution image processing. With scalability as a key consideration, the encoder processes each image a single time, producing high-dimensional embeddings. These embeddings are pre-computed and stored, allowing for reuse across multiple prompts, thereby enhancing efficiency and minimizing computational requirements by separating image encoding from the prompting stage.

The Prompt Encoder processes user-provided prompts, such as points, bounding boxes, masks and textual descriptions, into embeddings that align with the image embeddings, shown in Fig. 10.2. Positive and negative points highlight specific regions, bounding boxes define areas of interest, and mask prompts provide an

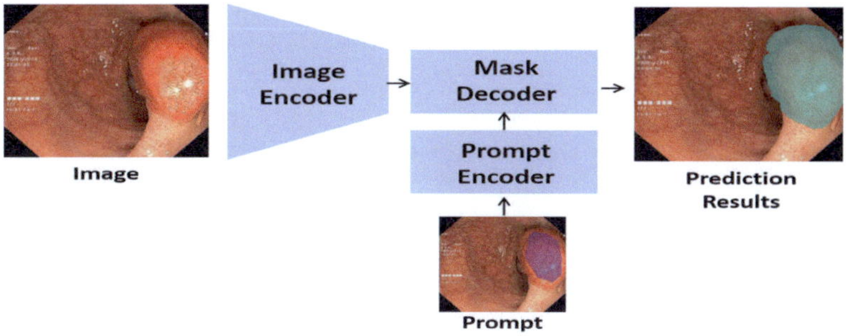

Fig. 10.1 The architecture of SAM. The image encoder, designed with a ViT transformer architecture, extracts image features. Both the lightweight prompt encoder and mask decoder, with minimal parameters, process prompts and generate segmentation masks, respectively

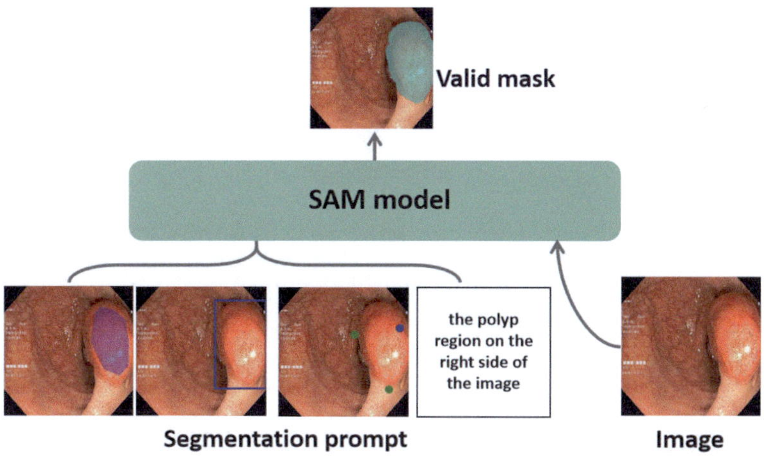

Fig. 10.2 SAM supports four types of prompts: points, bounding boxes, masks, and textual descriptions. These prompts guide the model in generating a valid segmentation mask from the input image

initial segmentation mask to guide further refinement. This lightweight component ensures that SAM can efficiently process user inputs in real time, enabling flexible and interactive segmentation across a variety of scenarios.

The Mask Decoder combines the image and prompt embeddings to generate high-resolution segmentation masks. It employs a transformer-based design to refine segmentation results iteratively based on user-provided inputs, supporting dynamic interaction. By leveraging pre-computed embeddings, the decoder ensures quick responses, which is essential for real-time interactive segmentation tasks.

Researchers [10] have conducted extensive evaluations of SAM's generalization performance on medical images. As shown in Fig. 10.3, the study assessed SAM's performance under three types of prompts—bounding box, point, and semantic mask

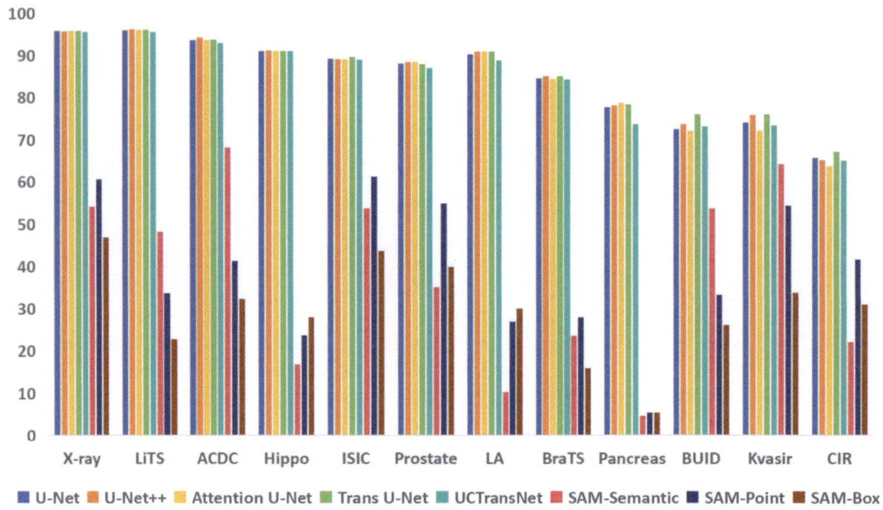

Fig. 10.3 Performance comparison of SAM and deep learning models across 12 medical image datasets [10]

prompts, using 12 publicly available datasets spanning multiple anatomical regions. The evaluation, using Dice as the metric, compared SAM with five mainstream medical image segmentation methods, including U-Net, U-Net++, Attention U-Net, Trans U-Net, and UCTransNet. The datasets covered various anatomical regions, such as the heart, brain, liver, lungs, and skin, and included six medical imaging modalities: endoscopy, dermoscopy, CT, MRI, X-ray, and ultrasound. The results demonstrated that, across all datasets, the original SAM significantly underperformed compared to deep learning-based medical image segmentation models. Visualization results, as shown in Fig. 10.4, provide further insights. From left to right, the figure shows the original image, the image mask, and SAM's output under the bounding box prompt. It is evident that while SAM can roughly identify the target region with the bounding box prompt, it struggles to provide satisfactory results due to the relatively indistinct boundaries in medical images (e.g., the boundary between polyps and normal tissues or between different organs).

These findings highlight the necessity of adapting SAM's parameters to account for the unique features of medical images, as the original model is not well-suited to handle the complexities of this domain.

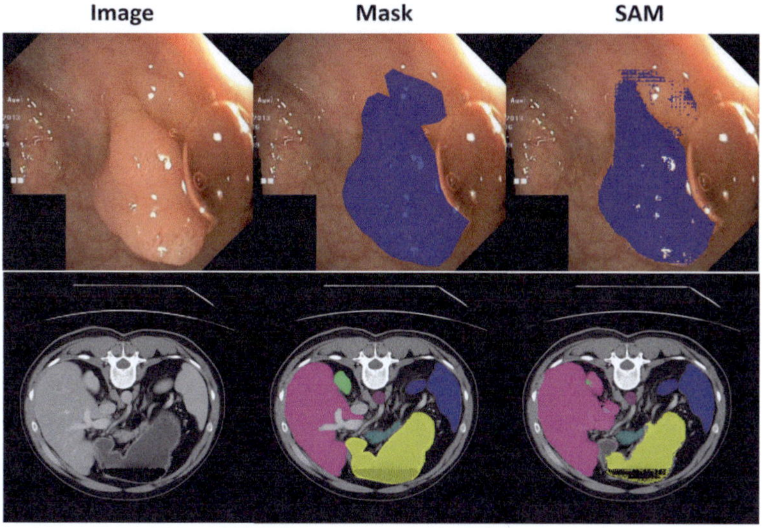

Fig. 10.4 The visualization results of SAM performance in medical image

10.3 Parameter Efficient Fine-Tuning Methods for SAM

10.3.1 Decoder-Only Fine-Tuning

In SAM, more than 90% of the parameters are concentrated in the Image Encoder. To address the challenges of adapting SAM for medical image segmentation tasks, MedSAM [11] proposes freezing the parameters of the Image Encoder and Prompt Encoder during fine-tuning, focusing exclusively on adjusting the parameters within the Mask Decoder, as illustrated in Fig. 10.5. Fine-tuning the Mask Decoder is particularly important because it directly generates the segmentation masks, which are crucial for accurately delineating regions of interest in medical images. By focusing on this component, the model can better align the output masks with the intricate and domain-specific features of medical images, such as tumors, lesions, or organs, without requiring extensive changes to the pre-trained encoders. We refer to this approach as Decoder-Only Fine-Tuning (DOFT). This method aims to enhance SAM's adaptability to medical image segmentation while significantly reducing the computational burden associated with training.

10.3.2 Low-Rank Adaptation Fine-Tuning

Low-Rank Adaptation (LoRA) [12] provides an efficient approach for fine-tuning large pre-trained models by introducing trainable low-rank matrices while keeping

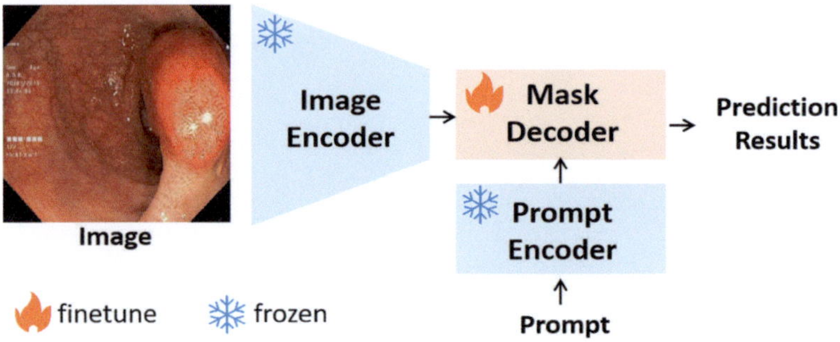

Fig. 10.5 Decoder-only fine-tuning framework for SAM adaptation

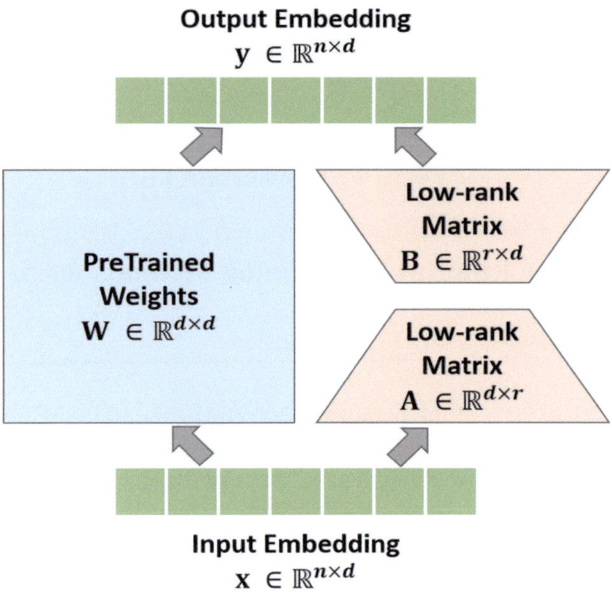

Fig. 10.6 Illustration of the low-rank adaptation (LoRA) mechanism

the original model weights frozen. As illustrated in Fig. 10.6, LoRA augments the pre-trained weight matrix $W \in \mathbb{R}^{d\times d}$ with a task-specific update $\Delta\mathbf{W} = \mathbf{A} \cdot \mathbf{B}$, where $\mathbf{A} \in \mathbb{R}^{d\times r}$ and $\mathbf{B} \in \mathbb{R}^{r\times d}$ are low-rank matrices, and $r \ll d$.

The transformation of the input embedding $\mathbf{x} \in \mathbb{R}^{n\times d}$ into the output embedding $\mathbf{y} \in \mathbb{R}^{n\times d}$ originally occurs through the operation Eq. (10.1):

$$\mathbf{y} = \mathbf{W} \cdot \mathbf{x} \tag{10.1}$$

With LoRA, the transformation incorporates the low-rank adaptation, resulting in Eq. (10.2):

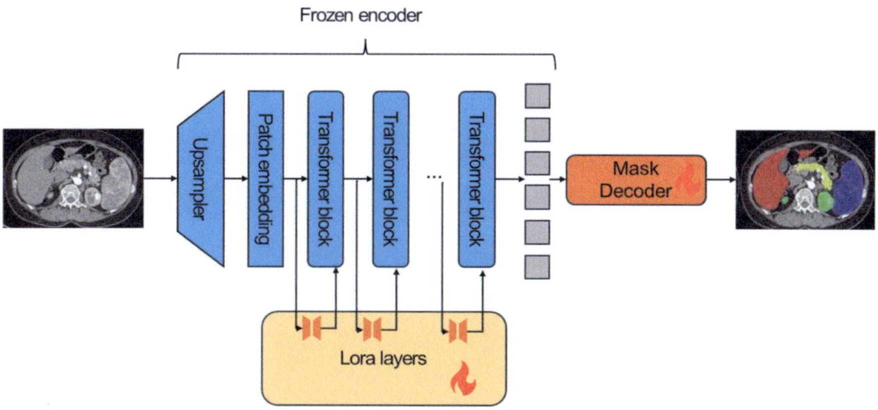

Fig. 10.7 The pipeline of SAMed using LoRA. The flame symbol indicates trainable components

$$\mathbf{y} = (\mathbf{W} + \Delta\mathbf{W}) \cdot \mathbf{x} = \mathbf{W} \cdot \mathbf{x} + \mathbf{A} \cdot \mathbf{B} \cdot \mathbf{x} \tag{10.2}$$

By parameterizing $\Delta\mathbf{W}$ as $\mathbf{A} \cdot \mathbf{B}$, LoRA significantly reduces the number of trainable parameters, enabling efficient fine-tuning for task-specific adaptations while preserving the pre-trained model's generalization capabilities.

SAMed [13] integrates the LoRA technique into SAM to enable effective fine-tuning for medical imaging tasks. The model architecture is shown in Fig. 10.7. Instead of fine-tuning all parameters, SAMed adopts a parameter-efficient strategy by freezing the pre-trained Image Encoder and introducing LoRA layers into each transformer block to enhance its adaptability to medical image features. Additionally, the Mask Decoder is fine-tuned to improve segmentation performance, while the Prompt Encoder is optimized with default embeddings to better align with task-specific requirements.

This approach leverages the remarkable segmentation capabilities of SAM and applies low-rank optimization to adapt it to medical image datasets efficiently. By updating only a small fraction of parameters, SAMed achieves precise semantic segmentation of medical images while maintaining computational efficiency, making it a robust solution for domain-specific applications.

10.3.3 Adapter-Based Adaptation Fine-Tuning

10.3.3.1 Medical SAM Adapter

An adapter is a lightweight, trainable module integrated into a pre-trained network to facilitate task-specific adaptation without modifying the original model's weights. These adapters introduce a lightweight, trainable module that allows for efficient

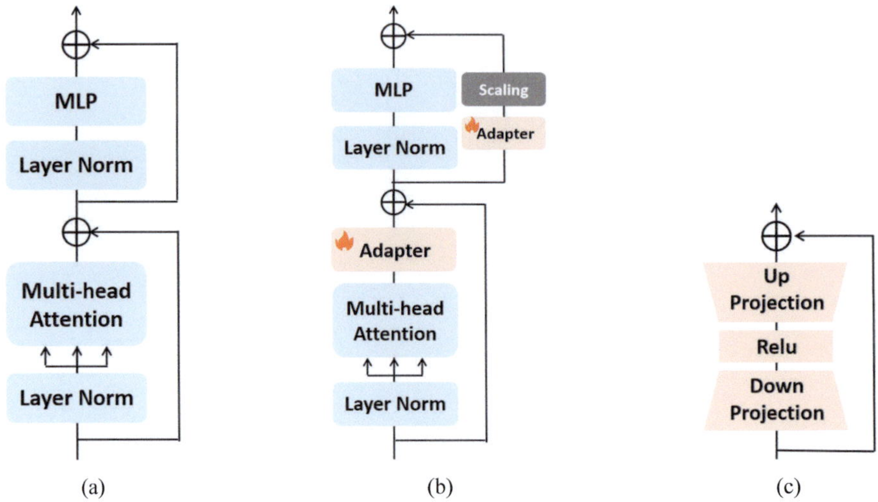

Fig. 10.8 Explanation of adapter methods used in MSA. **a** Standard self-attention block without adapters; **b** modified self-attention block with adapters, the flame symbol indicates trainable components; **c** adapter bottleneck structure

adaptation to specific downstream tasks without modifying the pre-trained weights of the original model.

MSA [14] integrates adapter-based techniques into the fine-tuning of SAM for medical tasks. Specifically, MSA introduces additional adapter structures into every self-attention block of SAM's transformer-based Image Encoder, shown in Fig. 10.8. In the original SAM structure (Fig. 10.8a), input tokens first undergo Layer Normalization, followed by Multi-Head Attention, and the resulting features are added to the input through a residual connection. This is followed by another Layer Normalization and an MLP module, with the output again added to the input via a second residual connection. MSA enhances this structure (Fig. 10.8b) by adding adapters after the Multi-Head Attention module and in the MLP residual path. Additionally, a scale factor s is applied to the embeddings after the second adapter. The internal structure of the adapter (Fig. 10.8c) follows a bottleneck design, where the input embedding is first compressed into a lower dimension using an MLP layer (Down-Projection), passed through a ReLU activation for non-linearity, and then expanded back to its original dimension with another MLP layer (Up-Projection).

10.3.3.2 Spatial Prior Adapter

The Spatial Prior Adapter(SPA) [15] offers an alternative adapter design by incorporating convolutional features extracted by a CNN into SAM's Image Encoder. Foundation models like SAM are predominantly designed with Transformer-based architectures. While Transformers excel at capturing long-range dependencies, CNNs are

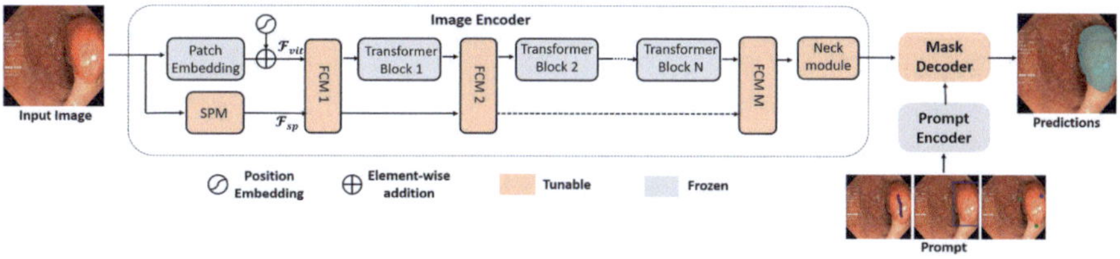

(a) The training framework of proposed SPA.

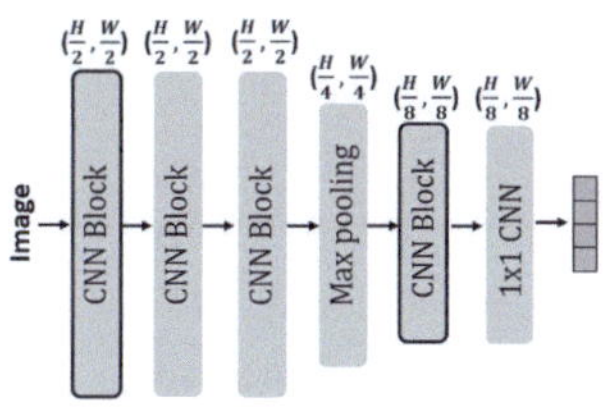

(b) The structure of SPM.

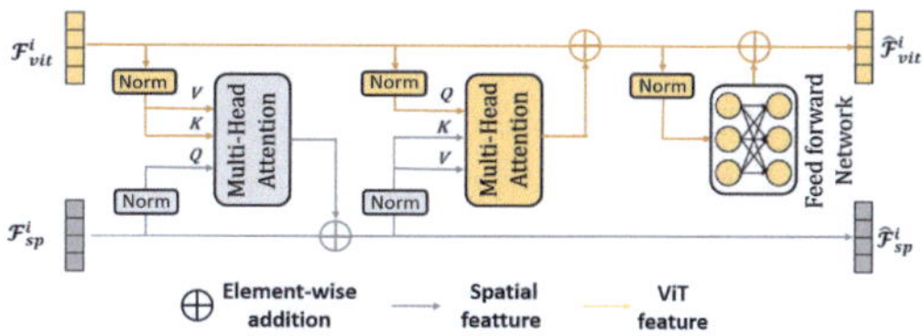

(c) The structure of FCM.

Fig. 10.9 The overview of SPA. a During fine-tuning, we fixed the parameters of prompt encoder, patch embedding, and all transformer block, only fine-tuning the neck module of the image encoder, the mask decoder, and the SPA to enhance downstream task performance. **b** The SPM extracts spatial prior information from the image through multiple convolutional layers, reducing the feature size to 1/8 of the original image. **c** The FCM facilitates interaction between the spatial prior features and transformer features through two cross-attention computations and one residual feed-forward network

particularly effective at extracting local features. This distinction highlights the complementary strengths of the two architectures. Numerous studies have demonstrated that combining CNN and Transformer architectures can significantly enhance performance in image recognition tasks. By integrating CNN-based spatial priors through the adapter, the Spatial Prior Adapter enables SAM to leverage both global contextual information from Transformers and detailed local features from CNNs, leading to improved adaptability for image-based tasks.

This method enhances efficiency by freezing the prompt encoder, patch embedding layer, and all transformer block components within the image encoder. The training process focuses on fine-tuning the Neck module of the image encoder, the mask decoder, and the Spatial Prior Adapter (SPA), as illustrated in Fig. 10.9a. SPA consists of two components: the Spatial Prior Module (SPM) and the Feature Communication Module (FCM). SPM, composed of multiple convolutional layers, extracts image-related spatial prior features directly from the input image. These features are then integrated with the transformer-based features from SAM's image encoder through FCM, which facilitates effective interaction between the spatial prior features and the transformer features. For instance, in the ViT-B model, the parameters associated with SPA constitute only 3% of the original image encoder's parameters, emphasizing the approach's focus on parameter-efficient fine-tuning.

More specifically, the SPM, as depicted in Fig. 10.9b, is designed using a CNN-based architecture. Each CNN block comprises a 3×3 convolution, batch normalization, and a ReLU activation function. The module includes four CNN blocks: the first block employs a stride of 2, halving the spatial dimensions, while the second and third blocks use a stride of 1, preserving the existing dimensions. Afterward, a max-pooling layer is applied to further reduce the spatial dimensions by half. The fourth CNN block, also with a stride of 2, reduces the feature map size to 1/8 of the original image dimensions. Finally, a 1×1 convolution projects the feature map to D dimensions, ensuring compatibility with SAM's image encoder. In summary, the SPM extracts spatial prior features, represented as $\mathcal{F}_{sp}^{0}$, which are later integrated with features from SAM's image encoder through the Feature FCM.

The FCM integrates the Transformer block features with the spatial prior features extracted by the SPM, as illustrated in Fig. 10.9c. Let $\mathcal{F}_{vit}^{i}$ and $\mathcal{F}_{sp}^{i}$ represent the Transformer and spatial prior features input to the i-th FCM, where $\mathcal{F}_{sp}^{0}$ is the output from the SPM. Each FCM performs three operations: Cross-attention from $\mathcal{F}_{sp}^{i}$ (as queries) to $\mathcal{F}_{vit}^{i}$, as shown in Eq. (10.3). Cross-attention from $\mathcal{F}_{vit}^{i}$ (as queries) to $\hat{\mathcal{F}}_{vit}^{i}$, as shown in Eq. (10.4). Residual feed-forward network (FFN) computation for $\mathcal{F}_{vit}^{i}$, as shown in Eq. (10.5). Here, $norm(\cdot)$ denotes LayerNorm [16].

The FFN contains two fully connected layers with an activation function. In this model, the standard multi-head cross-attention is replaced with deformable attention [17], which focuses on the most relevant data regions. This reduces computational complexity while enhancing performance.

$$\hat{\mathcal{F}}_{sp}^{i} = \mathcal{F}_{sp}^{i} + Attention(norm(\mathcal{F}_{sp}^{i}), norm(\mathcal{F}_{vit}^{i})) \tag{10.3}$$

$$\hat{\mathcal{F}}_{vit}^{i} = \mathcal{F}_{vit}^{i} + Attention(norm(\mathcal{F}_{vit}^{i}), norm(\hat{\mathcal{F}}_{sp}^{i})) \tag{10.4}$$

$$\hat{\mathcal{F}}_{vit}^{i} = \hat{\mathcal{F}}_{vit}^{i} + FFN(norm(\hat{\mathcal{F}}_{vit}^{i})) \tag{10.5}$$

10.4 Performance Validation

10.4.1 Dataset Description

To compare the performance differences of various fine-tuning methods for SAM, we conducted experiments on two distinct medical imaging datasets. The PROMISE12 dataset [18] consists of 50 transversal T2-weighted MRI images of the prostate from various centers. For preprocessing, MRI slice intensity values were clipped to the range between the 0.5 and 99.5 percentiles, normalized to [0, 1], and augmented with random flips, brightness and contrast adjustments, as well as resizing to 1024×1024. The Kvasir dataset [19] contains 1,000 gastrointestinal endoscopy images of polyps. Images were normalized using mean values [0.485, 0.456, 0.406] and standard deviations [0.229, 0.224, 0.225], resized to 1024×1024, and augmented with random flips, color jitter, and affine transformations inspired by [20]. Both datasets were split into 80–20% partitions for training and validation.

10.4.2 Implement Details

We conducted experiments on the Kvasir and PROMISE12 datasets using bounding box prompts. The training process employed the AdamW optimizer with cosine decay for learning rate scheduling. For both datasets, the initial learning rate was set to 0.0001, with a weight decay of 0.001. The model was fine-tuned for 30 epochs with a batch size of 2. All experiments were executed using PyTorch on a single NVIDIA RTX 3090 GPU, and model performance was evaluated using Dice, IoU, and HD95 metrics.

10.4.3 Results and Comprehensive Analysis

Under Bounding Box Prompts, we compared the original, un-tuned SAM with several fine-tuning methods, including MedSAM [11], MSA [14], SAMed [13], ViT-adapter [21], and SAM-adapter [22], as shown in Tables 10.1 and 10.2. These tables detail the method name, the number of parameters requiring fine-tuning, and the performance on the Kvasir and PROMISE12 datasets. All fine-tuning methods demonstrated improved accuracy compared to the original SAM model, highlighting the critical role of fine-tuning for medical datasets. Among these, MedSAM exhibited the lowest accuracy, as it exclusively fine-tunes the Mask Decoder parameters. While it required the fewest parameters (4.06 M), its limited performance underscores the importance of enhancing the image encoder's feature extraction capability for domain-specific tasks.

Several methods achieved a strong balance between parameter efficiency and performance. For example, SPA showed substantial improvements over SAM on the Kvasir dataset, with gains of 5.39% in Dice, 8.32% in IoU, and a reduction of

Table 10.1 Performance comparison of original SAM with other SAM fine-tuning methods on PROMISE12 datasets under bounding box prompts

Method	Grad. Param. (M)	PROMISE12		
		Dice (%)	IoU (%)	HD95
SAM [5]	–	$89.91 \pm 0.64^{**}$	$82.33 \pm 1.01^{**}$	$7.37 \pm 0.26^{**}$
MedSAM [11]	4.06	$92.46 \pm 0.11^{**}$	$86.23 \pm 0.17^{**}$	$4.69 \pm 0.10^{**}$
MSA [14]	11.16	$93.66 \pm 0.08^{**}$	$88.32 \pm 0.18^{**}$	$4.20 \pm 0.03^{**}$
SAMed [13]	4.21	$93.47 \pm 0.31^{**}$	$87.94 \pm 0.52^{**}$	$4.38 \pm 0.25^{**}$
ViT-adapter [21]	12.36	$93.77 \pm 0.18^{**}$	$88.48 \pm 0.28^{**}$	$4.24 \pm 0.24^{**}$
SAM-adapter [22]	4.12	$93.29 \pm 0.34^{**}$	$87.63 \pm 0.56^{**}$	$4.89 \pm 0.61^{**}$
SPA [15]	6.88	$\mathbf{94.29 \pm 0.21}$	$\mathbf{89.39 \pm 0.39}$	$\mathbf{3.92 \pm 0.26}$

* and ** denote statistically significant differences compared to SPA with *: $p < 0.05$, **: $p < 0.01$.
Note: Bold represents the highest performance

Table 10.2 Performance comparison of original SAM with other SAM fine-tuning methods on Kvasir dataset under bounding box prompts

Method	MGrad. Param. (M)	Kvasir		
		Dice(%)	IoU(%)	HD95
SAM [5]	–	$90.96 \pm 0.45^{**}$	$84.77 \pm 0.44^{**}$	$24.53 \pm 0.21^{**}$
MedSAM [11]	4.06	$95.20 \pm 0.27^{**}$	$91.06 \pm 0.44^{**}$	$12.77 \pm 0.12^{**}$
MSA [14]	11.16	$96.08 \pm 0.02^{**}$	$92.64 \pm 0.06^{**}$	$10.02 \pm 0.17^{*}$
SAMed [13]	4.21	$95.89 \pm 0.03^{**}$	$92.32 \pm 0.09^{**}$	$10.59 \pm 0.09^{**}$
ViT-adapter [21]	12.36	$95.85 \pm 0.12^{**}$	$92.21 \pm 0.21^{**}$	$11.91 \pm 0.13^{**}$
SAM-adapter [22]	4.12	$95.77 \pm 0.08^{**}$	$92.06 \pm 0.13^{**}$	$12.32 \pm 0.77^{**}$
SPA [15]	6.88	**96.35 ± 0.04**	**93.09 ± 0.10**	**9.06 ± 0.17**

* and ** denote statistically significant differences compared to SPA with *: $p < 0.05$, **: $p < 0.01$.
Note: Bold represents the highest performance

15.47 in HD95. It also outperformed the competitive MSA method with increases of 0.27% in Dice, 0.45% in IoU, and a reduction of 0.96 in HD95. On the PROMISE12 dataset, SPA exceeded ViT-adapter with gains of 0.52% in Dice, 0.91% in IoU, and a reduction of 0.32 in HD95. To ensure the significance of these improvements, T-tests were performed for all comparisons, with statistically significant differences marked in the table using superscripts * and **. SPA achieved competitive results on both datasets while requiring 6.88M parameters for fine-tuning, approximately 7% of the total model parameters.

10.5 Summary

Foundation models, such as SAM in computer vision, excel in diverse tasks by leveraging prompt-based learning and extensive training on large datasets, offering broad applicability across multiple domains. However, adapting these models to domain-specific tasks presents challenges due to feature disparities and domain shifts, particularly in specialized fields like medical imaging. Fine-tuning these foundation models provides an efficient solution, optimizing performance without the need for extensive computational resources or full retraining.

This chapter focuses on the Segment Anything Model (SAM) as a foundational model for medical imaging and explores techniques such as adapters and low-rank fine-tuning. These methods aim to address the unique characteristics of medical imaging, enhancing SAM's adaptability and effectiveness in this critical field.

References

1. Devlin, J., Chang, M.-W., et al. (2018). *BERT: Pre-training of deep bidirectional transformers for language understanding*. arXiv preprint arXiv:1810.04805
2. R OpenAI. (2023). *GPT-4 technical report* (Vol. 2, p. 13). arxiv: 2303.08774
3. Wang, X., Zhang, X., Cao, Y., Wang, W., Shen, C., & Huang, T. (2023). *SegGPT: segmenting everything in context*. arXiv preprint arXiv:2304.03284
4. Zou, X., Yang, J., Zhang, H., Li, F., Li, L., Gao, J., & Lee, Y. J. (2023). *Segment everything everywhere all at once*. arXiv preprint arXiv:2304.06718
5. Kirillov, A., Mintun, E., Ravi, N., Mao, H., Rolland, C., Gustafson, L., Xiao, T., Whitehead, S., Berg, A. C., Lo, W.-Y., et al. (2023). *Segment anything*. arXiv preprint arXiv:2304.02643
6. Deng, R., Cui, C., Liu, Q., Yao, T., Remedios, L. W., Bao, S., Landman, B. A., Wheless, L. E., Coburn, L. A., Wilson, K. T., et al. (2023). *Segment anything model (SAM) for digital pathology: Assess zero-shot segmentation on whole slide imaging*. arXiv preprint arXiv:2304.04155
7. Hu, C., & Li, X. (2023). *When SAM meets medical images: An investigation of segment anything model (SAM) on multi-phase liver tumor segmentation*. arXiv preprint arXiv:2304.08506
8. Dosovitskiy, A., Beyer, L., Kolesnikov, A., Weissenborn, D., Zhai, X., Unterthiner, T., Dehghani, M., Minderer, M., Heigold, G., Gelly, S., Uszkoreit, J., & Houlsby, N. (2021). An image is worth 16 × 16 words: Transformers for image recognition at scale. In *ICLR, 2021*.
9. He, K., Chen, X., Xie, S., Li, Y., Dollár, P., & Girshick, R. (2021). *Masked autoencoders are scalable vision learners*. arXiv:2111.06377
10. He, S., Bao, R., Li, J., Stout, J., Bjornerud, A., Grant, P. E., & Ou, Y. (2023). *Computer-vision benchmark segment-anything model (SAM) in medical images: Accuracy in 12 datasets*. arXiv preprint arXiv:2304.09324
11. Ma, J., & Wang, B. (2023). *Segment anything in medical images*. arXiv preprint arXiv:2304.12306
12. Hu, E., Shen, Y., Wallis, P., Allen-Zhu, Z., Li, Y., Wang, S., & Chen, W. (2022). Lora: Low-rank adaptation of large language models. In *ICLR, 2022*.
13. Zhang, K., & Liu, D. (2023). *Customized segment anything model for medical image segmentation*. arXiv preprint arXiv:2304.13785
14. Wu, J., Fu, R., Fang, H., Liu, Y., Wang, Z., Xu, Y., Jin, Y., & Arbel, T. (2023). *Medical SAM adapter: Adapting segment anything model for medical image segmentation*. arXiv preprint arXiv:2304.12620
15. Hu, J., Li, Y., Jain, R. K., Lin, L., & Chen, Y. (2025). SPA: Leveraging the SAM with spatial priors adapter for enhanced medical image segmentation. *IEEE Journal of Biomedical and Health Informatics*
16. Ba, J. L., Kiros, J. R., & Hinton, G. E. (2016). *Layer normalization*. arXiv preprint arXiv:1607.06450
17. Zhu, X., Su, W., Lu, L., Li, B., Wang, X., & Dai, J. (2020). *Deformable DETR: Deformable transformers for end-to-end object detection*. arXiv preprint arXiv:2010.04159
18. Litjens, G., Toth, R., Van De Ven, W., Hoeks, C., Kerkstra, S., Van Ginneken, B., Vincent, G., Guillard, G., Birbeck, N., Zhang, J., et al. (2014). Evaluation of prostate segmentation algorithms for MRI: The promise12 challenge. *Medical Image Analysis*, *18*(2), 359–373.
19. Pogorelov, K., Randel, K. R., Griwodz, C., Eskeland, S. L., de Lange, T., Johansen, D., Spampinato, C., Dang-Nguyen, D.-T., Lux, M., Schmidt, P. T., et al. (2017). Kvasir: A multi-class image dataset for computer aided gastrointestinal disease detection. In *Proceedings of the 8th ACM on Multimedia Systems Conference* (pp. 164–169).
20. Sanderson, E., & Matuszewski, B. J. (2022). FCN-transformer feature fusion for polyp segmentation. In *Annual Conference on Medical Image Understanding and Analysis*. Springer International Publishing.

21. Chen, Z., Duan, Y., Wang, W., He, J., Lu, T., Dai, J., & Qiao, Y. (2022). *Vision transformer adapter for dense predictions*. arXiv preprint arXiv:2205.08534
22. Chen, T., Zhu, L., Ding, C., Cao, R., Zhang, S., Wang, Y., Li, Z., Sun, L., Mao, P., & Zang, Y. (2023). *Sam fails to segment anything?–SAM-adapter: Adapting SAM in underperformed scenes: Camouflage, shadow, and more*. arXiv preprint arXiv:2304.09148

MIX
Papier aus verantwortungsvollen Quellen
Paper from responsible sources
FSC® C105338

If you have any concerns about our products,
you can contact us on
ProductSafety@springernature.com

In case Publisher is established outside the EU,
the EU authorized representative is:
Springer Nature Customer Service Center GmbH
Europaplatz 3, 69115 Heidelberg, Germany

Printed by Libri Plureos GmbH
in Hamburg, Germany